Clay Swelling *and* Colloid Stability

Clay Swelling *and* Colloid Stability

Martin Smalley

CRC Press
Taylor & Francis Group
Boca Raton London New York

CRC Press is an imprint of the
Taylor & Francis Group, an **informa** business
A TAYLOR & FRANCIS BOOK

First published in 2006 by Taylor & Francis Group

Published in 2019 by CRC Press
Taylor & Francis Group
6000 Broken Sound Parkway NW, Suite 300
Boca Raton, FL 33487-2742

First issued in paperback 2019

ISBN-13: 978-0-367-45374-9 (pbk)
ISBN-13: 978-0-8493-8079-2 (hbk)

Visit the Taylor & Francis Web site at
http://www.taylorandfrancis.com

and the CRC Press Web site at
http://www.crcpress.com

Library of Congress Card Number 2005046699

Library of Congress Cataloging-in-Publication Data

Smalley, Martin V.
Clay swelling and colloid stability / Martin V. Smalley.
p. cm.
Includes bibliographical references and index.
ISBN 0-8493-8079-0 (alk. paper)
1. Colloids--Stability. 2. Swelling soils. 3. Clay. I. Title.

QD549.S54 2006
541'.345--dc22 2005046699

Dedication

This book is dedicated to my parents, Gordon Smalley (1918–1994) and Barbara Joan Smalley (1922–2002).

Acknowledgments

I would like to thank the following scientists for reading parts of the manuscript: Professor R.K. Thomas, FRS (Physical Chemistry Laboratory, Oxford University); Professor I.S. Sogami (Kyoto Sangyo University, Japan); Professor N. Ise (emeritus professor, Kyoto University, Japan); Professor K.S. Schmitz (University of Missouri-Kansas City); Professor J. Swenson (Chalmers University of Technology, Sweden); Dr. R.J. Crawford (Unilever Research); Dr. S.M. King (ISIS Facility, Rutherford Appleton Laboratory); Dr. N.T. Skipper (University College London); and Dr. H.L.M. Bohm (Corus Steel). Their comments and criticisms were greatly appreciated.

Author

Martin Smalley worked as a professional research scientist from 1980 to 2000, with the exception of two years when he trained and worked as a physics teacher. In the late 1980s, after spells in theoretical physics at the Universities of Erlangen-Nürnberg and Oxford, he started investigating clays at the Physical Chemistry Laboratory, Oxford University, and the Department of Polymer Chemistry, Kyoto University. He became interested in polymers on an ERATO (Exploratory Research for Advanced Technology) project in Japan in the mid-1990s, and pursued his studies of colloids and polymers as a lecturer in the Department of Physics and Astronomy, University College London. This book summarizes the 36 papers (21 experimental, 15 theoretical) he has written on clays and colloids.

Preface

The function of a preface in a science book seems to be to spoil all the fun for readers by telling them what is to come. If you feel you need a guide, some kind of overview, before setting out, then please continue with the next paragraph. If you enjoy the thrill of the chase, I would advise you to skip to the beginning of Chapter 1 now, and read how in 1986, just before my 30th birthday, I received a box of yellowy brown translucent crystals.

For those still with me, in Chapter 1 I give a brief overview of clay chemistry and mineral structure and introduce the n-butylammonium-substituted Eucatex vermiculite system. Our experimental investigations, mainly by neutron scattering, of the three-component clay-salt-water system composed of n-butylammonium vermiculite, n-butylammonium chloride and water form the cornerstone of the book. Earlier workers had discovered that the swelling of n-butylammonium vermiculite crystals was nearly perfectly one-dimensional, taking place perpendicular to the plane of the silicate layers, and homogeneous in the sense that if a crystal with a repeat distance along the swelling axis (d-value) of 20 Å swelled sixfold macroscopically in the crystal-to-gel transition, the d-value in the gel phase was 120 Å. They pointed out that this means that the n-butylammonium vermiculite system could be used as a model system for testing theories of colloid stability. They also knew that the results of their studies could not be explained by the established theory of colloid stability, the DLVO (Derjaguin-Landau-Verwey-Overbeek) theory.

The main problem that the early workers identified with the DLVO theory was that it could not explain how the d-value varied with the electrolyte concentration c. The observed behavior was that d was inversely proportional to the square root of c, whereas the DLVO theory predicted a much more rapid variation. The origin of this fault is that DLVO theory explains the stable interplate separation in the gel phase as arising from a balance between an electrostatic repulsion, which does depend on c as the inverse square root, and a van der Waals attraction that is more or less independent of the salt concentration. I started studying the gels by neutron diffraction and extended the earlier results to a much wider range of salt concentrations. There was no way the DLVO theory could be made to fit the data, even when the van der Waals force was introduced with an adjustable parameter.

A definite prediction of DLVO theory is that charge-stabilized colloids can only be kinetically, as opposed to thermodynamically, stable. The theory does not mean anything at all if we cannot identify the crystalline clay state ($d \approx 20$ Å) with the primary minimum and the clay gel state ($d \approx 100$ to 1000 Å) with the secondary minimum in a well-defined model experimental system. We were therefore amazed to discover a reversible phase transition of clear thermodynamic character in the n-butylammonium vermiculite system, both with respect to temperature T and pressure P. These results rock the foundations of colloid science to their roots and are described in detail in the final part of Chapter 1. We obtained all the thermodynamic parameters for the transition by a combination of calorimetry and neutron scattering measurements as a function of *T* and *P*. Early on, we pointed out that the existence of this reversible phase transition is a counterexample to DLVO theory.

Toward the end of one of my early neutron scattering experiments at the Institut Laue Langevin (ILL) in Grenoble, I went up to the ILL library to take a few photocopies of raw diffraction patterns for my coworkers. A chunky volume of the *Journal of Chemical Physics* was open to the first page of an article by Sogami and Ise on electrostatic interactions in macroionic solutions. It does not take much concentration to photocopy a few printouts, and I scanned the abstract. My eye was immediately taken by two sentences in the middle of the abstract, as follows. "However, the Gibbs pair potential leads us to *repulsion* at small interparticle separations and *attraction* at large distances, creating a secondary minimum (with a potential valley deeper than the thermal energy for spherical colloidal particles). Without taking refuge in a van der Waals attraction the theory evidently substantiates the experimental fact recently reported, namely the existence of Coulombic intermacroion attraction through the intermediary of counterions" [Sogami and Ise's italics].

The main part of Chapter 2 is devoted to my adaptation of Sogami-Ise theory to plate macroions. I begin by proving that, within the linearization approximation, both DLVO and Sogami theory give identical expressions for the Helmholtz free energy of the system. However, the thermodynamic processes of the macroionic particles develop so as to minimize the Gibbs free energy of the system. It is the long-range attractive tail of the effective pair interaction included in the expression for the Gibbs free energy in Sogami-Ise theory that is interpreted as promoting the gel formation in the n-butylammonium vermiculite system. I prove that for plates, the interparticle spacing should be a constant number of Debye screening lengths, in qualitative agreement with the results on the n-butylammonium vermiculite gels and the general behavior of charge-stabilized colloids. Since the origin of the secondary minimum is clearly a result of long-range coulombic attraction in the new theory, the theory is named the coulombic attraction theory from this point on. I demonstrate that the coulombic attraction theory accounts naturally for the reversible phase transition in the n-butylammonium vermiculite system.

Theories of interparticle interactions make definite predictions for force-distance curves. The first part of Chapter 3 is devoted to an exhaustive investigation of the effect of uniaxial stress on n-butylammonium vermiculite gels. We studied the effect of compression along the axis of the one-dimensional colloid at eight different salt concentrations in a range between 0.001 and 0.2 M. This factor of $\times$ 200 in salt concentration is essential in obtaining a global picture of the interaction. Many objections to the coulombic attraction theory based on force-distance curves taken at one poorly controlled salt concentration are dealt with later in the chapter, where I give a full analysis of the force-distance curves in the new theory and in DLVO theory. The latter are not as those represented by a purely exponentially decaying function, as is often taken to be the case, but include the van der Waals force.

The main experimental result of Chapter 3 is that the effective surface potential of the particles is constant with respect to the electrolyte concentration, its average value being approximately 70 mV across the $\times 200$ range in salt concentration. Many objections to the coulombic attraction theory, dealt with in detail in Chapter 7, have been based on the mistaken idea that, for some reason, the Nernst equation should apply to the surface potential of colloidal particles. This comes about as a result of treating a single inhomogeneous colloidal phase (the gel phase, in the case of the n-butylammonium vermiculite system) as a two-phase system composed of an electrode (solid) and an electrolyte solution (liquid). The Nernst equation states that we should expect a variation of about 60 mV per decade of concentration of potential-determining ions. This is clearly not the case. Instead, by taking the constancy of the surface potential and the predictions of the coulombic attraction theory as starting points, one opens up the possibility of a new theoretical discovery, as described in Chapter 4.

Membrane equilibria underlie all dialysis experiments and many biological phenomena, especially at the level of cell membranes. The basis for their physico-chemical description was laid down by Donnan in 1924, and the Donnan equilibrium is well known in biochemistry. It describes how electrolyte is distributed between two regions separated by a membrane when a macroion (charged colloidal particle) is present in one region but not the other. Salt is always expelled from the macroion-rich region into the external fluid. I proved that the Donnan equilibrium is a limiting case, the noninteracting case, of a wider description for interacting macroions. I predicted that for the model n-butylammonium vermiculite system, the salt should be distributed at a constant ratio between the two phases, irrespective of the salt concentration, and that this ratio should be equal to 2.8, a result markedly different from the prediction given by the Donnan equilibrium.

As you can imagine, I would not be writing this book if we had not obtained excellent agreement between the new theory and experiment, as described in detail in Chapter 5. In Chapter 5, we give our results on the salt-fractionation effect together with the effect on the swelling obtained by varying the clay concentration r. We have by this stage investigated the three-component system composed of n-butylammonium vermiculite, n-butylammonium chloride and water as function of five variables: the clay concentration r, the salt concentration c, temperature T, hydrostatic pressure P, and uniaxial stress along the swelling axis p.

The considerations of Chapters 1 to 5 show that the coulombic attraction theory is well adapted to explain the existence, extent and properties of the two-phase region of colloid stability. It was, however, desirable to place the theory on a firmer footing in statistical thermodynamics. Sogami, Shinohara and I took a major step in this direction when we calculated the Helmholtz free energy of the plate system exactly in mean field theory. The original calculations were based on two plates immersed in an infinite container, but we have recently calculated the Helmholtz free energy exactly for two plates in a finite container. The calculations and results are presented in detail in Chapter 6, together with our most recent results for the Gibbs free energy in an exact representation that includes the nonlinear effects. We now have a complete intellectual solution to the problem, and the origin of the long-range attraction is clearly revealed to be an "electric rope" binding the plates together at large separations.

Both the adiabatic Gibbs and Helmholtz potentials give rise to qualitatively the same behavior when the nonlinear Poisson-Boltzmann equation is solved exactly. They are expressed as the sum of the contributions from pairs of surfaces consisting of an osmotic part and an electric part. The osmotic pressure arises from the random motions of the small ions. On the other hand, the electric part of the adiabatic potential represents the interaction mediated not by material entities, but by the electric field produced by the equilibrium distribution of the small ions and the plate charges. This is the electric rope that binds the plates together at long range. By attracting counterions and repelling co-ions, the fixed surface charges of the plates create asymmetric distributions of small ions both in number and in charge. From these asymmetries, the plates receive two kinds of reaction. First, the excess of small ions in number induces an osmotic repulsion between the plates. Second, in the intermediate cloud of small ions, the counterions, dominating co-ions, give rise to an excess of charge opposite to that of the plates, and this induces an effective electric attraction between the plates. It is a delicate balance of these two diametrically opposed effects that leads to the osmotic and electric stability attained at the minimum of the adiabatic potential.

When you propose a new theory, you face opposition. Several criticisms of the coulombic attraction theory and their refutations are described in Chapter 7, which begins with a reconsideration of the derivation of various thermodynamic quantities in the original Debye-Hückel theory of simple electrolyte solutions. Although this chapter is of more historical than practical interest, many workers in colloid science are still interpreting their data in terms of DLVO theory, so it is necessary to answer objections to the new theory and demonstrate its logical coherence in the face of attack. As ever in the academic world, it is better to be attacked than to be ignored. Many workers in the field simply ignore the existence of the coulombic attraction theory, whether applied to plates or to spheres. I make a digression into the world of spherical particle interactions to undermine this limited viewpoint. From this point on, I regard questions about the validity of the coulombic attraction theory as closed.

Theories of colloid stability based on electrostatics go way back beyond the DLVO theory, to the Gouy-Chapman theory of the electrical double layer proposed in the early 1910s and the Stern theory of counterion condensation proposed in 1924. There was much weighty speculation about the counterion distribution around colloidal particles throughout the 20th century, but nobody succeeded in measuring it until our work in 1997. This work is described in detail in Chapter 8.

We first used isotope substitution in diffuse neutron scattering measurements to determine the distribution of water molecules and counterions (n-butylammonium ions) around the clay layers in the gel state, and obtained a unique picture of a "dressed" macroion in solution. We obtained a structure in which the naked clay plate of 10 Å thickness was extended out to about 35 Å by layers of water molecules and counterions. The dressed macroion has exactly two layers of water molecules coating the clay layers; these layers are 6 Å thick on both sides, extending the effective clay plate out to 22 Å, before any counterions at all are found. This is in direct contradiction to the Stern layer picture, widely held in colloid science, that counterions are in direct contact with the particles. A large proportion of the counterions, about one-half, were found in a layer 4 Å thick on both sides, extending the effective clay plate out to 30 Å, and there was also evidence for another layer of partially ordered water molecules on the outside of the counterion layer, giving a block of a thickness of at least 35 Å. This is an entirely new picture of the environment of a colloidal particle in solution.

Attempting to pursue the discovery of the structure around the macroion in solution, we went on to measure the complete counterion distribution as a function of distance along the swelling axis in a clay gel, as described in Chapter 9. It is fair to describe this as the first determination of the complete ion distribution inside a colloid. For technical reasons, we were unable to carry out the experiment on the n-butylammonium system; instead, we used its sister system, the n-propylammonium Eucatex system. We used a completely different kind of analysis than that employed in Chapter 8 by trying to understand the structure from a crystallographic point of view. We used *in situ* substitution to obtain a *d*-value of exactly 43.6 Å for gels prepared with both hydrogenated n-propylammonium ions and deuterated n-propylammonium ions. This enabled us to determine the complete scattering-length-density profile along the swelling axis, on the understanding that we were averaging over a lot of liquid-like motion in the interlayer region. First, we found that the surface was covered by exactly two layers of water molecules. Since this structure had now been obtained by an inverse Monte Carlo routine on a single-particle distribution function after its discovery by Fourier transformation and the pair-correlation method, it must surely be regarded as proven, and may turn out to be widespread in aqueous colloids. Second, we again found local maxima in the counterion distribution immediately outside the two water layers. Third, we found that the remaining counterions are broadly distributed around the center of the interlayer region. Of course, if the dressed macroion thickness of 35 Å is reproduced even at such low *d*-values, in a gel with a *d*-value of 43.6 Å, there is less than 10 Å between the effective particles, so it is hardly surprising that the rest of the counterions are smeared out around the center. It remains an experimental challenge to determine the complete counterion distribution for larger interparticle separations.

Chapter 10 is our final chapter on the three-component clay-salt-water system. All the results in Chapters 1 to 9 are for the clay plates in liquid water. In Chapter 10 we describe freezing experiments on the clay gels. Upon freezing of the water, the colloidally swollen gel phase collapsed into the crystalline phase. The two phase transitions of gel-to-crystal and water-to-ice appeared to occur simultaneously. As with the phase transition in liquid water, when heating a gel causes collapse into the crystalline phase, the gel-to-crystal phase transition around 0°C was observed to be reversible, the gel phase always being recovered upon warming through the freezing point of water. This phenomenon was observed throughout a wide range of clay and salt concentrations. Such a reversible phase transition between swollen and collapsed clay mineral phases may be important in the weathering of rocks in freezing cycles.

A completely different set of phenomena are observed when the samples are frozen rapidly. If we quench a sample in liquid water at a rate of 1000 degrees Kelvin per second, we do not see the thermodynamic phase transitions described above. Indeed, it is an underlying assumption of electron microscopy experiments on aqueous systems that we should more or less preserve the room-temperature structure by rapid freezing. We carried out an electron microscopy study of n-butylammonium vermiculites, and the beautiful micrographs we obtained are shown in the second half of Chapter 10. It was noteworthy that the *d*-values obtained from the electron microscopy experiments were definitely different from those obtained by *in situ* neutron scattering studies of the gel phase, with the electron microscopy results being greater by a factor of between one and one-half and two, the discrepancy being outside the range of experimental error. The neutron results are certain to be more accurate, so the results cast doubt on the idea that the freeze-fracture technique preserves the room-temperature structure.

A huge advantage of characterizing the three-component system so thoroughly, both with regard to the general characteristics of the swelling and the subtle information that can be obtained from neutron scattering experiments with isotope substitution, was that it enabled us to unravel the mechanisms of clay-polymer interactions by adding polymers to the model system under carefully controlled conditions. I started out studying the effects of adding polymers to the n-butylammonium vermiculite system in the mid-1990s, and our preliminary results on the effects of adding poly (ethylene oxide) (PEO) and poly (vinyl methyl ether) (PVME) to our model clay colloid are described in Chapter 11. The polymer-stabilized colloids are the other great class of colloid, alongside the charge-stabilized colloids, and the vermiculite-PEO system is described in detail in the following Chapters 12 and 13 as a model for studying interactions in them. From a more practical point of view, the importance and potential of the four-component clay-polymer-salt-water system in agricultural and industrial applications could hardly be overstated, being the central problem of soil science.

We discovered that the generally accepted mechanism of depletion flocculation was unable to explain our results even qualitatively. Instead, we were able to propose a new mechanism for bridging flocculation, as described in Chapter 12. Using isotope substitution on the polymer chains in neutron scattering experiments, we were able to determine the distribution of polymer segments in the interlayer region, the first time such a profile has been obtained between colloidal particles. Putting it all together with the old results on the effect of uniaxial stress on the gels, we were able to propose a quantitative model for the bridging mechanism, with a drawing force of 0.6 pN (pico-Newtons) per polymer bridge. The *a priori* calculation of the absolute value of the drawing force constitutes a new theoretical challenge in colloid and polymer science.

In Chapter 13 we describe our most sophisticated neutron scattering experiments, in which we used isotope substitution on both the polymer chains and the counterions, enabling us to obtain the two distributions under identical conditions. This gave us a uniquely detailed picture of the polymer segments in the interlayer region. Remarkably, the addition of polymer does not affect the ion distribution. Polymer segments do come into direct contact with the clay plates, displacing water molecules in the layer directly adjacent to the surface oxygens of the vermiculite layers. Otherwise, however, the dressed macroion structure is relatively unaffected, with the majority of the counterions still situated in a well-defined layer about 11 to

15 Å from the center of the clay plates. Outside of the "trains" of polymer segments situated about 5 to 8 Å from the center of the clay plates, the region up to about 18 Å is relatively depleted of polymer segments, giving us a picture of some stretched "stringy" bits poking through the rest of the dressed macroion, which still retains the water layers on either side of the butylammonium ion layer. We discovered that there was some kind of random coil, comprising approximately half the segments, in the middle between the vermiculite layers. For the sake of example, let us say that a polymer with a molecular weight of 10,000 is just large enough to form one bridge between the clay layers. We discovered that the only difference in the interlayer structure of a polymer with a molecular weight of 1 million is that it forms 100 identical bridges, with the number of bridges per unit amount of polymer the same in both cases.

The widely invoked model to explain why interparticle separations between colloidal particles decrease when a large polymer is introduced into the system is known as depletion flocculation. This is basically an equilibrium osmotic exclusion model, in which the reduction in *d*-value is driven by an osmotic pressure of polymer molecules excluded from the interparticle region. I have not sought controversy in my research career. When I started adding polymers to clay gels, I did not set out thinking, "I must show that the depletion flocculation mechanism is wrong," but I realized that the constancy of the *d*-value with respect to molecular weight at a fixed volume fraction rules out the depletion flocculation mechanism. Instead, the bridging mechanism gives a coherent explanation of all the available data. I am not seeking an argument, but looking for a paradigm shift.

Likewise, I did not set out on my studies of the charge-stabilized gels thinking, "I must disprove DLVO theory." I just discovered that the established theory did not fit our experimental results even qualitatively. I mention luck and serendipity several times in this book, and I was genuinely lucky to see the Sogami-Ise paper around the same time I was studying Verwey and Overbeek's book, *Theory of the Stability of Lyophobic Colloids*. I was therefore able to compare them on an equal basis, and found that the Sogami-Ise theory adapted to plate macroions fitted our clay swelling results like a glove on a hand. I came to the natural conclusion that the new theory was the correct one, and all my subsequent theoretical and experimental studies have confirmed this conclusion. Again, I am not looking for an argument; I am looking for a paradigm shift.

This is an account of my own research work and conclusions in the field. I do not wish to slight other workers, and I have paid homage to the pioneers in the study of both clay swelling and colloid stability. I hope that this book will contribute significantly to the understanding of these subjects.

Table of Contents

1 The n-Butylammonium Vermiculite System

In 1986, just before my 30th birthday, I received a box of yellowy brown translucent crystals. These vermiculite crystals came from Brazil, from a mine in Eucatex. The place sounds exotic, and the dark crystals had a fetching bronzy luster. In fact, they looked like one in a standard guide to minerals and had the perfect basal cleavage described there [1]. This means that the vermiculites are layer crystals.

I was lucky to receive some jars of sodium-substituted Eucatex vermiculite and n-butylammonium-substituted Eucatex vermiculite. These synthetic systems had been obtained by cation exchange on the raw minerals. Such cation exchange plays a major role in clay chemistry, and the process is described in detail in standard books on clay colloids, like that of van Olphen [2]. Some years later, I was able to obtain the following chemical formula for the dry sodium Eucatex vermiculite [3]

$$Si_{6.13}\,Mg_{5.44}\,Al_{1.65}\,Fe_{0.50}\,Ti_{0.13}\,Ca_{0.13}\,Cr_{0.01}\,K_{0.01}\,O_{20}\,(OH)_4\,Na_{1.29}$$

This is a typical formula for a vermiculite. If we put all the elements in their formal oxidation states, the condition for charge neutrality of the mineral is satisfied if the transition metal elements are in their higher oxidation states.

To make the discussion of clay chemistry simpler, let us consider the idealized magnesiosilicate, $Si_{(8-x)}Mg_6Al_xO_{20}(OH)_4M_x$, where M is a monovalent cation and x represents the amount of trivalent substitution for tetravalent silicon. The prototype mineral is talc, $Si_8Mg_6O_{20}(OH)_4$, in which the magnesiosilicate layers with $x = 0$ are obviously charge neutral, with Si^{4+} and Mg^{2+} as formal ionic charges. The flaky layer mineral talc is well known as an ingredient of talcum powder. These neutral layers are held together by weak, short-range van der Waals forces, so the crystals shear very easily in the a,b-plane, the plane of the layers. These individual layers are about 10 Å thick along the c-axis of the mineral, which is perpendicular to the clay plates.

When there is substitution of a trivalent element (aluminum in our idealized example) into the silicon atomic positions (which occur in two layers equidistant from the central layer of magnesium atoms), the layer as a whole acquires a net negative charge (equal to x per structural unit in the formula above), which is probably mainly located on the surface oxygen atoms. This negative charge is balanced by interlayer cations (x univalent cations in our example). We will not enter into the vast mineralogy of the magnesiosilicates as x varies from 0 to 2, with the highest charge corresponding to a member of the mica group. Instead, we focus on the vermiculites, in the narrower range between 1.2 and 1.6, where the minerals can

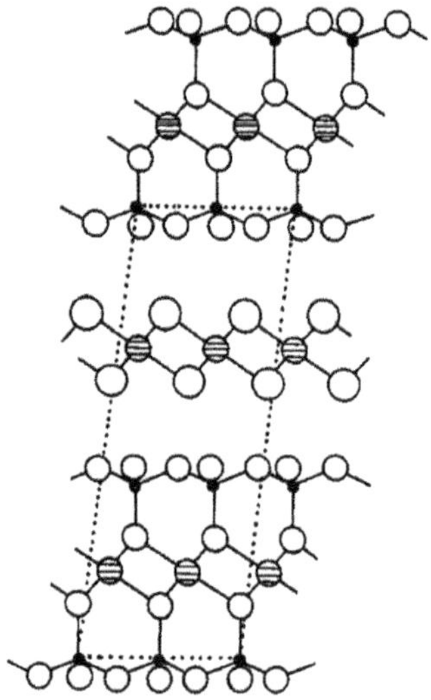

FIGURE 1.1 Schematic illustration of the structure of a crystalline vermiculite. Within the clay plates, the open circles represent oxygen, the closed circles silicon, and the shaded circles magnesium. The oxygen-oxygen separation of the surface layers is about 7 Å. In the interlayer region, the open circles represent water molecules and the shaded circles univalent cations. The dotted lines show the unit cell.

exhibit the property of macroscopic swelling. A schematic illustration of the structure of a crystalline vermiculite is shown in Figure 1.1.

Let us consider the case $x = 1.3$, corresponding to the Eucatex mineral. When the raw crystals are mined, they are obtained in a dry form, with a variety of interlayer cations, mainly sodium, potassium and calcium. To obtain sodium vermiculite from these, the crystals are soaked in hot (50°C) molar sodium chloride, with many exchanges of solution. Sodium ions and water molecules gradually force their way in between the layers, displacing the interlayer cations without altering the layer composition. The progress of the exchange can be monitored by X-ray diffraction, the repeat distance along the c-axis (the *d*-value) of the pure sodium form being 14.9 Å. The process is a slow one because the crystals have dimensions on the order of a centimeter in the a,b-plane and diffusion into the middle is slow, but pure sodium vermiculite patterns are obtained after about a year [4].

The sodium vermiculite obtained from the cation exchange is in a wet form, with approximately six water molecules per sodium ion and a c-axis spacing of approximately 15 Å. The complete structure of such a sodium vermiculite has been obtained [5] and is shown in Figure 1.2. There is nothing surprising in the crystallographic findings. The sodium ions are each surrounded octahedrally by six water molecules, and water molecules are found adjacent to the surface oxygen atoms of the negatively charged clay layers. It is easy to dry the mineral out by heating, and the dehydrated form has a c-axis *d*-value of approximately 12 Å. We can gain a feeling for the thickness of the clay plates themselves by noting that the layer

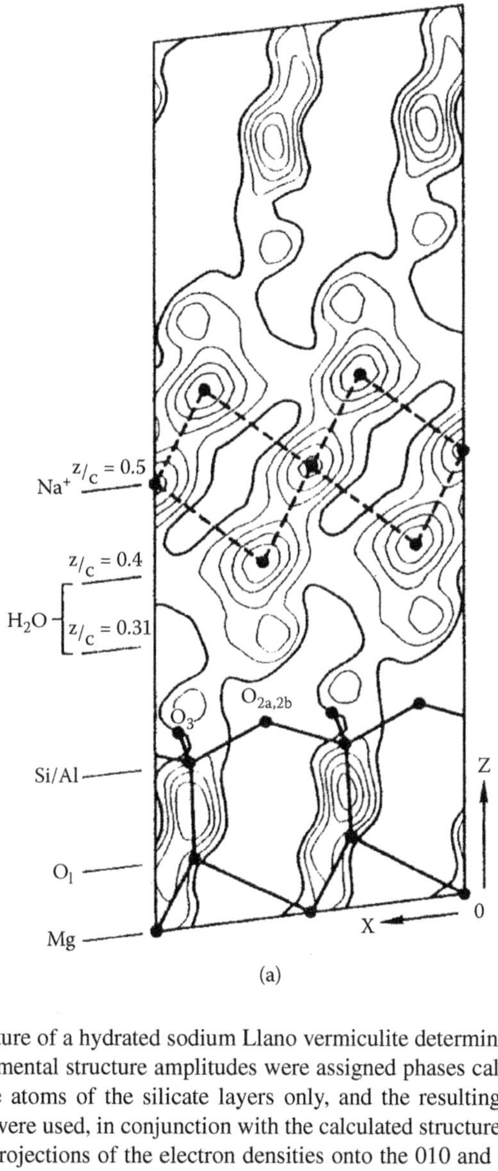

(a)

FIGURE 1.2 Structure of a hydrated sodium Llano vermiculite determined by X-ray diffraction [5]. The experimental structure amplitudes were assigned phases calculated on the basis of scattering by the atoms of the silicate layers only, and the resulting observed structure factors (Fo values) were used, in conjunction with the calculated structure factors (Fc values), to compute Fo-Fc projections of the electron densities onto the 010 and the 100 faces of the unit cell, shown in the parts (a) and (b), respectively. That the interlayer cations and water molecules are in octahedral coordination accords with these Fourier projections. (Reproduced with kind permission of the Clay Minerals Society, from Slade, P.G., Stone, P.A., and Radoslovich, E.W., *Clays Clay Min.*, 33, 51, 1985.)

thickness in talc is approximately 10 Å. The dry minerals are not considered further here, save to note that the unit cell weight of the structural formula given above for the dry sodium Eucatex vermiculite is 807. Therefore, 807 g of this clay contains 1.29 cations, which corresponds to a cation-exchange capacity (cec) of 160 meq/100 g

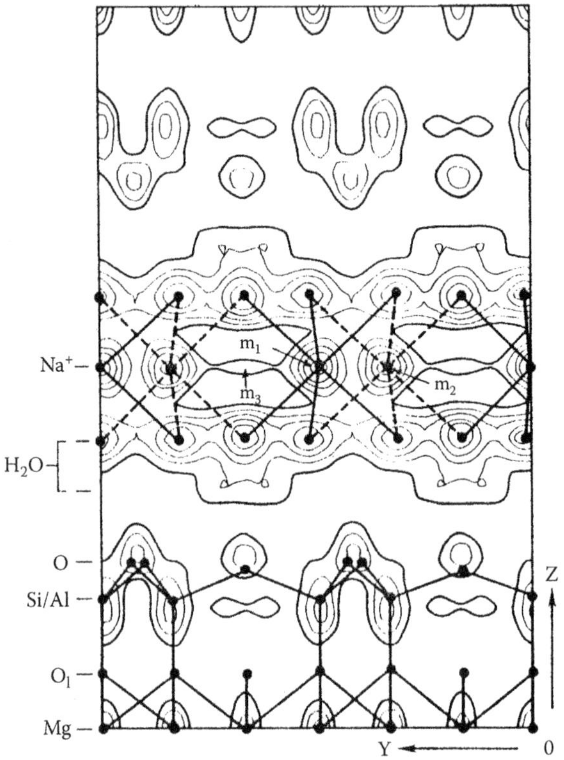

(b)

FIGURE 1.2 (Continued).

for the mineral, where meq are milliequivalents; the units are common in clay science.

The wet sodium vermiculite (d = 15 Å) does not swell further along the c-axis. Swelling minerals can be obtained by further substitution of the sodium ions with an interesting set of other cations, namely lithium [6], n-butylammonium [7], n-propylammonium [8], ornithine [9], lysine and maybe others. The so-called osmotic swelling of the n-alkylammonium vermiculites was first reported in the early 1960s. The paper by Garrett and Walker [8] is a fine example of descriptive science. These authors investigated the series of alkylammonium ions $C_nH_{2n+1}NH_3^+$ for the interlayer univalent cation M^+, for values of n from 1 to 8. Such synthetic crystals are easily obtained from the wet sodium vermiculite form by soaking in hot (50°C) molar alkylammonium chloride solutions, with regular changes of solution. When this process was performed for the Eucatex crystals by Humes [4], he used atomic absorption spectroscopy to measure the amount of sodium ions displaced into the external solution by each exchange, and thereby determined the cec to be 160 meq/100 g by standard chemical analysis, corresponding to 1.3 cations per unit cell. Garrett and Walker [8] investigated four vermiculites in the range from x = 1.2 to x = 1.6. For the Kenya

FIGURE 1.3 Butylammonium vermiculite (Kenya) crystals before and after swelling in water (lateral dimensions of the crystals approximately 2.5 × 2.5 mm). (Reproduced with kind permission of the Clay Minerals Society, from Garrett, W.G. and Walker, G.F., *Clays Clay Min.*, 9, 557, 1962.)

vermiculite, with $x = 1.3$, they found the remarkable phenomenon illustrated in Figure 1.3 when $n = 4$, namely for n-butylammonium vermiculite. The swelling also developed to a lesser extent in the n-propylammonium series. The reason why such macroscopic swelling only occurs for $n = 3$ and $n = 4$ remains a mystery.

In principle, it would be desirable to study the swelling for many different vermiculites and many different cations. However, as we shall see, the study of even one swelling system becomes a complicated many-variable problem. The swelling seems to be most pronounced and homogeneous for vermiculites with $x = 1.3$ and $M = C_4H_9NH_3$, so we choose to investigate this ideal clay colloid system as rigorously as possible and focus our studies on the Eucatex minerals in hand.

We started out by studying the macroscopic swelling with a traveling microscope. Garrett and Walker [8] had reported that single crystals swell anisotropically, perpendicular to the plane of the silicate layers, leading to increases in volume of as much as 4000% (see Figure 1.3). The crystals seemed to cleave naturally into slivers about 0.1 to 1 mm thick, easily measurable with a micrometer screw gauge. Soaking crystals about 5 × 5 mm in cross section in large excesses of dilute solutions of n-butylammonium chloride produced semitransparent yellow gel stacks with clearly recognizable outlines. The heights of these stacks, between a millimeter and several centimeters, could easily be read using a traveling microscope, with the samples mounted in a constant-temperature bath. By dividing the gel height by the crystal thickness, the macroscopic swellings for salt concentrations of 0.1, 0.01 and 0.001 M were found to be about 6-, 16- and 32-fold, respectively as illustrated schematically in Fig. 1.4. These findings were broadly in agreement with the preliminary results reported a quarter of a century previously [7, 8].

Humes [4] had already studied the n-butylammonium Eucatex crystals by X-ray diffraction. At relative humidities (RH) somewhat less than 100%, the n-butylammonium vermiculite exists in two stable crystalline phases having c-axis spacings of about

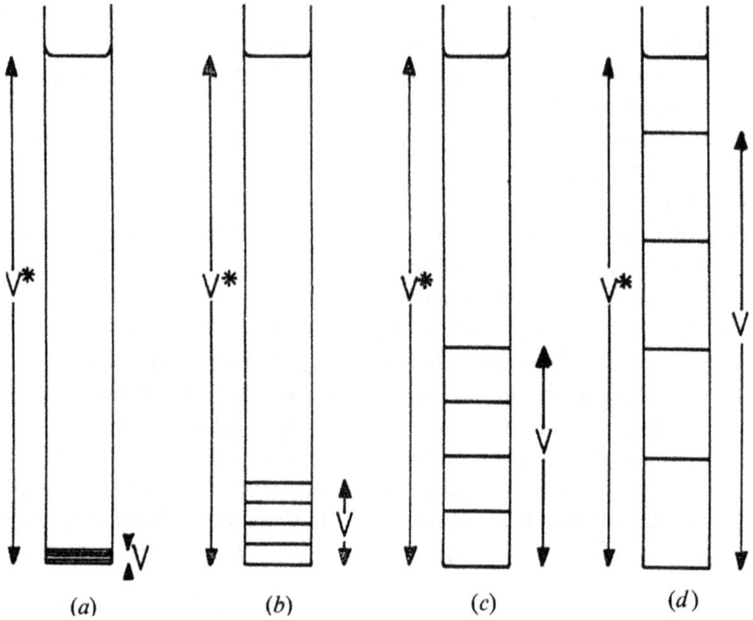

FIGURE 1.4 Schematic illustration of the swelling of n-butylammonium vermiculite. (a) shows the unexpanded in a 1.0 M butylammonium chloride solution. (b), (c) and (d) show the gels formed in 0.1 M, 0.01 M and 0.001 M solutions, respectively. Note how the extent of swelling is suppressed by an increase in the salt concentration. V represents the volume occupied by the clay, with V* the volume of the whole condensed matter system (clay plus excess soaking solution.)

15 Å and 19 Å and a number of poorly characterized, metastable intermediate phases. The metastable intermediate phases correspond to different states of hydration of the counterion or surface of the clay plate, and changes between them are brought about by changes in RH. For a sample soaking in 1 M n-butylammonium chloride, the equilibrium structure is a phase with $d = 19.4$ Å. Again, this is in broad agreement with the results of Garrett and Walker [8], who found spacings between 19.0 and 19.2 Å for the four n-butylammonium vermiculites they studied.

It is of prior importance to determine whether or not the swelling is homogeneous. It is one thing to say that a crystal with $d \approx 20$ Å expands sixfold along the c-axis in an 0.1 M solution. It is quite another to say that the c-axis spacing in the swollen gel phase is 120 Å under these conditions. There could, for example, be unexpanded crystalline regions separated by huge aqueous regions or gels with a wide range of spacings. This question had also been decided, at least for spacings up to about 200 Å, by X-ray experiments in the 1960s [10, 11]. Rausell-Colom's paper [11] on a small-angle X-ray diffraction study of the swelling of butylammonium vermiculite is particularly clear. It is obvious that when the swelling occurs the crystalline phase disappears completely, and the gel has a well-defined interlayer spacing whose value corresponds well with the extent of macroscopic swelling at the given salt concentration. We studied the swelling by neutron diffraction. Because of the longer wavelengths of cold neutrons, the use of small-angle neutron scattering (SANS) permits the measurement of larger spacings than small-angle X-ray scattering (SAXS). We were therefore able to extend the results of the early X-ray studies to a much wider range of salt concentrations.

Most of our earlier diffraction experiments [12, 13] were carried out on the D16 long-wavelength neutron diffractometer at the Institut Laue-Langevin (ILL), Grenoble. The incident wavelength was 4.52 Å with a wavelength spread $\Delta\lambda/\lambda$ of about 2%, which is suitable for studying c-axis spacings in the range 2 to 350 Å. For larger spacings, the D17 and D11 small-angle diffractometers at ILL were used. Spacings up to about 650 Å were measured on the D17 diffractometer, which used incident wavelengths in the range 10 to 30 Å. The D11 instrument offers wavelengths between 2 and 20 Å and is equipped with a multidetector and collimators that can be positioned at distances of 2, 5, 20 and 40 m from the sample, allowing measurements of c-axis spacings up to 2000 Å. A fuller description of the three instruments is given in reference [14].

The samples for the diffraction experiments were prepared as follows. A crystal of the n-butylammonium vermiculite was cut to a thickness of about 0.5 mm and an area of about 5 × 5 mm and soaked in the appropriate solution of protonated n-butylammonium chloride in D_2O. D_2O was used because it gave a lower incoherent-scattering background than H_2O. The gel was allowed to come to its equilibrium swelling distance for 48 hours in a cold room at a temperature of 7°C. A slice of this swollen gel about 1 mm thick was then transferred to a quartz cell of internal dimensions 30 × 5 × 2 mm. The remainder of the quartz cell was filled with some of the original solution. The cell was sealed with Parafilm to prevent loss of solution by evaporation, clamped into an aluminum block, as shown in Figure 1.5, and

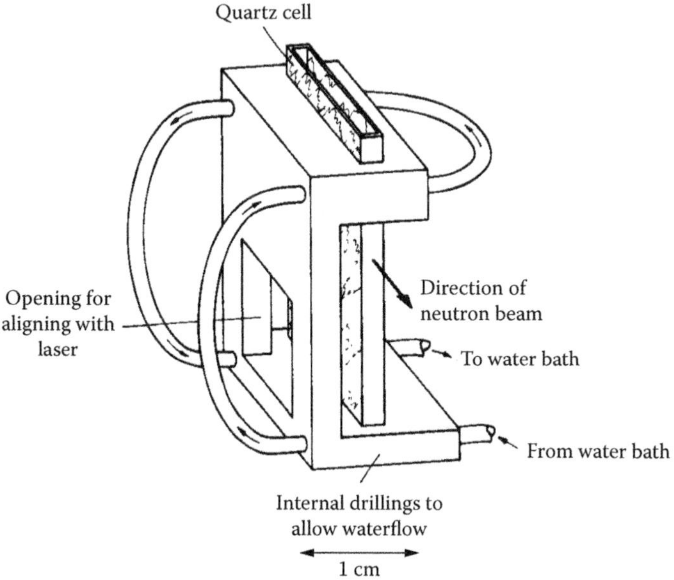

FIGURE 1.5 Arrangement for controlling the temperature of the quartz sample cell in the D16 experiments.

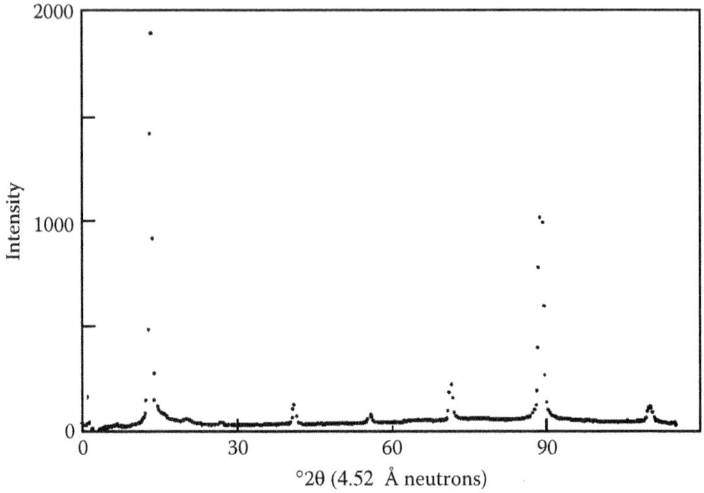

FIGURE 1.6 Neutron diffraction pattern of the fully hydrated crystalline phase of n-butylammonium vermiculite measured in reflection geometry; $\lambda = 4.52$ Å.

allowed to equilibrate at the required temperature. The temperature of the block was controlled by circulating water from a bath whose temperature was maintained at ±0.2°C. The aluminum block was mounted on a standard goniometer, and the gel was aligned accurately in the neutron beam by means of a laser.

The diffraction pattern of the crystalline 19.4 Å phase, for which the first-order c-axis peak is at a scattering angle 2θ of 13.4° for an incident wavelength of 4.52 Å, is shown in Figure 1.6. The range of the D16 instrument was sufficient to show the first seven orders of the $00l$ reflection. Only this series of peaks was observed, because the experiment was done in reflection with the momentum transfer perpendicular to the plane of the crystal, and because the mosaic spread of the crystal was small. For the sample whose pattern is shown in Figure 1.6, the mosaic spread was determined by a rocking curve. The rocking curve was measured by rotating the sample about its vertical axis (angle ω), while 2θ was fixed at the peak of the 001 reflection. The angular width of the rocking curve is an accurate measure of the spread in orientation of the crystallites making up the crystal. Here, the full width at half height of the rocking curve, shown in Figure 1.7, was found to be 5.3° in ω. This value is therefore the mean spread in orientation of the platelets in the crystalline vermiculite and was typical of all the crystals examined.

Upon swelling of the vermiculite, the diffraction pattern changed completely, with the peaks associated with the crystalline phase disappearing and new, broader peaks appearing at low angles. The pattern of the 0.1 M sample at 6°C (Figure 1.8) shows an intense first-order peak corresponding to a spacing of 122 Å, a weaker second-order peak having the same spacing and a very weak third-order peak, not clearly evident on the scale of Figure 1.8. The pattern is shown with a 2θ scale up to 20° to demonstrate the absence of the very intense first-order peak of the crystalline phase. The scale of Figure 1.8 is considerably expanded compared with that

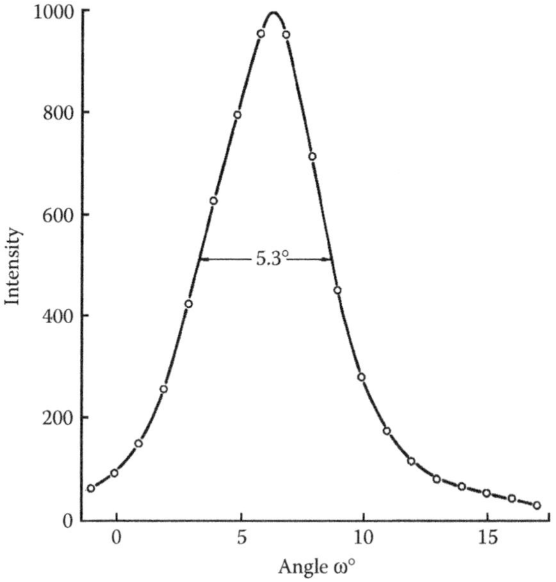

FIGURE 1.7 Rocking curve on the 001 peak of a fully hydrated crystalline n-butylammonium vermiculite; $d = 19.4$ Å.

of Figure 1.6, the low-angle pattern of the gel phase being much weaker than the high-angle pattern of the crystalline phase.

The microscopic expansion from 19.4 Å to 122 Å matches the sixfold macroscopic expansion almost exactly, confirming the assertions of other authors [7, 8, 10, 11, 15] that the expansion in this material is perfectly homogeneous.

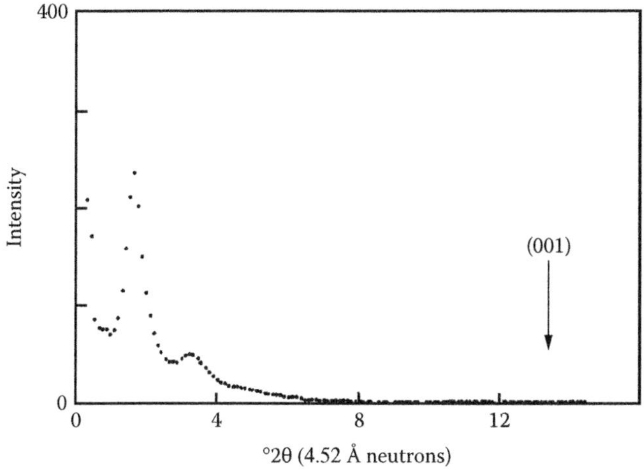

FIGURE 1.8 Low-angle diffraction pattern of n-butylammonium vermiculite at $T = 6$°C and an external salt concentration of $c = 0.1$ M.

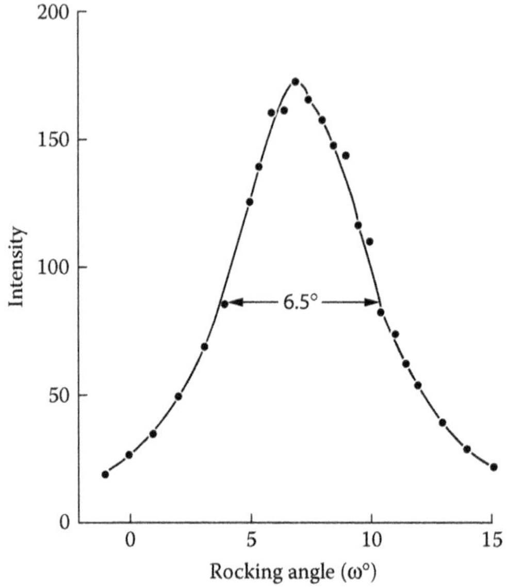

FIGURE 1.9 Rocking curve on the 001 peak of the swollen gel phase of n-butylammonium vermiculite at T = 6°C and an external salt concentration of c = 0.1 M; d = 122 Å.

What is perhaps also remarkable is that the orientation of the platelets was largely retained in the expansion, as demonstrated by the rocking curve on the first-order peak of the low-angle pattern, shown in Figure 1.9. The full width at half height was 6.5°, scarcely different from that of the crystalline material. It is clear that the clay plates remain parallel to a high degree in the swollen gels, and the spacings are sufficiently large that the system behaves as a one-dimensional colloid. This makes the system ideal for testing theories of colloid stability. Obviously, the first test of any theory is to explain the interlayer spacings as a function of salt concentration. Garrett and Walker's results [8], obtained via the macroscopic swelling, are reproduced in Figure 1.10. It is clear that d is inversely proportional to the square root of the salt concentration c. We obtained similar results from our microscopic SANS experiments.

At an external concentration of $c = 10^{-2}$ M, the macroscopic swelling indicated a d-value of about 330 Å. The low-angle diffraction pattern obtained on the D17 instrument using an incident wavelength of 17 Å (Figure 1.11) had first- and second-order peaks also corresponding to a spacing of 330 Å. This spacing was also measured on the D16 instrument. Some arguments had been raised about the possible contribution of specular reflection from either the clay or the cell to such diffraction patterns [16]. At these low angles, specular reflection may be confused with diffraction peaks; however, we were able to use the mosaic spread of the crystals to rotate the sample until contributions from specular reflection, which were indeed observed, vanished. Figure 1.11 corresponds to just such an orientation. The neutron experiment, under these conditions, had much higher resolution than a laboratory-based X-ray diffraction instrument, and no difficulties were encountered due to experimental

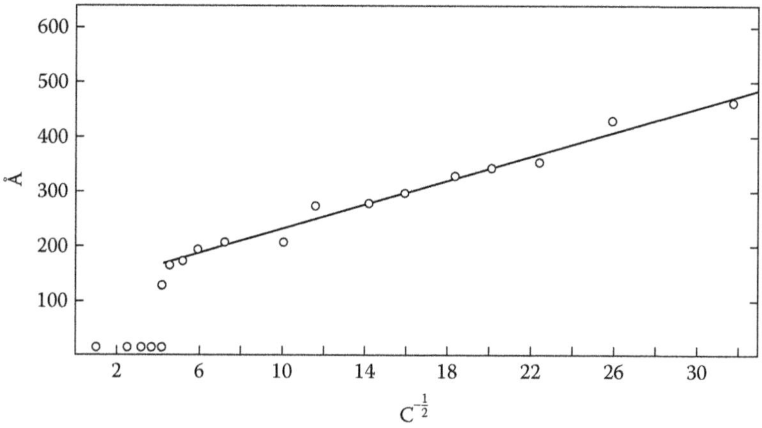

FIGURE 1.10 Relationship between interlayer distance of butylammonium vermiculite and concentration of butylammonium chloride in external solution (*c* expressed in units of M). (Reproduced with kind permission of the Clay Minerals Society, from Garrett, W.G. and Walker, G.F., *Clays Clay Min.*, 9, 557, 1962.)

artifacts in this range of *d*-values. Similar experiments at 10^{-3} M gave a *d*-value of 620 Å. This c-axis spacing was also reproduced on the D11 instrument, which allowed spacings as large as 900 Å to be observed for a 5×10^{-4} M solution.

The complete range of spacings studied is shown in Figure 1.12, together with the prediction from the established theory of colloid stability, the DLVO (Derjaguin-Landau-Verwey-Overbeek) theory. The situation for DLVO theory is even worse than it looks from the figure, as an adjustable parameter has been used to fit the data in the middle of the concentration range, at c = 0.01 M. The whole shape of the curve is qualitatively wrong.

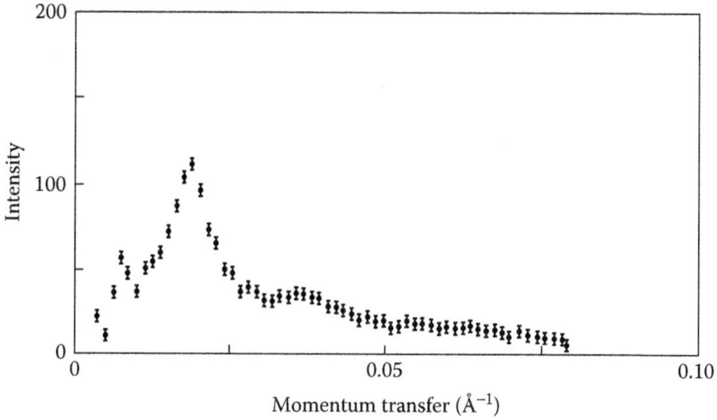

FIGURE 1.11 Low-angle diffraction pattern of the swollen gel phase of n-butylammonium vermiculite at T = 6°C and an external salt concentration of c = 0.01 M; λ = 17 Å.

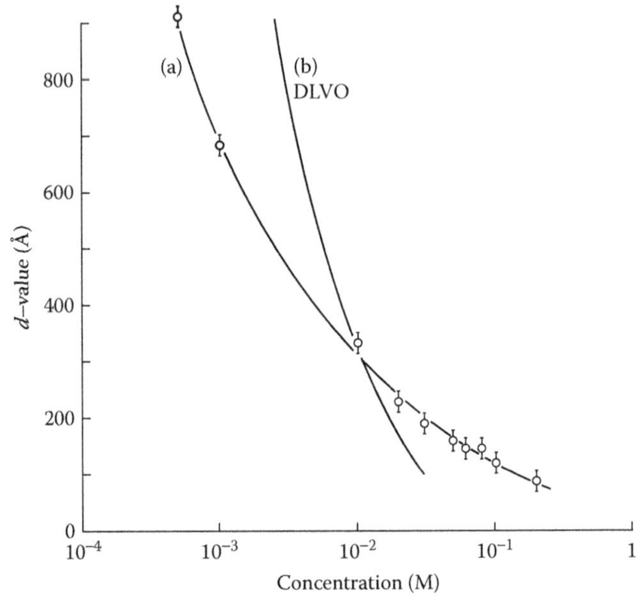

FIGURE 1.12 Comparison of (a) experimentally observed and (b) calculated *d*-values using DLVO theory, with an adjustable parameter (proportional to the Hamaker constant) chosen to fit the data at $c = 0.01$ M

When there is a huge, qualitative difference between experiment and theory, it is always easy to explain the difficulty away by saying that there is something anomalous with the experimental system. I was obviously therefore concerned that somehow something was "odd" about the n-butylammonium vermiculite system. Is the behavior $d \propto c^{-1/2}$ an aberration in an obscure branch of clay science, or is it an example of a more general phenomenon? The answer is clearly stated, though in a slightly more subtle form, in the standard book on colloid science by Hunter [17]. The statement reads, "The secondary minimum that occurs at larger distances ($\approx 7\kappa^{-1}$) is responsible for a number of important effects in colloidal suspensions." It refers to the position of the secondary minimum as being at a constant number of Debye screening lengths, where κ^{-1} is the said Debye length. In other words, the long-range interaction between charged colloidal particles leads to a minimum in the effective pair potential that tends to localize the particles at a constant number of Debye lengths. To anyone who knows the Debye-Hückel theory of simple electrolyte solutions, the following relationship is acceptable without further ado.

$$\kappa^2 \propto \frac{I}{\varepsilon T} \tag{1.1}$$

where κ is the inverse Debye length, I is the ionic strength, ε is the dielectric constant of the medium and T is the temperature. For water at a constant temperature, this relationship reduces to $\kappa^2 \propto I$, and the ionic strength is proportional to the salt concentration, so $\kappa^2 \propto c$. The Debye length is thus inversely proportional to the square root of c, as $\kappa^{-1} \propto c^{-1/2}$. So, $d \propto c^{-1/2}$ and $d \propto \kappa^{-1}$ (Hunter [17] quotes $7\kappa^{-1}$ as the typical separation) are the same relationship. The n-butylammonium vermiculite gels behave as a *typical* colloidal system in exhibiting a secondary minimum at a few Debye lengths. The reason for the discrepancy lies not in the model experimental system, but in a fundamental flaw in DLVO theory.

How was the theoretical DLVO curve in Figure 1.12 obtained? The DLVO model [18, 19] postulates that the appropriate thermodynamic potential energy of interaction between two parallel flat plates can be described in terms of two components: a repulsive term V_R, resulting from the overlap of electrical double layers, and an attractive van der Waals interaction, V_A. It also assumes that these interactions are additive, so that the total potential energy can be written as

$$V = V_R + V_A \tag{1.2}$$

Both the Boer-Hamaker treatment [20, 21] and the macroscopic approach of Lifschitz [22] yield an inverse cubic decay of the attractive van der Waals force,

$$\frac{dV_A}{dx} = \frac{A'}{x^3} \tag{1.3}$$

for small separations x. At large separations, retardation reduces the attractive force, resulting in Equation 1.4

$$\frac{dV_A}{dx} = \frac{A}{x^4} \tag{1.4}$$

The force of attraction between the two plates is now inversely proportional to the fourth power of the separation. Because the present investigation deals with separations between 100 and 1000 Å, Equation 1.4 is applicable. A is considered to be an arbitrary constant, because it cannot be calculated *a priori*. The double-layer repulsion, V_R, at large distances is given by Equation 1.5

$$V_R \propto \frac{1}{\kappa}\exp(-\kappa x) \tag{1.5}$$

where κ is the inverse Debye screening length given by

$$\kappa = (8\pi\lambda_B n_0)^{1/2} \tag{1.6}$$

where n_0 is the number density of a uni-univalent (n-butylammonium chloride in the model system) electrolyte solution and λ_B is the Bjerrum length given by

$$\lambda_B = \frac{e^2}{\varepsilon kT} \tag{1.7}$$

where e is the charge of the electron, k is the Boltzmann constant, and the other symbols are as defined for Equation 1.1. For water at room temperature, $\lambda_B \approx 7$ Å, and for monovalent ions in aqueous solution at 25°C

$$\kappa^2 = 0.107c \tag{1.8}$$

where κ has the units of Å^{-1} and c has units of moles per liter (M).

From the sharpness of the phase transition described below, the observed d-values should clearly be treated as equilibrium separations of the plates; thus the repulsive and attractive forces given by Equation 1.5 and Equation 1.4 must balance. Writing

$$\frac{dV_R}{dx} = B\exp(-\kappa x) \tag{1.9}$$

where B is a constant determined from electrostatic theory, we have

$$B\exp(-\kappa x_{eq}) = \frac{A}{x_{eq}^4} \tag{1.10}$$

where x_{eq} is the equilibrium separation. Equating this with d and writing $A/B = C$ gives

$$d^4 \exp(-\kappa d) = C \tag{1.11}$$

Theory and experiment can now be compared by choosing C such that $d = 320$ Å for $\kappa = 0.032$ Å^{-1}, that is, for a salt concentration of 10^{-2} M, so that the results agree in the middle of the concentration range studied. This gives $C = 4.26 \times 10^5$ Å^4; the curve

$$d^4 \exp(-\kappa d) = 4.26 \times 10^5 \text{ Å}^4 \tag{1.12}$$

is plotted together with the experimental curve in Figure 1.12. This diagram shows that the DLVO theory does not account for the variation of d with c. The same conclusion is reached for both retarded and nonretarded van der Waals attractions.

In DLVO theory, the secondary minimum can only be created by the van der Waals force, which is essentially independent of the salt concentration across the concentration range 0.001 M $< c <$ 0.1 M. This force has to be balanced with a force that decays exponentially as a function of κ, which means that it decays by a factor exp(−10) across this range. The unhappy consequence of this prediction is that the position of the secondary minimum, and therefore the interlayer d value, varies very rapidly as a function of κ, in contradiction to the experimental results. A further unhappy consequence of this balance is that it always produces a primary minimum much deeper than the secondary minimum. The full, standard DLVO thermodynamic potential energy curve, which also includes a very-short-range Born repulsion, is shown in Figure 1.13 [23]. It is therefore a definite prediction of DLVO theory that charge-stabilized colloids can only be kinetically, as opposed to thermodynamically, stable. The theory does not mean anything at all if we cannot identify the crystalline

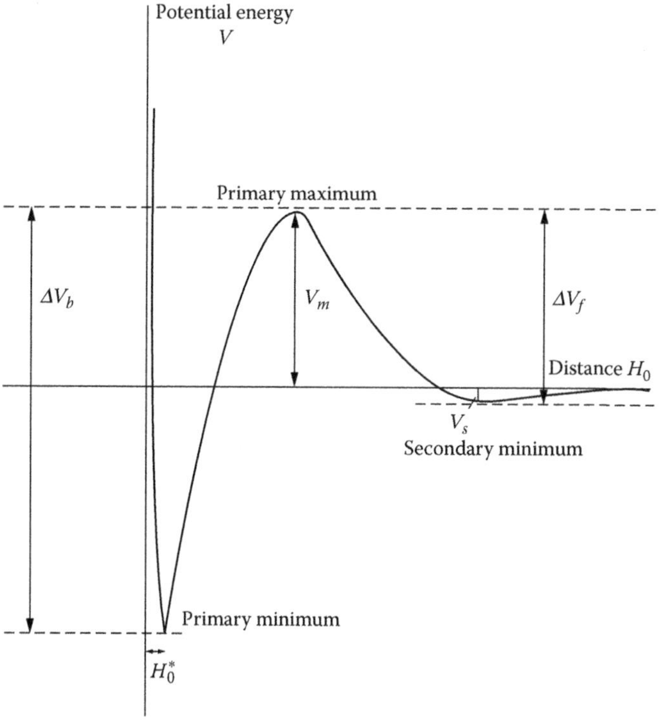

FIGURE 1.13 General form of the curve of potential energy against distance of surface separation for interaction between two particles, according to DLVO theory. (Taken from Barclay, L., Harrington, A., and Ottewill, R.H., *Kolloid-Z. u. Z. Polymere*, 250, 655, 1972. With permission.)

clay state ($d \approx 10$ Å) with the primary minimum and the clay gel state ($d \approx 100$ to 1000 Å) with the secondary minimum in a well-defined model experimental system. We were therefore amazed to discover a reversible phase transition of clear thermodynamic character in the n-butylammonium vermiculite system, both with respect to temperature T and pressure P. These results rock the foundations of colloid science to their roots, and they are now described in detail.

First, at atmospheric pressure, heating the osmotically swollen samples at a well-defined temperature brought about a transition to the crystalline phase. For example, for the sample in an external solution concentration of 0.1 M, no change was detected in the diffraction pattern upon heating to 13.5°C, but at 14.5°C the low-angle pattern shown in Figure 1.8 had collapsed into a broad envelope of scattering, and the pattern of the crystalline phase reappeared. Further scans at 20, 25 and 30°C were identical to that at 14.5°C, showing that the phase change was complete in the interval between 13.5°C and 14.5°C. The transition was shown to be reversible by cycling the temperature between 10 and 20°C. The d-value was always found to be in the range 120 to 125Å at <13.5°C, and the full diffraction pattern of the crystalline phase was always recovered at >14.5°C. There is clearly a reversible phase transition at T_c = 14°C for c = 0.1 M.

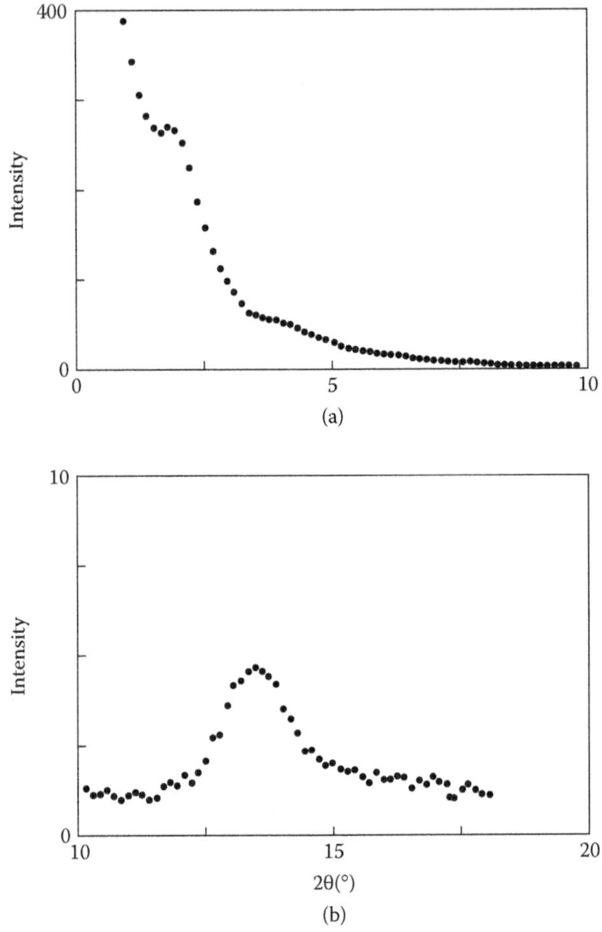

FIGURE 1.14 Neutron diffraction patterns of n-butylammonium vermiculite close to the crystalline-gel phase transition at $T = 14°C$ and an external salt concentration of $c = 0.1$ M: (a) low-angle region, (b) wide-angle region

At temperatures close to the phase transition, some variation of the c-axis spacing and a small coexistence region of the gel and crystalline phases were found. Figure 1.14 shows the pattern obtained at 14.1°C. The low-angle diffraction pattern (Figure 1.14a) corresponds to a spacing of 140 Å, and the presence of the crystalline phase is shown by the presence of the first-order peak at $2\theta = 13.4°$ (Figure 1.14b). The intensity of this peak, however, was about one-quarter of the value observed at higher temperatures. Close to the center of the temperature range of the phase transition (14°C), the vermiculite took on a silvery sheen. The sheen always appeared at the phase transition and could therefore be used as a means of following the phase transition experimentally.

For the sample at $c = 0.01$ M, the temperature was changed in approximately 2°C steps between 22 and 34°C. The scattering patterns obtained on D17 are shown in

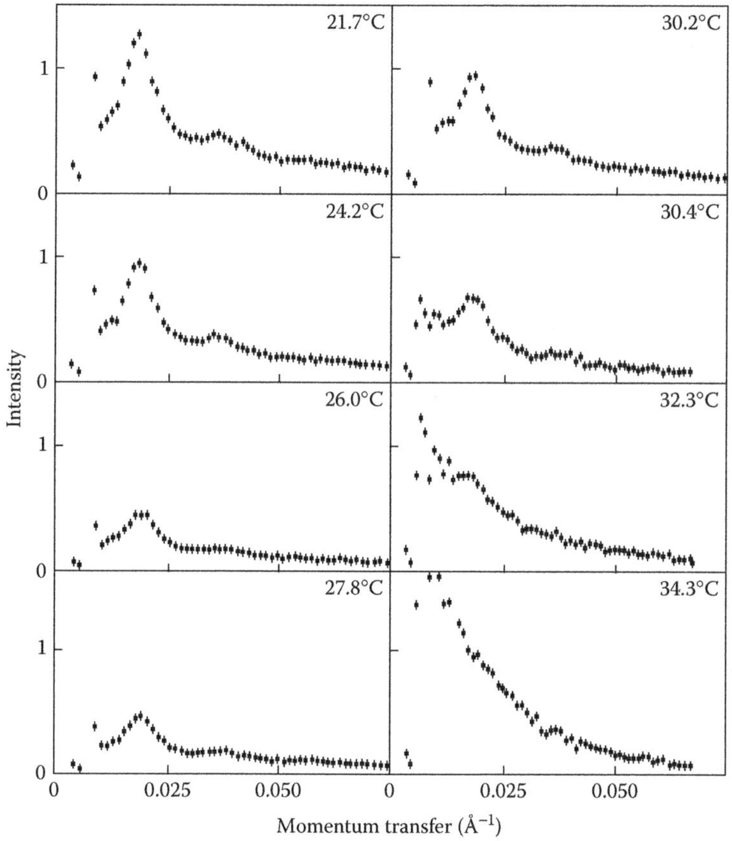

FIGURE 1.15 Temperature-induced phase transition between the gel and crystalline phases of n-butylammonium vermiculite at an external salt concentration of $c = 0.01$ M; $\lambda = 17$ Å.

Figure 1.15. The intensity of the first-order peak varied with temperature, but its position was unchanged up to a temperature of 30°C. At 32°C it appeared at a slightly lower angle, just as was observed for the 0.1 M sample at 14°C, and at 34°C it was obliterated by a broad envelope of small-angle scattering. Parallel experiments on the D16 instrument showed that the pattern of the crystalline phase was just visible at 33.5°C, but not at 32.6°C. At 35°C the sample was fully crystalline. There is a reversible phase transition at $T_c = 33$°C for $c = 0.01$ M. Similar experiments at $c = 0.001$ M gave a transition temperature of 45°C, verified also by both the change in the low-angle pattern and the appearance of the crystalline diffraction pattern using the D16 and D17 instruments. The phase diagram for the transition is shown in Figure 1.16.

Having determined the position of the phase boundary by neutron scattering experiments, we performed direct calorimetric measurements on the system. The calorimeter and its operation are described in reference [24]. A known weight of n-butylammonium vermiculite (about 1.5 g) was sealed in a calorimetric sample cell together with a known weight (about 3 g) of the appropriate n-butylammonium

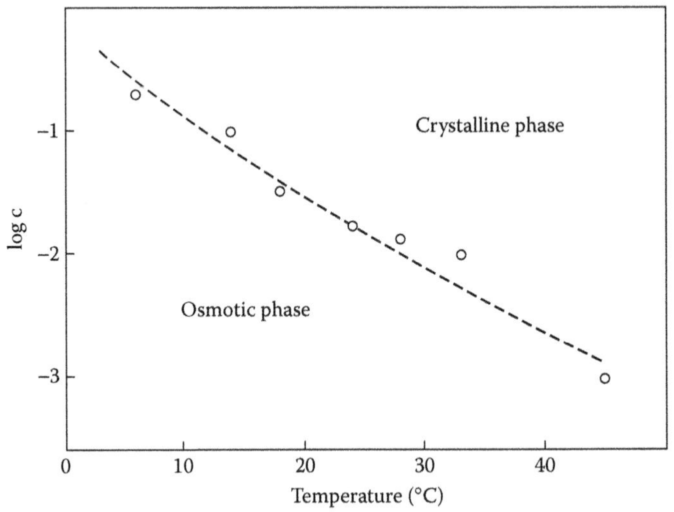

FIGURE 1.16 Temperature-salt concentration (*T-c*) phase diagram of n-butylammonium vermiculite.

chloride solution. The remaining space (about 0.5 cm^3) was filled with air at atmospheric pressure. The experiment thus effectively gave the heat capacity at constant pressure. The first results [12], at an effective salt concentration slightly greater than 0.1 M, are shown in Figure 1.17. The heat capacity across the swelling transition

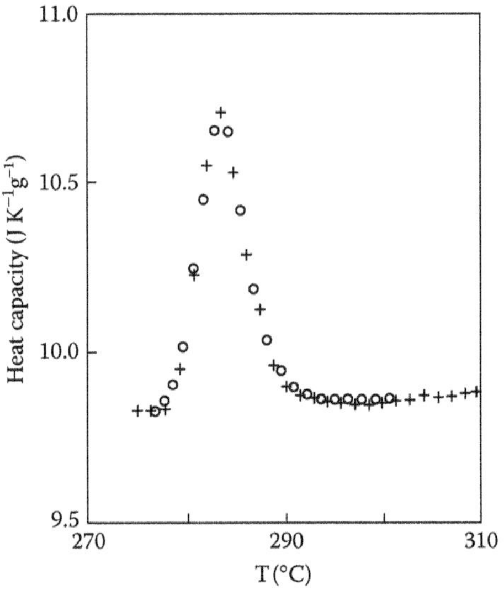

FIGURE 1.17 Heat capacity across the swelling transition of n-butylammonium vermiculite (per gram of crystalline material) in a 0.1 M solution of n-butylammonium chloride.

is given per gram of dry n-butylammonium vermiculite in Figure 1.17. Points are shown for two series of measurements, and the good agreement between them confirms the reproducibility of the swelling transition. A peak in the heat capacity curve was noted at 10.1°C, consistent with neutron scattering results giving T_c = 14°C for c = 0.1 M and T_c = 6°C for c = 0.2 M [13].

The enthalpy, ΔH, and entropy, ΔS, of the transition were calculated from the heat capacity curve by integration of the excess heat capacity, ΔC_p, and $\Delta C_p/T$, respectively. The normal heat capacities for the calculation were determined by interpolation of the heat capacities outside the transition region using the linear function

$$\Delta C_p/\mathrm{JK^{-1}g^{-1}} = 9.438 + 0.00140T/\mathrm{K}$$

The numerical coefficients were determined by the method of least squares. The following results were obtained:

$$\Delta H = 5.2 \pm 0.1\ \mathrm{Jg^{-1}}$$

$$\Delta S = 0.0183 \pm 0.0004\ \mathrm{JK^{-1}g^{-1}}$$

We determined further thermodynamic parameters for the reversible phase transition between the crystalline (high temperature) and gel (low temperature) phases by studying the effect of hydrostatic pressure on the phase transition at c = 0.1 M. A swollen gel was loaded into a quartz cell as described above, and on this occasion the quartz cell was contained in a Zircal pressure cell designed for pressures up to 2 kbar [25].

At ambient pressure, at an effective salt concentration slightly greater than 0.1 M and at 9°C, the sample was in the swollen gel state. At this temperature, three orders of the 00l reflection were observed corresponding to a d-value of 120 Å. Upon warming the sample to 11°C, the gel underwent the phase transition to the crystalline phase having a d-value of 19.4 Å. This transition was monitored, as before, by the growth of the first-order reflection of the crystalline phase at a value of the momentum transfer Q of 0.325 Å^{-1}. The momentum transfer is defined as

$$Q = \frac{4\pi \sin\theta}{\lambda} \tag{1.13}$$

where 2θ is the scattering angle and λ is the incident wavelength of the neutrons.

The initial measurement of the effect of hydrostatic pressure was made at 20°C, 9°C above the transition temperature to the crystalline phase at atmospheric pressure. With reference to the behavior of the first-order peak of the crystal at Q = 0.325 Å^{-1}, no significant change in this peak was noted as the hydrostatic pressure was increased to 500 bar. Above this pressure, however, the intensity of the peak decreased continuously as the pressure was increased. The diffraction patterns over the range 500 to 1100 bar, shown in Figure 1.18, indicate that the crystalline phase disappeared at about 1050 bar. The integrated intensities of the peak after subtraction of a flat background are given in Table 1.1. No further change was noted in this region of the diffraction pattern as the pressure was increased further to 2000 bar.

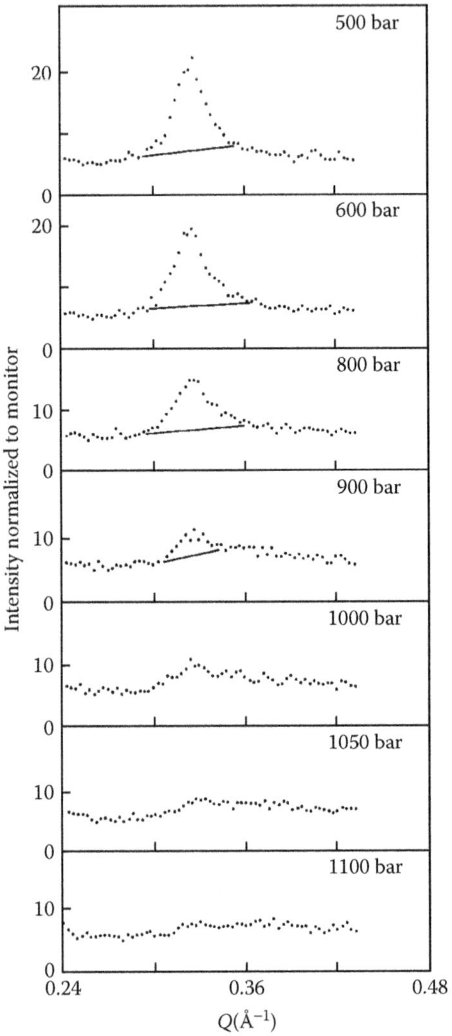

FIGURE 1.18 Effect of hydrostatic pressure on the 001 reflection of the crystalline phase of n-butylammonium vermiculite at $T = 20°C$, $c = 0.1$ M.

No change in the small-angle scattering pattern was detected as the pressure was increased to 800 bar. The patterns obtained between 800 and 1300 bar are shown in Figure 1.19. Only slight changes occurred between 800 and 1000 bar, but at 1050 bar the first-order reflection of the swollen gel phase appeared. At 1300 bar, first, second and third orders of this reflection were observed; the 1300-bar pattern is shown on an expanded scale in Figure 1.20. The integrated intensities, positions of the reflection, and the corresponding c-axis spacings are given in Table 1.2. As the pressure was increased further to 2000 bar, the c-axis spacing decreased by only about 1 Å per 300 bar, and the intensity of the first-order peak increased steadily over the whole range.

TABLE 1.1
Intensity of the 001 Reflection of the Crystalline Phase of n-Butylammonium Vermiculite in 0.1-M n-Butylammonium Chloride

Pressure (bar)	Intensity (arbitrary units)
500	146
600	141
800	121
900	115
1000	81
1050	26
1100	≈0

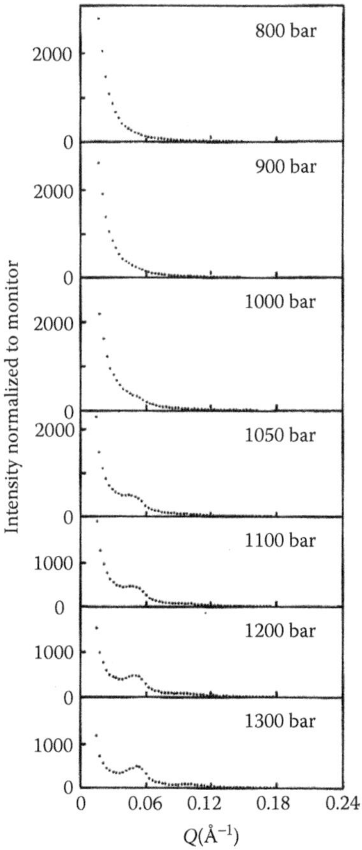

FIGURE 1.19 Effect of hydrostatic pressure on the low-angle diffraction pattern of n-butylammonium vermiculite at $T = 20°C$, $c = 0.1$ M.

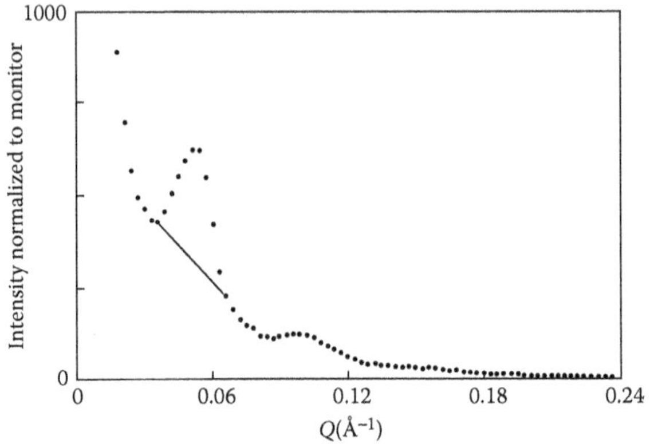

FIGURE 1.20 Small-angle scattering pattern of the gel phase of n-butylammonium vermiculite at P = 1300 bar, T = 20°C, c = 0.1 M.

Thus, at 20°C, no crystalline material was detected at pressures >1100 bar, and no gel phase was detected at pressures <1000 bar. A narrow coexistence region at about 1050 bar was found, which we define as the transition pressure at this temperature. The transition was found to be reversible as the pressure was cycled between 500 and 1500 bar, just as it had previously been found to be reversible with respect to changes of temperature. The rate of equilibration was surprisingly rapid, on the order of minutes. The whole pressure cycle was repeated at different temperatures, and nine transition pressures were obtained as a function of T. The resulting P-T phase diagram of the 0.1 M n-butylammonium vermiculite system is shown in Figure 1.21. This is an equilibrium phase diagram. The swollen gel phase is the thermodynamically stable low-temperature, high-pressure phase.

Because the application of hydrostatic pressure caused the vermiculite to swell to its gel phase, the total volume of the gel phase was less than that of the crystalline phase plus the appropriate amount of solution, even though the gel phase itself

TABLE 1.2
Intensity of the 001 Reflection from the Swollen Gel Phase of n-Butylammonium Vermiculite in 0.1M n-Butylammonium Chloride

Pressure (bar)	Intensity of 001 Reflection	Position of Intensity Maximum, Q_{max} (Å^{-1})	Layer Distance, d (Å)
1000	—	—	—
1050	501	0.0498	126
1100	665	0.0510	123
1200	889	0.0514	122
1300	1357	0.0516	122
1400	1542	0.0513	122
1600	1663	0.0521	121
2000	1788	0.0524	120

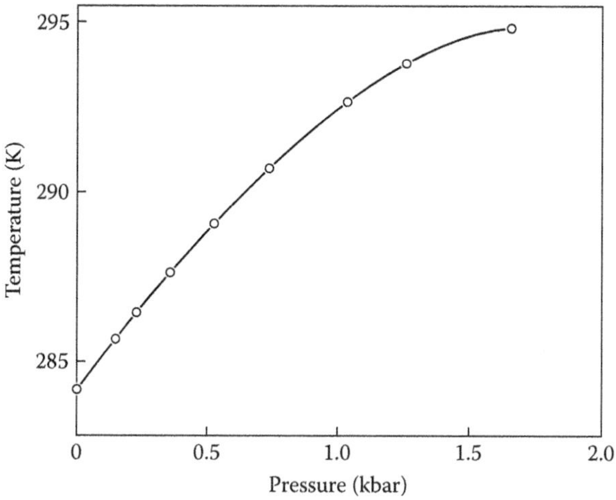

FIGURE 1.21 Phase diagram for swelling of n-butylammonium vermiculite in water as a function of temperature T and pressure P, at $c = 0.1$ M. The swollen gel phase lies in the lower half of the diagram; the gel phase is the high-density, low-temperature phase.

represents a sixfold expansion of the crystalline phase. We were able to determine the contraction from the gradient dP/dT along the phase boundary in Figure 1.21 by applying the standard thermodynamic equation

$$\frac{dP}{dT} = \frac{\Delta S}{\Delta V} \tag{1.14}$$

with the entropy $\Delta S = 0.0183$ JK^{-1}g^{-1} (g^{-1} means per gram of dry clay) already known from the calorimetric measurements. The slope dP/dT was approximately constant at low pressures and equal to 85 atm K^{-1}, or 8.5×10^6 J m^{-3} K^{-1}. The combination of the two values gave a volume change ΔV for the transition of -2.15 cm^3 per kilogram of dry clay. If this contraction is converted to a change in the water volume from bulk solution to between the clay plates, it is equivalent to a fractional decrease of slightly less than 0.1%. Every thermodynamic parameter for the transition has been determined.

You cannot have faith in experimental science and uphold the view that experiment is the sole judge of scientific truth without rejecting the DLVO theory on the basis of the experiments described in this chapter. However, it is much easier to say that an old theory is wrong than it is to say what the right theory is. Fortunately, at the time of our discoveries, Sogami and Ise had pointed the way to a correct understanding of these phenomena [26, 27].

REFERENCES

1. Hamilton, W.R., Woolley, A.R., and Bishop, A.C., *The Country Life Guide to Minerals, Rocks and Fossils,* 12th ed, Country Life Books, Middlesex, U.K., 1984.
2. van Olphen, H., *An Introduction to Clay Colloid Chemistry,* 2nd ed., Wiley, New York, 1977.

3. Smalley, M.V., Jinnai, H., Hashimoto, T., and Koizumi, S., *Clays Clay Min.,* 45, 745, 1997.
4. Humes, R.P., *Interparticle Forces in Clay Minerals,* D. Phil. thesis, Oxford University, Oxford, U.K., 1985.
5. Slade, P.G., Stone, P.A., and Radoslovich, E.W., *Clays Clay Min.,* 33, 51, 1985.
6. Walker, G.F. and Milne, A., *Trans. 4th Int. Congr. Soil Sci.,* 2, 62, 1950.
7. Walker, G.F., *Nature,* 187, 312, 1960.
8. Garrett, W.G. and Walker, G.F., *Clays Clay Min.,* 9, 557, 1962.
9. Rausell-Colom, J.A., Saez-Auñón, J., and Pons, C.H., *Clay Min.,* 24, 459, 1989.
10. Norrish, K. and Rausell-Colom, J.A., *Clays Clay Min.,* 10, 123, 1963.
11. Rausell-Colom, J.A., *Trans. Faraday Soc.,* 60, 190, 1964.
12. Smalley, M.V., Thomas, R.K., Braganza, L.F., and Matsuo, T., *Clays Clay Min.,* 37, 474, 1989.
13. Braganza, L.F., Crawford, R.J., Smalley, M.V., and Thomas, R.K., *Clays Clay Min.,* 38, 90, 1990.
14. Institut Laue-Langevin, Neutron Research Facilities at the High Flux Reactor, Institut Laue-Langevin, Grenoble, France, 1986.
15. Viani, B.E., Roth, C.B., and Low, P.F., *Clays Clay Min.,* 33, 244, 1985.
16. Simonton, T.C., *Clays Clay Min.,* 33, 472, 1985.
17. Hunter, R.J., *Foundations of Colloid Science,* Clarendon Press, Oxford, U.K., 1987.
18. Derjaguin, B.V. and Landau, L., *Acta Physicochimica,* 14, 633, 1941.
19. Verwey, E.J.W. and Overbeek, J.Th.G., *Theory of the Stability of Lyophobic Colloids,* Elsevier, Amsterdam, 1948.
20. de Boer, J.H., *Trans. Faraday Soc.,* 32, 10, 1936.
21. Hamaker, H.C., *Physica,* 4, 1058, 1937.
22. Lifschitz, E.M., *Zh. Eksp. Teor. Fiz.,* 29, 94, 1955.
23. Barclay, L., Harrington, A., and Ottewill, R.H., *Kolloid-Z. u. Z. Polymere,* 250, 655, 1972.
24. Matsuo, T. and Suga, H., *Thermochimica Acta,* 88, 149, 1985.
25. Braganza, L.F. and Worcester, D.L., *Biochemistry,* 25, 2591, 1986.
26. Sogami, I., *Phys. Lett. A,* 96, 199, 1983.
27. Sogami, I. and Ise, N., *J. Chem. Phys.,* 81, 6320, 1984.

2 The Coulombic Attraction Theory of Colloid Stability

During one of the ILL (Institut Laue Langevin, Grenoble) experiments described in Chapter 1, I came across an article on electrostatic interactions in macroionic solutions by Sogami and Ise [1]. My eye was immediately taken by two sentences in the middle of the abstract, as follows. "However, the Gibbs pair potential leads us to *repulsion* at small interparticle separations and *attraction* at large distances, creating a secondary minimum (with a potential valley deeper than the thermal energy for spherical colloidal particles). Without taking refuge in a van der Waals attraction the theory evidently substantiates the experimental fact recently reported, namely the existence of Coulombic intermacroion attraction through the intermediary of counterions [Sogami and Ise's italics]."

I was already aware that the van der Waals attraction was unable to explain either of the two main features of the secondary minimum in the n-butylammonium vermiculite system, the position of the minimum with respect to salt concentration, or its reversible phase transition to the primary minimum state, so the idea of a theory dispensing with this force in creating the secondary minimum was obviously an appealing one. After assimilating the paper, I then studied Verwey and Overbeek's book [2] and started doing some calculations of my own. The bulk of Verwey and Overbeek's book deals with the interaction between charged macroionic plates, whereas the Sogami and Ise paper dealt with charged macroionic spheres as the example system. It seemed a good starting point to adapt the new general theory on the electrostatic interaction in macroionic solutions to plate particles. In particular, I wanted to calculate the Helmholtz free energy of the interaction of plate macroions from the Sogami and Ise (SI) formalism to compare it with the DLVO (Derjaguin-Landau-Verwey-Overbeek) result. This calculation is readily achieved by transposing paragraph 4 of Sogami and Ise [1] to one dimension as follows.

We presume that the electric charge of the plate particle is distributed uniformly on the surface. Thus, if the nth plate, whose center of mass is situated at X_n, has the thickness $2a_n$, the charge-distribution function is given by

$$\rho_n = \delta(|X - X_n| - a_n)$$

With the Fourier transform of this function, $\rho_n(k)$, introduced by

$$\rho_n(k) = \int \exp(-ikX)\delta\,|X - X_n| - a_n)dX$$

$$= \exp[-ik(X_n + a)] + \exp[-ik(X_n - a)]$$

the effective electrostatic potential between a pair of plates at X_m and X_n is expressed in the form

$$U_{mn}^{E} = \frac{2e^2}{\varepsilon} Z_m Z_n \int \frac{k^2}{(k^2+\kappa^2)^2} \rho_n(k)\rho_m(-k)dk$$

A residue calculation gives

$$U_{mn}^{E} = \frac{2\pi e^2}{\varepsilon} Z_m Z_n \exp(-\kappa X_{mn})\{[1+\cosh(2a\kappa)]\left(\frac{1}{\kappa} - X_{mn}\right) + 2a\sinh(2a\kappa)\} \quad (2.1)$$

where $X_{mn} = |X_m - X_n|$. For plates with like charge, $Z_m Z_n > 0$, there arises an attractive interaction in addition to the repulsive screened interaction, whose origin is self-evident. The quasistatic configuration of the plates leads to an attractive force between them. Likewise, the integral representation of the Helmholtz pair potential

$$U_{mn}^{F} = \frac{2e^2}{\varepsilon} Z_m Z_n \int \frac{1}{k^2+\kappa^2} \rho_n(k)\rho_m(-k)dk$$

leads to

$$U_{mn}^{F} = \frac{4\pi e^2}{\varepsilon} Z_m Z_n \frac{1}{\kappa} \exp(-\kappa X_{mn})[1+\cosh(2a\kappa)] \quad (2.2)$$

U_{mn}^{F} turns out to be purely repulsive for all plate distances. The entropy associated with the thermal motion of the simple ions works to cancel out completely the effective attraction in the electric pair potential U_{mn}^{E}.

Because this result has been obtained by solving a generalized Poisson-Boltzmann equation with the linearization approximation, it is necessary to compare it with the DLVO theory in the limit where the Debye approximation holds. In this case, Verwey and Overbeek [2], working in cgs (centimeter-gram-second) units, derived the following approximate equation for the repulsive potential:

$$V_R = \frac{64nkT}{\kappa} \gamma^2 \exp(-2\kappa d) \quad (2.3)$$

where κ is the inverse Debye screening length, $2d$ is the plate separation, kT is the thermal energy (in ergs), n is the number density of simple ions (in cm^{-3}) and the parameter γ is defined by

$$\gamma = \frac{e^{z/2} - 1}{e^{z/2} + 1}$$

where $z = z_i e\psi_0/kT$ is the ratio of the surface electrostatic energy to the thermal energy. We work with univalent ions — such that $z = e\psi_0/kT$ — to make the calculation simpler, but the following discussion is easily generalized. Within the Debye approximation, namely for $z < 1$, we have $\gamma^2 = (1/16)z^2$, which transforms Equation 2.3 into

$$V_R = \frac{4nkT}{\kappa} z^2 \exp(-\kappa X_{mn}) \tag{2.4}$$

where we have written $X_{mn} = 2d$ to bring the plate separation into the new notation. Because we are working within the Debye approximation, we take the linearized relation between the surface potential and the surface charge, which is (in cgs units) $\sigma_0 = (\varepsilon\kappa/4\pi)\psi_0$, where ψ_0 is the surface potential and σ_0 is the surface charge. We then have

$$z^2 = \left(\frac{e\psi_0}{kT}\right)^2 = \frac{16\pi^2}{\varepsilon^2\kappa^2(kT)^2}\sigma_0^2$$

substitution of which into Equation 2.4 gives

$$V_R = \frac{64\pi^2 ne^2\sigma_0^2}{kT\varepsilon^2\kappa^2}\frac{e^{-\kappa X}}{\kappa} \tag{2.5}$$

into which we now insert (again in cgs units) $\kappa^2 = 8\pi ne^2/\varepsilon kT$, from which

$$V_R = \frac{8\pi}{\varepsilon}\sigma_0^2\frac{e^{-\kappa X}}{\kappa} \tag{2.6}$$

Now σ_0 is the surface charge density, such that $\sigma_0 = Z_n e$, where Z_n is the valency of the nth plate in number of charges per unit area, so this expression becomes

$$V_R = \frac{8\pi e^2}{\varepsilon} Z_m Z_n \frac{e^{-\kappa X}}{\kappa} \tag{2.7}$$

which is to be compared with Equation 2.2. Because in the Debye approximation $a\kappa << 1$, $\cosh(2a\kappa) = 1$, the term in square brackets in Equation 2.2 is equal to 2, and the new theory gives

$$U_{mn}^F = \frac{8\pi e^2}{\varepsilon} Z_m Z_n \frac{e^{-\kappa X}}{\kappa}$$

proving that the calculations give identical results for the Helmholtz free energy in this limit.

Both theories are based on the calculation of the electrical contribution to the free energy of the region bounded by the macroions. In both theories it is assumed that (a) the motion of the macroions is adiabatically cut off from that of the simple ions and (b) the distribution of simple ions is determined by the Boltzmann distribution. In both theories the total electrostatic energy of the solution is obtained by solving the Poisson-Boltzmann equation using the so-called primitive model, in which the solvent is treated as a dielectric continuum, and the Helmholtz free energy is calculated from this. Within the linearization approximation, the theories yield identical results for this quantity.

If you like, $V_R(\text{DLVO}) = U^{\text{F}}_{mn}$ (SI) for the Helmholtz free energy of the interaction for two flat plates is our baseline, a point at which everyone can agree. It was the next step in the SI formalism, the calculation of the Gibbs free energy, that sundered the colloid world. It may seem incredible to scientists outside the field that the basic thermodynamic relation

$$G = F + PV \tag{2.8}$$

between the Gibbs free energy G, and the Helmholtz free energy F, the pressure P and volume V, could cause a controversy that has persisted for 20 years. The problem arises from the interpretation of V in this equation. In DLVO theory, this V is identified with the volume of the whole condensed matter system. We label this volume V^*. In SI theory, V is identified as the region bounded by the macroions. These two volumes, V and V^*, are obviously different in the case of the n-butylammonium vermiculite gels, as shown in Figure 2.1b. As we have seen, the clay plates, by virtue of some attractive force between them (obviously not the van der Waals force), form gels that can exist in equilibrium with an excess of solution. This excess bath of solution can become arbitrarily large without essentially affecting the equilibrium separation of the clay plates. In these circumstances, the volume of the system V in which macroions and simple ions are confined (by their mutual interactions) must be definitely distinguished from the volume V^* of the macroionic

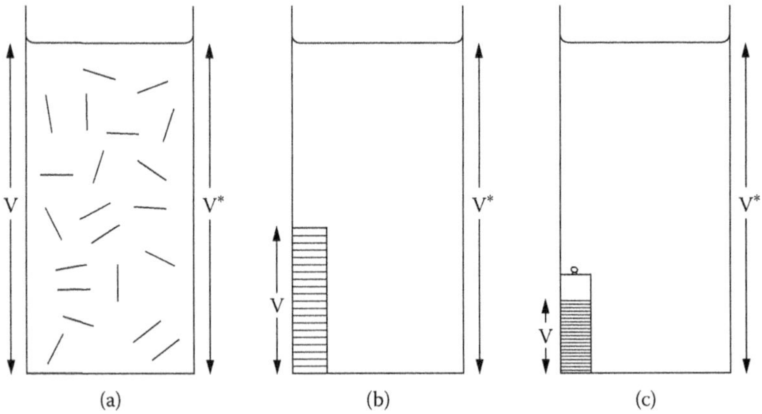

FIGURE 2.1 Schematic illustration of the difference between suspensions and gels. (a) Suspension: $V = V^*$. (b) Gel: $V \ll V^*$. (c) Gels are highly compressible: when a load is applied to a gel, $\Delta V \approx V$, $\Delta V^* = 0$.

solution plus the volume of the excess solution (confined by the container walls). The virtually incompressible volume V^* is irrelevant as the thermodynamic variable for describing the properties of the macroionic solution: the behavior of the system must be described in terms of the variable V, whose value is determined by the equilibrium condition (not set equal to V^* *a priori*) and which depends on the configuration of the macroions.

Because this point is so important to understand, two other situations in clay swelling have also been illustrated in Figure 2.1. First, in Figure 2.1a, a suspension of randomly oriented clay plates — the more common situation in clay swelling than that illustrated in Figure 2.1b — is shown for comparison. In this case, the macroionic particles are homogeneously distributed throughout the bulk of the solvent medium in which they are dispersed. Such a system is virtually incompressible, and the difference between F and G is not significant. Second, in Figure 2.1c, a case is shown in which a gel is compressed uniaxially along its swelling axis. Such experiments have been performed by many authors [3–6], and we performed extensive measurements of this type on the n-butylammonium vermiculite system [7, 8], in experiments to be described in Chapter 3. Here we have a completely different situation; the gels are highly compressible — not because the solvent itself is compressible, but because it is easy to squeeze solvent molecules out of the macroionic phase. In the process of compressing a gel by uniaxial stress (the difference between Figure 2.1b and Figure 2.1c is not exaggerated; it is easy to compress a clay gel to half its volume), ΔV is large and ΔV^* is negligible. Now V and V^* must definitely be distinguished in the relation

$$\Delta G = \Delta F + P\Delta V \tag{2.9}$$

If the Gibbs and Helmholtz free energies are assumed to be equal, then any work done on the gel would simply disappear, in violation of the first law of thermodynamics [9].

Ise and Sogami went to great pains in their recent book [10] to explain the equation of the state of the simple ion system in the form

$$PV = G - F = G^0 - F^0 + E \tag{2.10}$$

where G^0 and F^0 are the free energies of the corresponding uncharged system and E is the electrostatic energy. This leads immediately to the relation

$$P\Delta V = \Delta E \tag{2.11}$$

Mechanical work done on the macroionic phase is stored as electrostatic energy in that phase. So, in the system under consideration, the Gibbs, Helmholtz and electrostatic energies are related by

$$\Delta G = \Delta F + \Delta E \tag{2.12}$$

In particular,

$$U^{\mathrm{G}}_{mn} = U^{\mathrm{F}}_{mn} + U^{\mathrm{E}}_{mn} \tag{2.13}$$

Sogami and Ise [1] derived the same result by using the alternative definition of G, as the sum of the chemical potentials. I also adopted this approach as an independent check of the calculation and obtained [9]

$$G = 2(F - F^0) + \kappa^2 \frac{\partial F}{\partial(\kappa^2)} \tag{2.14}$$

and

$$U_{mn}^{\mathrm{G}} = \left(2 + \kappa^2 \frac{\partial}{\partial(\kappa^2)}\right) U_{mn}^{\mathrm{F}} \tag{2.15}$$

By applying the operator $(2 + \kappa^2\partial/\partial\kappa^2)$ to the right-hand side of Equation 2.2, or by taking the sum of Equation 2.1 and Equation 2.2, we derive

$$U_{mn}^{\mathrm{G}} = \frac{2\pi e^2}{\varepsilon} Z_m Z_n \exp(-\kappa X_{mn})\{[(1+\cosh(2a\kappa)]\left(\frac{3}{\kappa} - X_{mn}\right) + 2a\sinh(2a\kappa)\} \tag{2.16}$$

for the Gibbs pair potential of interaction between plate macroions.

Preliminary inspection of Equation 2.16 reveals that the Gibbs pair potential leads to repulsion at small interplate separations and attraction at large distances. Because it is U_{mn}^{G} and not U_{mn}^{F} that is the appropriate pair potential for describing the effective interaction between the mth and nth macroions in solution under isobaric conditions, the different analytic properties of U_{mn}^{G} and U_{mn}^{F} have profound implications for colloid science.

Let us first consider the limit of dilute electrolyte solutions, where $a\kappa < 1$. For clay layers, $a \approx 10$ Å, which implies $\kappa < 0.1$ Å^{-1}, i.e., electrolyte solutions less concentrated than 0.1 M. This is the region of interest. In this case,

$$U_{mn}^{\mathrm{G}} \cong \frac{4\pi e^2}{\varepsilon} Z_m Z_n \frac{\exp(-\kappa X_{mn})}{\kappa}(3 - \kappa X_{mn}) \tag{2.17}$$

To make things simpler, let us abbreviate the effective thermodynamic pair potential U_{mn}^{F} by V and the separation between the macroions m and n, X_{mn}, by X. Let us also assume that the macroions m and n have the same charge Z. Then, abbreviating the constant $(4\pi e^2/\varepsilon)$ by b, we have

$$V = bZ^2 \frac{\exp(-\kappa X)}{\kappa}(3 - \kappa X) \tag{2.18}$$

for the effective interaction potential. Taking the derivative with respect to X gives the force F_X as

$$F_X = -\frac{dV}{dX} = cZ^2 \exp(-\kappa X)(4 - \kappa X) \tag{2.19}$$

which is attractive for long-range separations ($\kappa X > 4$). The minimum position created by the thermodynamic and electrical (thermoelectric) interactions is given by

$$X_{min} = \frac{4}{\kappa} \tag{2.20}$$

Just as we equated X_{eq}, the position of the secondary minimum predicted by DLVO theory, with the equilibrium separation d in Chapter 1, so now we equate X_{min}, the position of the secondary minimum predicted here, with the interlayer d-value. Equation 2.21

$$d \cong \frac{4}{\kappa} \tag{2.21}$$

obviously states that the interparticle spacing should be a constant number of Debye screening lengths, in qualitative agreement with the general statement by Hunter [11] and the specific results on the n-butylammonium vermiculite gels. The predictions of the two theories are plotted together with the experimental results in Figure 2.2. It is clear that Figure 2.2 calls for a paradigm shift in colloid science.

Looking back on it, anyone could have picked out the correct qualitative form for the interaction potential V by combining two ideas, quite irrespective of the

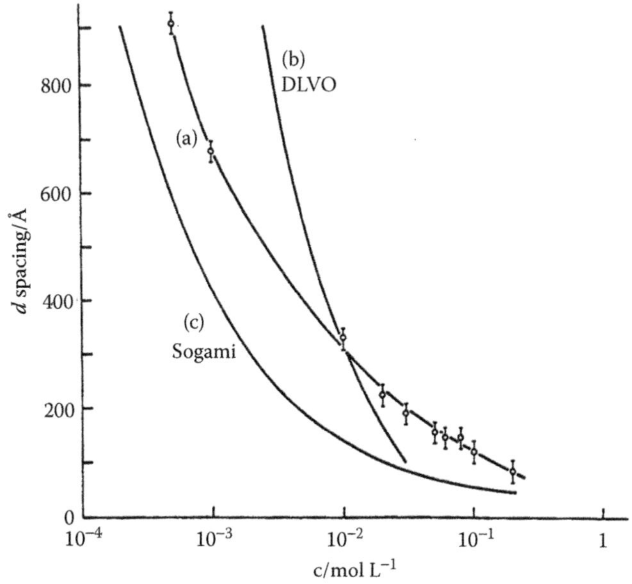

FIGURE 2.2 Comparison of (a) experimentally observed and (b) calculated d-values using DLVO theory and (c) the coulombic attraction theory. Curve (b) is obtained with an adjustable parameter (proportional to the Hamaker constant) chosen to fit the data at $c = 0.01$ M. Curve (c) contains no adjustable parameters.

SI formalism and the calculations based on it above. First, there is ample evidence that the medium-range osmotic repulsion of DLVO theory is basically correct [11]. A theory does not become textbook without being able to explain some phenomena well. Therefore, there must be something correct, at least to a leading approximation, about the repulsive potential

$$V_R = B\exp(-\kappa X) \tag{2.22}$$

of DLVO theory. (We name this purely repulsive potential a Yukawa potential to distinguish it from the DLVO potential given by Equation 1.2, which is the sum of the Yukawa potential and a van der Waals potential.) Second, the general knowledge, typified in Hunter [11] and exemplified in the n-butylammonium vermiculite gels, that the secondary minimum lies at a constant number of Debye screening lengths tells us unambiguously that *both* the repulsive and attractive forces must be electrical in origin. Finding a function that has a leading term expressed by Equation 2.22 and a minimum at a constant number of Debye lengths is not difficult. The function

$$V \propto \exp(-\kappa X)(c - \kappa X) \tag{2.23}$$

is the obvious choice. This is the *experimentally* correct function for long-range phenomena. It is no coincidence that this function agrees with the one derived from the SI formalism, as this provides the correct application of the laws of physics to the problem. Because the origin of the secondary minimum is clearly a result of long-range coulombic attraction in the new theory, the theory will be named the coulombic attraction theory from now on. The SI formalism is the version of this theory that calculates the Gibbs free energy from the linearized Poisson-Boltzmann (PB) equation.

So, we have seen that the interlayer spacings in the gels are in qualitative agreement with the new theory (the quantitative displacement of the two curves [a] and [c] in Figure 2.2 will be a major subject in Chapter 4). What about the reversible phase transition?

We begin by looking at a success of DLVO theory, namely the derivation of the Schulze-Hardy rule. The Schulze-Hardy rule refers to the "salting out" of colloids, namely collapse into the primary minimum when the electrolyte concentration is increased beyond a critical value. It states that the critical coagulation concentration is proportional to the sixth power of the valency of the counterion in the added electrolyte. In other words, for negatively charged colloidal particles like the clay plates, Ca^{2+} is $2^6 = 64$ times as effective at inducing coagulation as Na^+, and Al^{3+} is $3^6 = 729$ times as effective, at equivalent salt concentrations. Unlike flocculation into the secondary minimum (gel state in the n-butylammonium vermiculite system), coagulation into the primary minimum (crystalline state in the n-butylammonium vermiculite system) is a short-range phenomenon. DLVO theory correctly predicted the Schulze-Hardy rule by combining a short-range van der Waals potential with a Yukawa potential, as in Equation 2.1. It is clear that a total interaction potential V_T

combining V_A and an electric interaction is basically correct, so the obvious choice of function is to take

$$V = V_A + U_{mn}^G \tag{2.24}$$

where V_A is as in Equation 1.2 and U_{mn}^G is given by Equation 2.17. That a function of this type does indeed account for the Schulze-Hardy rule has recently been demonstrated by Ise and Sogami [10]. The way in which it accounts for the *reversible* phase transition in the n-butylammonium vermiculite system is illustrated schematically in Figure 2.3.

The introduction of the van der Waals potential in combination with a Yukawa potential produces a curve in which the primary minimum is *always* deeper than the secondary minimum. This must be so because the primary minimum state is that for which the particles have coalesced and the valency of the *n*th plate Z_n has dropped to zero: since $Z_n \to 0$ as $X_{mn} \to 2a$, $U_{mn}^F \to 0$ as $X_{mn} \to 2a$, and the van der Waals force introduced to explain the existence of the secondary minimum must be stronger at the primary minimum separation. The DLVO theory is therefore refuted by experiments that demonstrate the thermodynamic stability of charged colloidal dispersions [12, 13].

The thermodynamic stability of the colloidal gel state has a perfectly natural explanation in terms of the coulombic attraction theory. Just as $U_{mn}^F \to 0$ as $X_{mn} \to 2a$, so $U_{mn}^G \to 0$ as $X_{mn} \to 2a$ because $Z_n \to 0$ in this limit. The negative values of U_{mn}^G at large particle separations therefore represent a state of lower free energy than the crystal plus the appropriate amount of solution, as shown in Figure 2.3c. The addition of a short-range van der Waals potential to U_{mn}^G has the effect of bringing the primary minimum into balance with the secondary minimum, and a phase transition between the states becomes possible. In Figure 2.3c we have assigned the van der Waals potential its proper role in colloid science, namely as a negligible contribution at the secondary minimum separation, but as a constituent in understanding coagulation into the primary minimum [14].

Figure 2.3a is purely a sketch. The exact interaction potential between n-butylammonium-substituted clay plates (or other charged colloidal particles) in solution must incorporate many effects, such as the size of the small ions and the molecular degrees of freedom of the solvent, that are beyond the scope of either the coulombic attraction theory or DLVO theory. However, whatever the complicated functional dependence, the curve must comprise two states of equal thermodynamic potential. Somehow, the valleys in V_T, the total potential, must be of equal depth. As discussed previously, in Figure 2.3b we see that the DLVO theory can never account for this experimentally proved phenomenon.

It is clear from this chapter that the coulombic attraction theory potential is much better adapted to explain the experimental phenomena described in Chapter 1 than the DLVO theory potential (Equation 1.2). Of course, if you predict an interaction potential, you predict force-distance curves along the swelling axis. There have been a lot of arguments about how direct measurements of forces between spherical colloidal particles "refute" the coulombic attraction theory. Let us get the facts first. We now examine the experimental curves for the n-butylammonium vermiculite system.

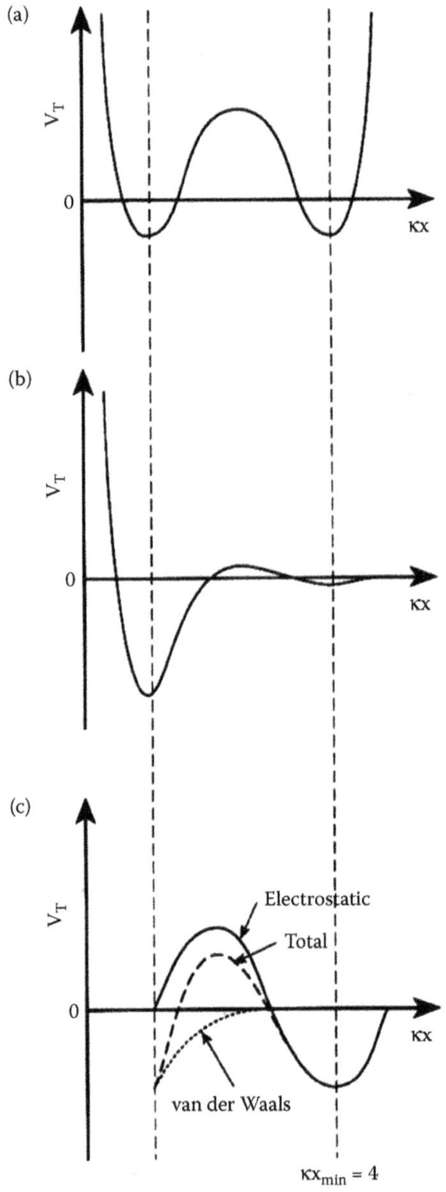

FIGURE 2.3 The phase transition in n-butylammonium vermiculite swelling is thermodynamic, so (a) the primary and secondary minima must be of equal depth, irrespective of the shape of the total interaction potential V_T. This cannot be accounted for by the DLVO potential, shown in (b). Part (c) shows V_T in the coulombic attraction theory (dashed curve), composed of an electrostatic part (solid curve) and a short-range potential (dotted curve) that includes the weak contribution from the van der Waals potential. The vertical dashed lines indicate the positions of the two minima.

REFERENCES

1. Sogami, I. and Ise, N., *J. Chem. Phys.,* 81, 6320, 1984.
2. Verwey, E.J.W. and Overbeek, J.Th.G., *Theory of the Stability of Lyophobic Colloids,* Elsevier, Amsterdam, 1948.
3. Rausell-Colom, J.A., *Trans. Faraday Soc.,* 60, 190, 1964.
4. Viani, B.E., Roth, C.B., and Low, P.F., *Clays Clay Min.,* 33, 244, 1985.
5. Low, P.F., *Langmuir,* 3, 18, 1987.
6. Rausell-Colom, J.A., Saez-Auñón, J., and Pons, C.H., *Clay Min.,* 24, 459, 1989.
7. Braganza, L.F., Crawford, R.J., Smalley, M.V., and Thomas, R.K., *Prog. Colloid Polym. Sci.,* 81, 232, 1990.
8. Crawford, R.J., Smalley, M.V., and Thomas, R.K., *Adv. Colloid Interface Sci.,* 34, 537, 1991.
9. Smalley, M.V., *Mol. Phys.,* 71, 1251, 1990.
10. Ise, N. and Sogami, I.S., *Structure Formation in Solution: Ionic Polymers and Colloidal Particles,* Springer Verlag, Berlin-Heidelberg-New York, 2005.
11. Hunter, R.J., *Foundations of Colloid Science,* Clarendon Press, Oxford, U.K., 1987.
12. Smalley, M.V., Thomas, R.K., Braganza, L.F., and Matsuo, T., *Clays Clay Min.,* 37, 474, 1989.
13. Braganza, L.F., Crawford, R.J., Smalley, M.V., and Thomas, R.K., *Clays Clay Min.,* 38, 90, 1990.
14. Smalley, M.V., *Prog. Colloid Polym. Sci.,* 97, 59, 1994.

3 Force-Distance Curves for Plate Macroions

We measured the force-distance curves for the n-butylammonium vermiculite gels on the D16 long-wavelength neutron diffractometer at the Institut Laue-Langevin (ILL), Grenoble, using the uniaxial pressure cell apparatus shown in Figure 3.1 [1]. A crucial feature of our data was that we obtained force-distance curves across the salt concentration range between 0.001 and 0.2 M. This factor of $\times 200$ in c is vitally important in testing theories of electrical interactions. Experiments carried out over a narrower range of salt concentrations are much less useful, and those carried out at just one salt concentration are more or less useless, as we shall see below.

For these experiments, some of which were also conducted on the ILL D17 small-angle diffractometer, crystals of n-butylammonium vermiculite exhibiting the fewest obvious structural defects were selected and trimmed to a rectangular cross section with a razor blade. This enabled the surface area, and therefore the applied pressure, to be measured accurately. The samples were then immersed in a dilute solution of n-butylammonium chloride of the desired concentration and allowed to swell freely. After equilibration for at least two days at 7°C, the swollen (or colloidal) gel phase samples were placed into the uniaxial pressure cell shown in Figure 3.1.

The sample was contained in the large outer quartz cell (internal dimensions 1×2 cm), its silica planes parallel to the narrow face. A quantity of the solution in which it had been soaking was pipetted into the cell. The sample rested on the bottom of and at the front of the cell. Behind it and contained within the larger cell was another quartz cell of external dimensions 0.6×0.6 cm, which was movable and served as the compression member. The entire clay gel was thus sandwiched between two quartz faces. The movable internal cell was supported by a cadmium-clad aluminum pillar that was, in turn, free to slide along steel rods that formed part of the slide assembly at the top of the apparatus. Turning the wheel at the back of the apparatus caused the pillar and attached cell to be pushed forward along its tracks, thus reducing the gap between the internal and external quartz cells at the front of the apparatus. With the sample in place, a uniaxial squeezing force was thus applied. The entire force on the sample was transmitted by the two vertical spring steel members that supported the slide assembly. Attached to each of these pieces was a resistance-type strain gauge that measured the extent of their bending motion as the force was applied. The changes in voltage across the strain gauges, caused by their changing resistance as they bent, was amplified and displayed. The device was calibrated by removing the outer quartz cell, laying the apparatus on its side, and loading known weights onto the inner cell. The response was linear up to an applied force of 7.5 N (7.5×10^5 dynes). The use of cgs (centimeter, gram, second)

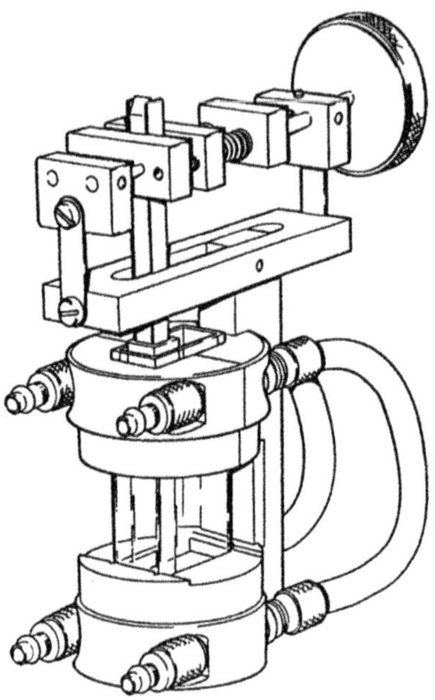

FIGURE 3.1 The uniaxial pressure cell apparatus used in the neutron-scattering experiments. The whole apparatus is 10 cm high.

units in this chapter facilitates comparison with the predictions of DLVO (Derjaguin-Landau-Verwey-Overbeek) theory [2, 3].

The whole apparatus, which was 10 cm tall, was mounted on a goniometer head, this being fixed to the θ-turntable of the diffractometer. This arrangement allowed the sample to be exactly located at the center of rotation of the diffractometer. The body of the cell holder was of aluminum, clad in cadmium foil. The sample and solution were cooled using the water circulation ports illustrated, the temperature within the cell being monitored by means of a metal junction thermometer.

When placing a gel between the faces of the quartz cells, it was important that the gel not be subjected to any initial squeezing force, particularly for the more dilute gels that are very easily compressed. In the absence of any applied force, the zero-stress d-value of the gel was obtained. Force was gradually brought to bear on the sample by moving the inner cell by means of the sliding mechanism and the control wheel. Having applied a certain force, time had to be allowed for equilibrium to be reached. During this period, the d-value was seen to decrease slightly, the force measurements decreasing concomitantly until equilibrium was attained. When both measurements were steady, they were recorded and the force was increased to obtain the next point. The experimental limits were reached at large applied forces, under which circumstances the gel becomes increasingly incompressible. Very gradual increases in force were then required to prevent the sample being ejected from between the constraining quartz faces. The time required for these very gradual increases in force became too great, and the experiment was terminated.

In our previous paper [4], we had reported the equilibrium *d*-value of n-butylammonium vermiculite gels to be 190 Å in a 0.03 M solution. This distance gives diffraction effects that lie in a convenient *Q*-range for study on the D16 diffractometer, and so this concentration of soaking solution was chosen for a preliminary study. A new batch of six samples had an average *d*-spacing of 174 Å (in the range between 167 and 183 Å) and an average mosaic spread of 5.8° (in the range between 4.7 and 6.8°). The mosaic spread of the crystalline phase is 5.3°, so the value of 5.8° for these gels confirmed that the swelling is nearly perfectly uniaxial up to a tenfold expansion of the original crystal.

It is apparent that there is some sample-to-sample variability in both the *d*-spacing and mosaic spread values, a consequence of the natural origin of the mineral, which may lead to slightly varying composition of the samples. We found the slightly larger values of the *d*-values exhibited by the older samples (the 1986 batch) to be a systematic variation. For the average *d*-values quoted in this chapter, which I denote $<x_{min}>$, I have taken an average over all the samples measured and have retained two significant figures of accuracy. In the case of the 0.03 M gels, for example, $<x_{min}>$ is 180 Å.

A series of small stresses were applied to a 0.03 M sample with an equilibrium *d*-value of 183 Å, reducing the *d*-value to 143 Å over a period of about four hours. A larger stress was then applied, and the gel was allowed to come to equilibrium with the applied stress for 15 hours. The *d*-value is plotted as a function of time in Figure 3.2. After 2 hours the *d*-value was 118 Å, and after 15 hours when the curve has obviously flattened out, the *d*-value was 115 Å. In the available neutron beam time, it was not possible to allow the gel to come fully to equilibrium with each applied stress. An equilibration time of two hours was chosen. After this time the voltage was always steady to two significant figures, and we may surmise that the error of approximately 3 Å introduced into the *d*-spacing is an absolute one, which does not affect the shape of the measured force-distance curve.

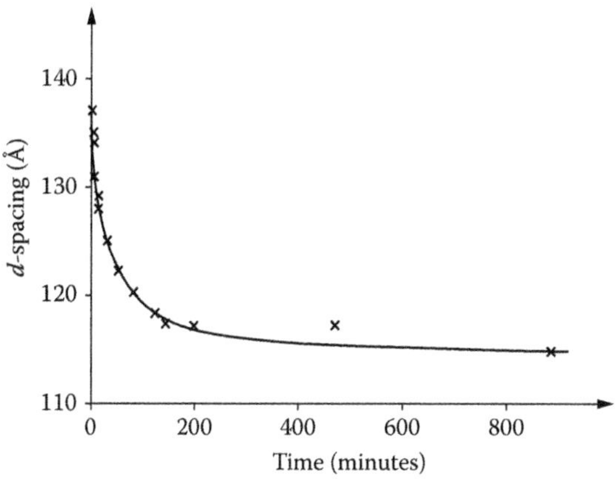

FIGURE 3.2 The *d*-value of a gel (c = 0.03 M) as a function of time after application of a uniaxial stress.

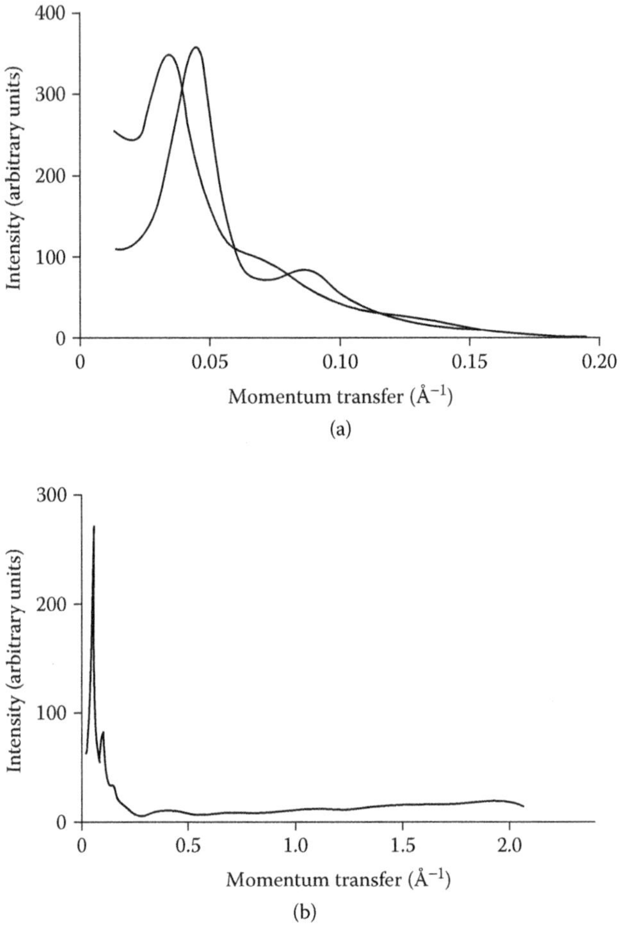

FIGURE 3.3 Observed diffraction traces from a gel at $c = 0.03$ M. (a) Small-angle scattering at zero stress ($d = 183$ Å) and at an applied stress of $p = 0.03$ atm ($d = 143$ Å). (b) Complete diffraction pattern at an applied stress of $p = 0.07$ atm. Incident wavelength $\lambda = 4.53$ Å.

The low-angle diffraction traces corresponding to the d-values of 183 Å and 143 Å are shown in Figure 3.3a. It is apparent that even with a small stress, in this case about 0.03 atmospheres, the diffraction effects become sharper, an effect that had been observed previously [5, 6]. As the stress is further increased, the one-dimensional colloid continues to contract along the swelling axis and the pattern sharpens still more. When the gel had come to equilibrium at a stress of about 0.07 atm, and the d-spacing was 117 Å, three orders of the 00l reflection were observed, which appear clearly in the complete diffraction pattern shown in Figure 3.3b. The broad maximum around $Q = 0.4$ Å^{-1} is not connected with the diffraction condition, but it is part of the diffuse scattering to be described in detail in Chapter 8

TABLE 3.1
Uniaxial Pressures and *d*-Spacings for a 0.01 M Gel

P (10^3 dynes/cm^2)	*d*-Spacing (Å)	ln *P*
3.5	260	1.26
3.5	245	1.26
3.5	230	1.26
7.0	216	1.95
10.5	204	2.35
21.1	184	3.05
24.6	168	3.20
59.7	147	4.09
66.7	137	4.20
98.3	130	4.59
140	122	4.94
218	113	5.38
334	103	5.81
362	99	5.89
534	92	6.28
688	85	6.53
850	80	6.74
1100	73	7.00
1310	67	7.18
1610	61	7.38

Note: The sample area was 0.21 cm^2, and the zero-stress d-spacing was 260 Å.

We now consider the *d*-values as a function of the applied uniaxial stress. Because our preliminary studies on the 0.03 M gels gave results that were slightly inconsistent with our subsequent measurements made at seven other salt concentrations (*vide infra*), we take as example a set of results obtained for a 0.01 M sample. These results are given in Table 3.1 and are plotted in Figure 3.4. The direct pressure–distance curve in Figure 3.4a is not very revealing, and the data have been replotted in the standard form of ln P vs. x in Figure 3.4b. The straight line in Figure 3.4b is the best least-squares fit to the linear relationship between ln P and x usually invoked. It appears to be an adequate fit and was corroborated by repeating the experiment with gels prepared by soaking a crystal from a different batch of samples in a 0.01 M solution. For the three least-squares fits obtained we have

$$\ln P = 9.29 - 0.034\,x$$
$$\ln P = 9.72 - 0.037\,x$$
$$\ln P = 8.26 - 0.025\,x$$

The agreement is reasonable and confirms the reproducibility of the properties of the gels. Some sample-to-sample variability was observed in the intercepts and gradients

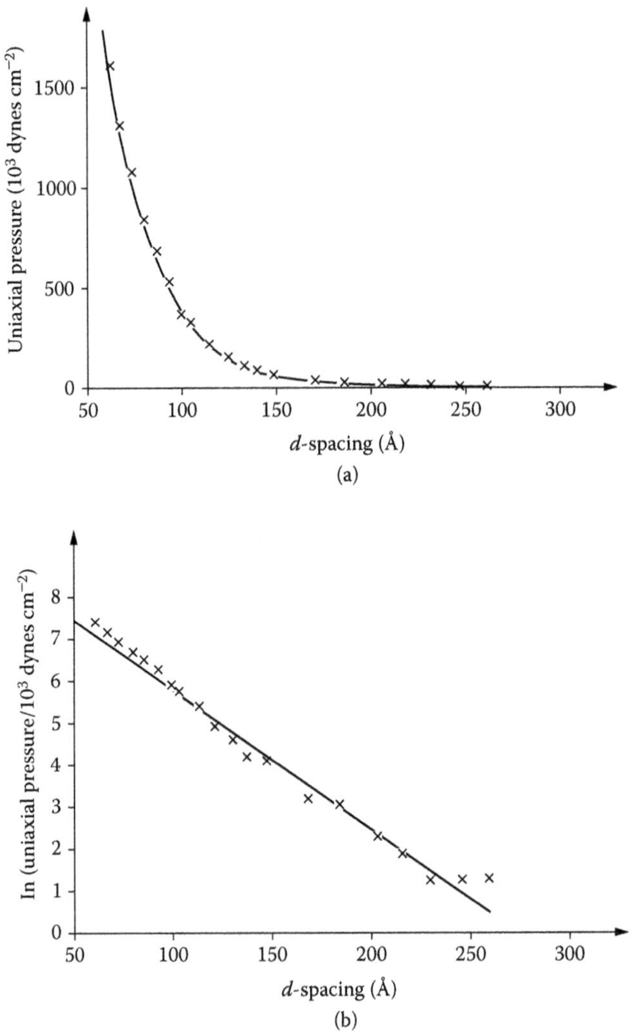

FIGURE 3.4 Pressure–distance data obtained from a gel at $c = 0.01$ M: (a) raw data and (b) standard plot of ln P vs. x.

obtained at one concentration, but each individual ln P vs. x plot was linear to the limit of the experimental accuracy, and all subsequent data are given in this form.

Pressure–distance information was obtained for a wide range of soaking solution concentrations. On one occasion we were able to use the D17 diffractometer to obtain the pressure–distance curve of a gel prepared in a 10^{-3} M soaking solution. This was particularly valuable, as we were able to obtain stress-strain data at unusually large spacings. Our unstressed 0.001 M sample gave a weak diffraction trace corresponding to a d-value of 550 Å, which is also the average value obtained at this concentration. As before, as soon as a stress was applied, the diffraction effect sharpened considerably,

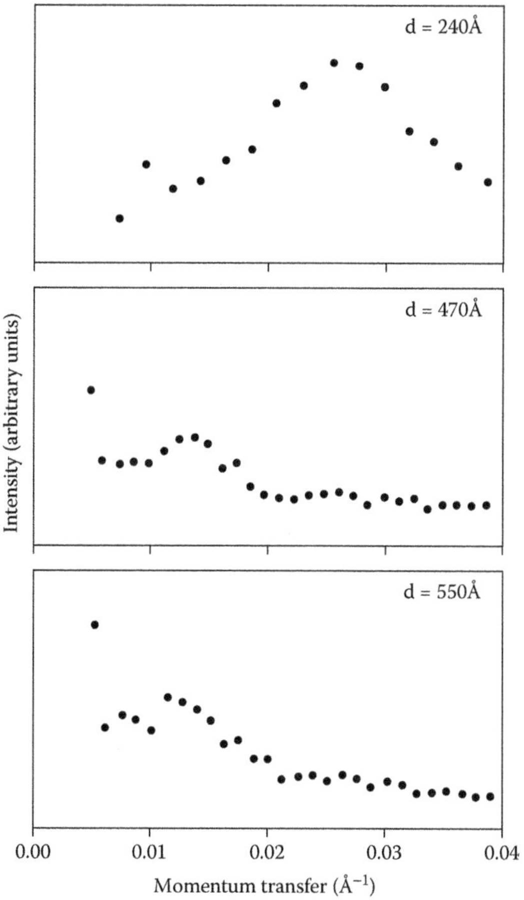

FIGURE 3.5 A selection of the diffraction traces observed in the D17 experiment. Incident wavelength $\lambda = 17.4$ Å.

and at higher stresses there was no difficulty in determining the *d*-value: the trace at $d = 240$ Å is given as an example in Figure 3.5, together with the patterns obtained at the first stress point ($d = 470$ Å) and at zero stress ($d = 550$ Å). Although there was only sufficient beam time to allow one hour for the sample to come to equilibrium with the applied stress in this particular experiment, scans every 15 minutes confirmed that the *d*-value had settled down well during this period. The results are given in Table 3.2, and the logarithmic plot is shown in Figure 3.6.

All other experiments were performed on the D16 diffractometer, yielding approximately 20 stress-strain curves. In each case, at least 6 data points, but usually around 15, were obtained, and all the plots of the logarithm of the pressure against the separation appeared linear to the eye. A selection of the data obtained at three concentrations is shown in Figure 3.7. A least-squares fit was made to each data set, and the negative gradients obtained by this method are given in Table 3.3. The intercepts of the plots are given in Table 3.4. These results contain information on

TABLE 3.2
Uniaxial Pressures and *d*-Spacings for a 0.001M Gel

P (10^3 dynes/cm^2)	*d*-Spacing (Å)	ln *P*
1.0	470	0.00
3.2	415	1.16
4.0	380	1.39
6.0	315	1.80
6.6	300	1.89
10.5	280	2.36
13.2	260	2.58
20.7	240	3.03
24.5	230	3.20
32.1	220	3.47
38.1	210	3.64
48.9	200	3.89
60.9	185	4.11
70.1	170	4.25
87.4	165	4.47

Note: The sample area was 0.20 cm^2, and the zero-stress *d*-spacing was 550 Å.

the surface potentials and charges on the clay plates. The derivation of the surface potentials (ψ_0) and surface charges (σ_0) given in Table 3.4 is explained below.

There are two leading features to Figure 3.7. The first is that the data points are quite scattered. The second is that a linear relationship between ln *P* and the *d*-value

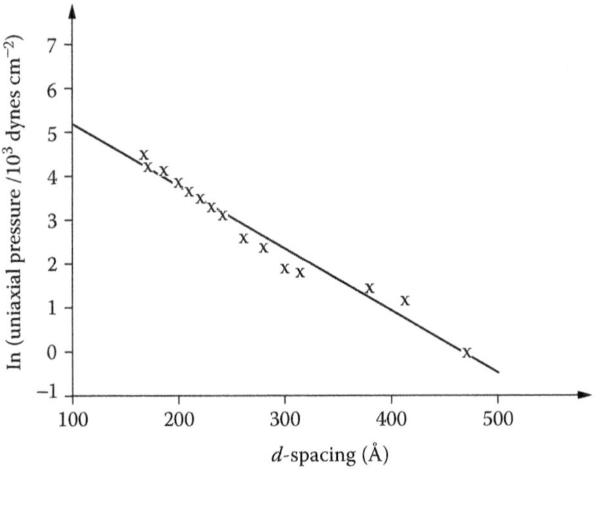

FIGURE 3.6 Plot of ln *P* vs. *x* for the D17 data. c = 0.001 M.

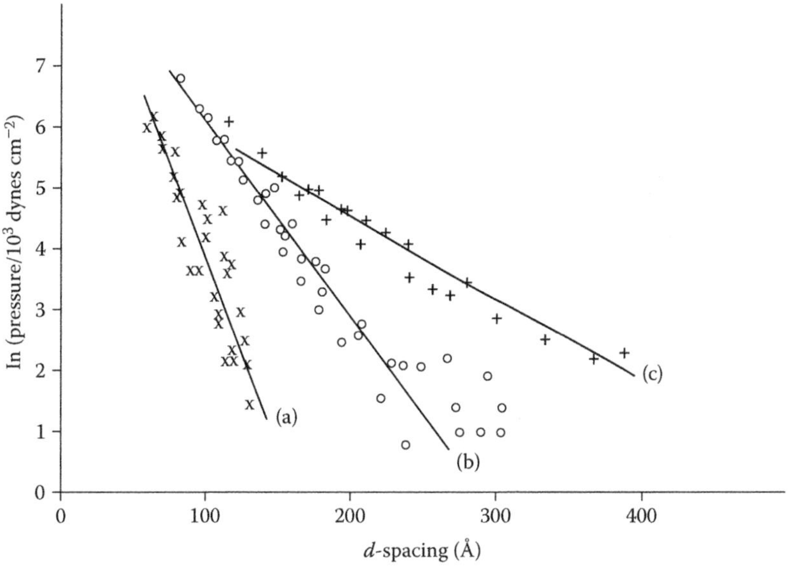

FIGURE 3.7 Plots of ln P vs. x for gels prepared in (a) 0.1 M, (b) 0.01 M and (c) 0.001 M soaking solutions.

holds to a good approximation. The existence of a linear relation between the logarithm of the swelling pressure P and the separation of the clay plates x is usually taken as evidence for the DLVO theory [5–8]. The reason for this is that the repulsive electrostatic force of DLVO theory is described by the equation

$$\frac{dV_R}{dx} = B\exp(-\kappa x) \tag{3.1}$$

TABLE 3.3
Averages of the Gradients of the Measured ln P vs. x Plots at Various Solution Concentrations

c_{ex} (mol dm^{-3})	Gradient 1 (Å^{-1})	Gradient 2 (Å^{-1})	Gradient 3 (Å^{-1})	Average Gradient (Å^{-1})
0.2	–0.1141	–0.0794	–0.0839	–0.0925
0.1	–0.0667	–0.0920	–0.0497	–0.0695
0.06	–0.0645	–0.0538	—	–0.0592
0.03	–0.0336	–0.0320	—	–0.0328
0.01	–0.0339	–0.0372	–0.0245	–0.0319
0.006	–0.0258	—	—	–0.0258
0.003	–0.0180	–0.0278	—	–0.0229
0.001	–0.0111	–0.0155	–0.0226	–0.0164

TABLE 3.4
Averages of the Intercepts of the ln *P* vs. *x* Plots at Various Solution Concentrations

c_{ex} (mol dm^{-3})	ln *B*	Surface Potential, ψ_0 (mV)	Surface Charge, σ (Å^{-2})	σ/σ_{max}
0.2	11.0	68	3.94×10^{-3}	0.28
0.1	11.1	95	2.96×10^{-3}	0.21
0.06	10.4	81	2.52×10^{-3}	0.18
0.03	8.1	45	1.40×10^{-3}	0.10
0.01	9.1	77	1.36×10^{-3}	0.097
0.006	7.2	37	1.10×10^{-3}	0.078
0.003	8.5	81	0.975×10^{-3}	0.069
0.001	7.3	62	0.699×10^{-3}	0.050

where V_R is the effective repulsive potential and *B* is a positive constant that can be calculated from electrostatic theory. Its value is given by [3]

$$B = 9.2 \times 10^8 \kappa^2 z^2 \tag{3.2}$$

where *B* is in dynes/cm^2, κ is in Å^{-1}, and z is the dimensionless ratio of the surface electrostatic energy to the thermal energy, defined by

$$z = \frac{e\psi_0}{kT} \tag{3.3}$$

where ψ_0 is the surface potential. If we fit the purely repulsive part of the DLVO potential (the Yukawa potential) to the data, the constant *B* in the equation

$$\ln P = \ln B - \kappa x \tag{3.4}$$

can be obtained from the intercepts of the plots given in Table 3.4, and the negative gradient can be identified with κ, the inverse Debye screening length.

Let us consider the combination of two general sets of facts about the n-butylammonium vermiculite system. First, the results on the phase transition and the position of the secondary minimum strongly support a coulombic origin for the long-range attraction that must exist to stabilize the gels, as described in Chapters 1 and 2. Second, the direct force-distance curves presented here in Chapter 3 show no obvious evidence for the long-range attraction (of whatever origin), but fit well to a Yukawa-type force, as in Equation 3.1. Do these experimental force-distance curves contradict the coulombic attraction theory? Emphatically not. For the sake of simplicity, let us take the $\kappa a \ll 1$ limit and the approximate coulombic attraction theory force

$$\frac{dV}{dx} = Z^2 \exp(-\kappa x)(4 - \kappa x) \tag{3.5}$$

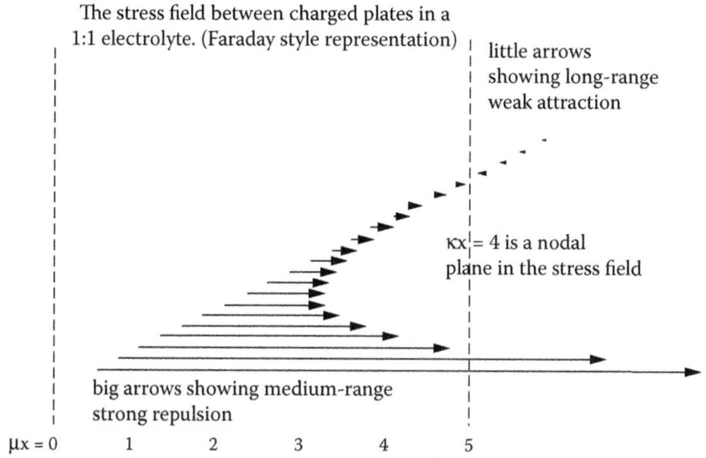

FIGURE 3.8 The stress field between charged plates in a 1:1 electrolyte (Faraday style representation).

To get a qualitative grasp of what is happening, I have plotted out the force field predicted by Equation 3.5 in a Faraday representation in Figure 3.8.

It is perfectly clear that if we start pushing the plates together, we will quickly encounter the medium-range strong repulsion because of the shallowness of the minimum. Experimentally, we are not able to distinguish between the pure Yukawa force of Equation 3.1 and the coulombic attraction theory force of Equation 3.5 because of the error bars in the data and reaching the limit of thermal noise. I will make this argument quantitative below. Here, let us be clear that approximately exponentially decaying (that is, exponential to the limit of experimental accuracy) force-distance curves are to be expected from the coulombic attraction theory. And such curves are emphatically not predicted by DLVO theory, as is widely assumed. This is only because the van der Waals force required to explain the existence of the secondary minimum in DLVO theory mysteriously disappears when the force-distance curves are considered. Let us consider the full DLVO force as

$$\frac{dV}{dx} = B\exp(-\kappa x) - \frac{A}{x^4} \tag{3.6}$$

At equilibrium

$$B\exp(-\kappa x_{\min}) = \frac{A}{x_{\min}^4} \tag{3.7}$$

Therefore

$$P = B\exp(-\kappa x)\left(\frac{1-\exp(-\kappa x_{\min})x_{\min}^4}{\exp(-\kappa x)x^4}\right) \tag{3.8}$$

Transforming to dimensionless variables $P' = P/B$ and $y = \kappa x$, we have

$$P' = \exp(-y)\left(1 - \frac{\exp(-y_{min})y_{min}^4}{\exp(-y)y^4}\right) \tag{3.9}$$

Thus

$$\ln P' = -y + \ln\left(1 - \exp(-y_{min})y_{min}^4 \frac{\exp(y)}{y^4}\right) \tag{3.10}$$

is the actual prediction of DLVO theory. This gives a linear relationship between ln P' and y (ln P and x) only if the second term is negligible. The obvious way to fix the Hamaker constant is to use it to fit the experimental fact that $y_{min} = 7$ ($\kappa x_{min} = 7$). We should therefore fit force-distance curves with the function

$$\ln P' = -y + \ln\left(1 - 2.2\frac{\exp(y)}{y^4}\right) \tag{3.11}$$

in DLVO theory, not with a pure Yukawa force. This function is plotted in Figure 3.9. As is clear from the figure, DLVO theory predicts collapse into the primary minimum at $y = 2.5$ or $y = 0.35y_{min}$. This is not only untrue for the results presented in Figure 3.7 here (we compressed a 10^{-3} M gel to $y = 0.2y_{min}$ without any crystallization, for example), it is also untrue of force-distance curves obtained by other authors for swollen clays [5–8] and likewise is untrue for the many force-distance curves obtained

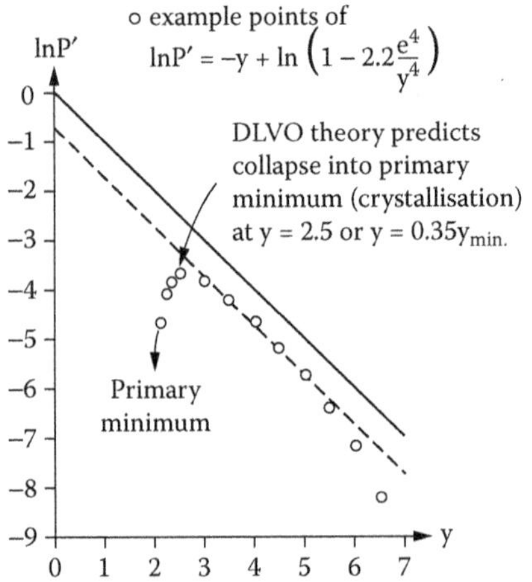

FIGURE 3.9 DLVO prediction for the force-distance curves. DLVO theory predicts collapse into the primary minimum at $y = 2.5$ or $y = 0.35y_{min}$. The straight line is for a pure repulsion.

with the mica force balance or atomic force microscope. Indeed, although the famous mica force balance experiments in the late 1970s [9] and early 1980s [10–12] were taken as evidence for DLVO theory, they were actually evidence for the following version of the DLVO potential

$$V_{\mathrm{DLVO}} = V_{\mathrm{R}} + \mathrm{NYSI} \tag{3.12}$$

where the term NYSI represents a "now you see it, now you don't" van der Waals potential. In using this strange nomenclature, I wish to emphasize the force of the argument; if you adjust the Hamaker constant to fit the position of the secondary minimum, you cannot account for the approximately linear ln P vs. x curves; *mutatis mutandis*, if you adjust the Hamaker constant to fit the approximately linear ln P vs. x curves (basically by setting it equal to zero), you cannot account for the position of the secondary minimum. It is only when the van der Waals force conveniently disappears that the DLVO potential reduces to a Yukawa potential. If DLVO theory is to account for the experimental fact that $y_{\mathrm{min}} = 7$ (of course, a constant number of Debye lengths is a natural prediction of the coulombic attraction theory), then the DLVO prediction for the force-distance curves, which includes the van der Waals force, is definitely false (see Figure 3.9).

Let us now analyze how the coulombic attraction theory accounts for the linear ln P vs. x plots observed in our model plate system and the linear ln P vs. r plots, where r is the radial distance, observed for colloidal spheres in laser tweezer experiments [13, 14]. I will occasionally make diversions into the field of spherical colloidal particles, the field for which the Sogami-Ise theory was originally developed. There were papers by Crocker and Grier in the 1990s [13, 14] that claimed to have refuted the Sogami-Ise theory by the measurement of linear ln P vs. r plots between unconfined charged colloidal spheres. The reason why our argument for plates applies in this instance to spheres is that the force-distance curves of both systems are dominated by the medium-range strong osmotic repulsion and a minimum at about 7 Debye lengths. The full analysis for plates is given by Smalley and Sogami [15], and a demonstration that the Sogami potential minimum for spheres lies at about 7 Debye lengths is given by Ise and Smalley [16]. Basically, to bring out the leading qualitative features for both systems, we can just choose the constant c in $V = B\exp(-\kappa x)(c - \kappa x)$ (see Equation 2.23) to give us a minimum at 7 Debye lengths. This function has been plotted as the curvilinear line in Figure 3.10.

The coulombic attraction theory prediction is the curvilinear line in Figure 3.10. We have added "error bars" to it at intervals of half-Debye screening lengths to represent the uncertainty inherent in any experimental determination of the force between colloidal particles, and have drawn a straight line through these simulated "experimental" points. It departs noticeably from the true curvilinear plot only at separations greater than 6 Debye lengths. The Yukawa prediction has been represented by the upper dotted line in Figure 3.10, adjusted to reproduce the same force at half the equilibrium separation (of course, with a purely repulsive potential, there is no equilibrium separation). It is straightforward to show that the experiments of Crocker and Grier [13, 14] are unable to distinguish between the Sogami and Yukawa potentials, as follows. (These authors, neglecting the NYSI (or V_{A}) term, refer to the Yukawa potential as the DLVO potential).

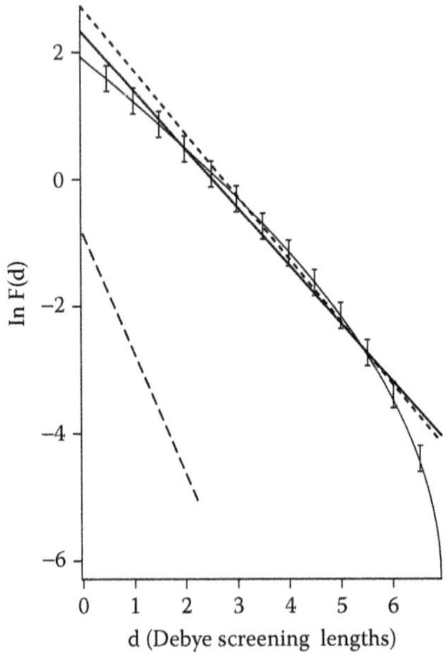

FIGURE 3.10 Force-distance curves. The curvilinear line is the prediction of the coulombic attraction theory. The straight line has been drawn through simulated error bars in the previous line to illustrate the approximately linear ln P vs. x (or ln P vs. r) plots expected from the coulombic attraction theory. The upper dotted line is the corresponding Yukawa prediction. If we scale the plot such that the 1996 data of Crocker and Grier [14] lies along the upper dotted line, the data for the same sample in their 1994 paper [13] is as represented by the lower dotted line.

Crocker and Grier used polystyrene latex spheres (Duke Scientific Catalog No. 5065A) in both their 1994 [13] and 1996 [14] papers. They used two-parameter fits to the data, expressed in terms of an effective surface charge Z^* and an effective Debye length κ^{-1}. No independent measurement of these parameters was made. In the 1996 paper, the Yukawa fit gave $Z^* = 5964$ and $\kappa^{-1} = 272$ nm. Let us now scale Figure 3.10 such that this fit lies along the upper dotted line. Once the scale has been fixed in this way, the Yukawa fit given in the 1994 paper, namely $Z^* = 1991$ and $\kappa^{-1} = 161$ nm, is represented by the lower dotted line. The idea that experiments on the same sample giving rise to the two dotted lines in Figure 3.10 can distinguish between the upper dotted line (Yukawa fit) and curvilinear plot (Sogami fit) seems implausible.

There are other major objections to Crocker and Grier's claim [13, 14], one of which we consider in Chapter 7. First, there is the timescale of the measurements. The laser tweezer experiments measure transient collisions, whereas equilibrium in these dispersions usually takes weeks or months to achieve [17]. Inspection of Figure 3.8 reveals that any transient experiment is going to be dominated by the medium-range strong repulsion. The minimum is so shallow that the true equilibrium nature of the forces takes weeks to reveal itself against the background of the Brownian motion. We also observed that when an n-butylammonium vermiculite gel had been compressed well inside its equilibrium position, it was very slow to respond. For example, the 0.03 M gel with an equilibrium *d*-spacing of 167 Å was compressed to a spacing of 85 Å over a period of 24 hours, at which time all applied stress was released. After three days its spacing had recovered to 140 Å, after several weeks to 160 Å, and after several months to 167 Å. Thus, although the gels do recover their original *d*-spacings, they re-expand more slowly than they come to equilibrium with an applied stress. The separations used in the laser tweezer experiments correspond to large applied stresses.

I may seem to be belaboring the point that the Crocker and Grier measurements do not distinguish between the Yukawa potential and the Sogami potential (in the same sense that you cannot measure the difference between the predictions for the force given by Newton's and Einstein's theories of gravitation with a couple of kilogram weights and a torsion balance), but the idea that the coulombic attraction theory had been "refuted" by these measurements spawned a spurious theory that tried to maintain a Yukawa potential for the pair interaction, with an effective attraction introduced by the interaction potential

$$V = V_{\mathrm{R}} + \mathrm{MMBI} \tag{3.13}$$

where MMBI is a "mysterious many-body interaction," known as a volume term, deemed too deep by the theoreticians for mere experimental scientists to understand (or use, apply, etc.) [18–22]. Again, in using a strange nomenclature, I wish to emphasize the force of an argument; a "volume term" that has no functional dependence on the separation of the particles is a fiction. *All* of the repulsive and attractive interactions are complicated many-body effects, and the result of these is an *effective* interaction that has a minimum separation at 7 Debye lengths. There is no hint of this important global fact in the volume-term theories, and the approach of references [18–22] is practically useless in colloid science.

The full potential studied in the new theory should really be written as

$$V = V_{\mathrm{R}} + \mathrm{MMBI} + \mathrm{NYSI} \tag{3.14}$$

because the van der Waals potential has a convenient way of disappearing for those who claim that Equation 3.4 supports DLVO theory. A classic example of this approach in the late 1990s was that of van Roij and Hansen [18, 19], who stated that "attraction mechanisms of electrostatic origin have been proposed, at the cost, though, of sacrificing the well-established DLVO potential." Their weighty considerations, applying the full armory of density functional theory, are summarized by Equation 3.13. This approach has been followed up by many authors [20–22], all aiming to explain the phase transitions observed in charged colloidal systems in terms of a linearized Poisson-Boltzmann equation with a state-dependent volume term. Recently, it has been shown that the prediction of a phase separation of a homogeneous suspension of spherical colloidal particles into dilute (gas) and dense (liquid) phases is a spurious result of the linearization scheme in these theories [23]. However, the least one can say about these theories is that there is an attempt to banish the van der Waals force from considerations of secondary minima phenomena, and one of the authors graciously stated that "one should credit Sogami and Ise with identifying the essential electrostatic origin of the effective attraction mechanism" [21].

In Chapter 2, we compared the predictions of the coulombic attraction theory and of the DLVO theory with the experimental results for the d vs. c curve for the n-butylammonium vermiculite system (see Figure 2.2). In the preceding paragraphs, we have compared both theoretical predictions with the corresponding force-distance curves and have concluded that the experimental results are consistent with either the Sogami potential or the Yukawa potential, but not with a DLVO potential that includes the van der Waals force. Let us now look at the experimental facts in greater detail, as they raise interesting questions about our interpretation of κ in macroionic solutions, and give us some vital information about the surface potential of the plates.

As all the ln P vs. x curves were linear within the limit of experimental accuracy, we could not do better than to fit these with the Yukawa force defined by Equation 3.1 and Equation 3.2. We consider first the gradients of the plots, given in Table 3.3. The value of κ in the external soaking solution is determined by Equation 3.15

$$\kappa^2 = 0.107 c_{ex} \tag{3.15}$$

where κ is in Å^{-1} and c_{ex}, the concentration of the external soaking solution, is in moles per liter. It is noteworthy that the κ values determined from the pressure–distance curves using Equation 3.4 are significantly different from those calculated from Equation 3.15. We label the average negative gradients of the least-squares fits as κ_{in}, referring to the internal kappa (inside the gel), as opposed to κ_{ex}, the value of the inverse Debye screening length in the external soaking solution. The corresponding values of these two variables are shown in columns 2 and 3 of Table 3.5. It appears that κ_{in} and κ_{ex} are approximately equal in the 0.01 M solution, but that κ_{in} is higher in the more dilute solutions and lower in the more concentrated solutions.

The value κ_{in} is plotted as a function of κ_{ex} in Figure 3.11. The straight line in Figure 3.11 represents $\kappa_{in} = \kappa_{ex}$, and the error bars on κ_{in} have been taken to be 10% of the average values obtained. The deviations from $\kappa_{in} = \kappa_{ex}$ are not caused by experimental uncertainties and are systematic. The relationship appears to be a linear

TABLE 3.5
Values of x_{min}, κ, and Concentration

c_{ex} (mol dm^{-3})	κ_{ex} (Å^{-1})	κ_{in} (Å^{-1})	$\langle x_{min} \rangle$	$\kappa_{in} x_{min}$
0.2	0.141	0.082	80	6.5
0.1	0.100	0.070	130	9.0
0.06	0.077	0.059	160	9.5
0.03	0.055	0.033	190	6.2
0.01	0.032	0.031	320	9.9
0.006	0.024	0.026	440	11.4
0.003	0.017	0.023	470	10.8
0.001	0.010	0.016	550	8.8

Note: Average value of $\kappa_{in} x_{min}$ = 9.0.

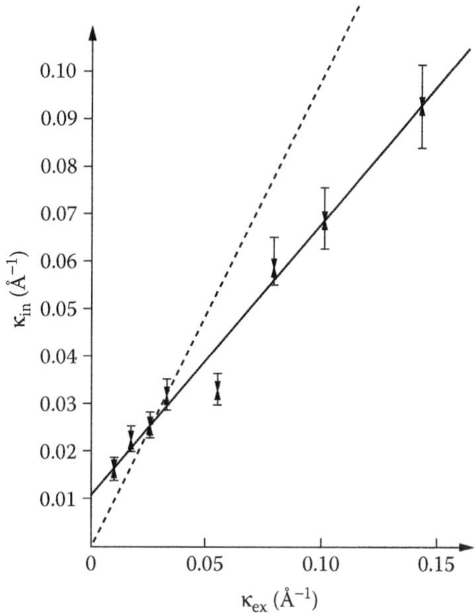

FIGURE 3.11 A plot of κ_{in} vs. κ_{ex}. The dotted line represents the relationship $\kappa_{in} = \kappa_{ex}$.

one; we should note that the one point that does not agree with such a linear relationship was obtained at $c_{ex} = 0.03$ M from our first two experiments with the pressure cell and may be the least reliable result. The slope of this line is equal to less than one-half; κ_{ex} varies by a factor of 14 across the concentration range studied, whereas κ_{in} varies by a factor of 5.6. We were unable to understand this result in terms of standard membrane equilibria [1].

Another interesting feature is revealed if we take the product $\kappa_{in}x_{min}$ at each of the eight salt concentrations studied. The results are shown in Table 3.5. The values of x_{min} given here differ slightly (though never by more than about 10%) from those given in Chapters 1 and 2 because we have taken averages over an increased number of measurements. There appears to be no systematic variation of $\kappa_{in}x_{min}$ across the wide range of concentrations studied (×200). Rather, it would appear that $\kappa_{in}x_{min}$ is approximately constant, its average value being equal to 9. This conclusion is confirmed by a plot of κ_{in} vs. $1/\langle x_{min}\rangle$, which is shown in Figure 3.12. The straight line in Figure 3.12 is a least-squares fit to a linear relationship. The error bars in $1/\langle x_{min}\rangle$ represent the spread of all observed *d*-values. The straight line passes through seven of these eight points, the exception again being the 0.03-M point, showing that $\kappa_{in}x_{min}$ is constant within the limits of experimental accuracy.

Before we can extract the surface potentials from the intercepts using Equation 3.2, we have to make a choice for κ. We thought it logical to use the values for κ determined by the slopes of the same plots, namely κ_{in}. This yields the values for ψ_0 shown in the third column of Table 3.4 given, as usual, in units of millivolts (mV). The scatter of values is large, but it appears that the surface potential is

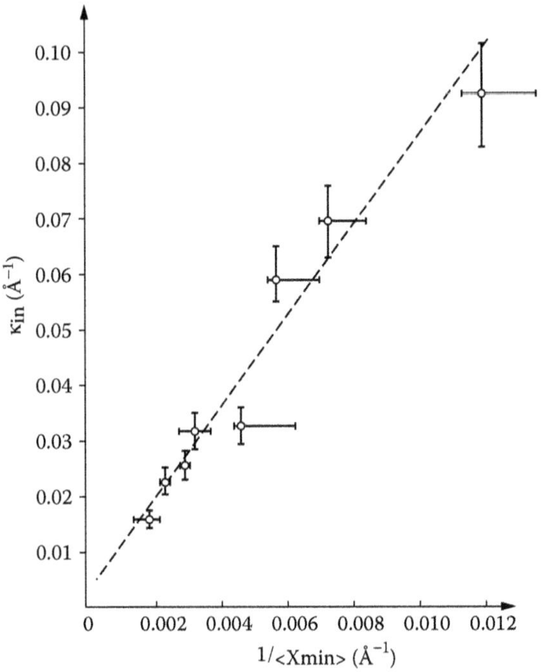

FIGURE 3.12 A plot of the internal kappa values κ_{in} obtained from the pressure–distance curves against $1/\langle x_{min}\rangle$, where $\langle x_{min}\rangle$ is the average observed *d*-spacing.

approximately constant across the whole concentration range, the average value being 70 mV in the pure-repulsion theory (Yukawa force fit). Low [8] has also found the surface potential to be independent of electrolyte concentration for the related smectite minerals, and our value of $\psi_0 \approx 70$ mV accords well with his result of $\psi_0 \approx 60$ mV for the smectites, given that vermiculites have a slightly greater "naked" charge density. The naked charge density refers to the plate charge if *all* the counterions were to desorb into the diffuse double layers when the clay plates go into solution, and Low uses σ_0 for this quantity, with σ_δ representing the charge density obtained from the measured surface potential ψ_δ. We have kept to the standard DLVO nomenclature for the surface potential, namely ψ_0; this corresponds to the physically measurable ψ_δ of Low's paper, where σ is obtained from ψ by the relation [2]

$$\sigma = \left[\frac{2n\varepsilon kT}{\pi}\right]^{1/2} \sinh\left(z/2\right) \tag{3.16}$$

where n is the number density of the ions in cm^{-3}, ε is the dielectric constant and the equation is written in cgs units. For $\psi_0 \approx 70$ mV, $z = 2.8$ and the value of the function $\sinh(z/2)$ is 1.9. Equation 3.16 then reduces to

$$\sigma = 4.26\times 10^{-2}\kappa \tag{3.17}$$

where σ is in units of number of charges per Å^2 and κ is in Å^{-1}. The approximate constancy of ψ_0 therefore implies a linear relation between the surface charge and κ. Because κ_{in} and κ_{ex} appear to be linearly related, this further implies that the surface charge increases with the square root of the concentration of the soaking solution. The premise that desorption into a diffuse double layer follows the square root of the concentration has been derived theoretically [24] from the original Gouy-Chapman theory [25–27] for a single diffuse double layer.

The fourth column of Table 3.4 has been calculated by inserting the eight κ_{in} values into Equation 3.17 using the smoothed value $\psi_0 \approx 70$ mV. The surface charge calculated from the cation-exchange capacity of the mineral on the assumption that all the counterions are desorbed into the diffuse double layers is 1.4×10^{-2} charges per Å^2. This represents the maximum possible surface charge in the system and is denoted as σ_{max}. This quantity is also known as the analytic surface charge, as it is obtained from chemical analysis. The right-hand column of Table 3.4 has been calculated by dividing the fourth column by σ_{max}. It represents the fraction of the n-butylammonium counterions that are present in the diffuse double layers. In the most dilute concentration studied, 10^{-3} M, the effective surface charge is approximately 5% of the naked charge. This number is also typical of charged spherical colloids in very dilute salt solutions [17]. In the most concentrated solution studied, 0.2 M, it is approximately 28%. The variation mirrors the 5.6-fold variation of κ_{in} across the concentration range because of the way we have used a smoothed potential in Equation 3.17. We will describe our attempts to measure the desorption of the n-butylammonium ions by diffuse neutron scattering in Chapter 8.

The approximate constancy of the surface potential over such a wide range of conditions is at first sight a surprising result. The chemical potential of the n-butylammonium ions must be equal throughout the gel phase and the external solution at equilibrium; in the external solution we can write

$$\mu_i = \mu_i^0 + RT \ln a \tag{3.18}$$

where a is the activity of the ions, or, less rigorously,

$$\mu_i = \mu_i^0 + RT \ln c \tag{3.19}$$

where c is their concentration. The fact that changing the concentration of the potential-determining ions by two decades in the external solution leads to no significant change in the surface potential, is certainly not the behavior we would therefore expect from classical chemistry if we treat the colloidal surface as a two-phase system of an electrode (solid) and an electrolyte solution (liquid) rather than as a single inhomogeneous (gel) phase. The Nernst equation states that we should expect a variation of about 60 mV per decade of concentration of potential-determining ions. This implies that if $\psi_0 \approx 70$ mV for a 0.1 M solution (see Table 3.4), it should be equal to 190 mV for a 0.001 M solution. This is clearly not the case. The fact that the surface potential is roughly constant with respect to the salt concentration in our model system is the experimental basis for the calculations of membrane equilibria described in Chapter 4.

REFERENCES

1. Crawford, R.J., Smalley, M.V., and Thomas, R.K., *Adv. Colloid Interface* Sci., 34, 537, 1991.
2. Derjaguin, B.V. and Landau, L., *Acta Physicochimica,* 14, 633, 1941.
3. Verwey, E.J.W. and Overbeek, J.Th.G., *Theory of the Stability of Lyophobic Colloids,* Elsevier, Amsterdam, 1948.
4. Braganza, L.F., Crawford, R.J., Smalley, M.V., and Thomas, R.K., *Clays Clay Min.,* 38, 90, 1990.
5. Rausell-Colom, J.A., *Trans. Faraday Soc.,* 60, 190, 1964.
6. Rausell-Colom, J.A., Saez-Auñón, J., and Pons, C.H., *Clay Min.,* 24, 459, 1989.
7. Viani, B.E., Roth, C.B., and Low, P.F., *Clays Clay Min.,* 33, 244, 1985.
8. Low, P.F., *Langmuir,* 3, 18, 1987.
9. Israelachvili, J.N. and Adams, G.E., *J. Chem. Soc. Faraday Trans.,* 1, 74, 975, 1978.
10. Pashley, R.M., *J. Colloid Interface Sci.,* 80, 153, 1981.
11. Pashley, R.M., *J. Colloid Interface Sci.,* 83, 531, 1981.
12. Pashley, R.M. and Israelachvili, J.N., *Colloids Surfaces,* 2, 169, 1981.
13. Crocker, J.C. and Grier, D.G., *Phys. Rev. Lett.,* 73, 352, 1994.
14. Crocker, J.C. and Grier, D.G., *Phys. Rev. Lett.,* 77, 1897, 1996.
15. Smalley, M.V. and Sogami, I.S., *Mol. Phys.,* 85, 869, 1995.
16. Ise, N. and Smalley, M.V., *Phys. Rev.* B, 50, 16722, 1994.
17. Ise, N. and Sogami, I.S., *Structure Formation in Solution: Ionic Polymers and Colloidal Particles,* Springer Verlag, Berlin-Heidelberg-New York, 2005.
18. van Roij, R. and Hansen, J-P., *Phys Rev. Lett.,* 79, 3082, 1997.
19. van Roij, R. and Hansen, J-P., *Prog. Colloid. Polym. Sci.,* 110, 50, 1998.
20. van Roij, R. and Evans, R., *J. Phys. Condens. Matter,* 11, 10047, 1999.
21. Warren, P.B., *J. Chem. Phys.,* 112, 4683, 2000.
22. Chan, D.Y.C., Linse, P., and Petris, S.N., *Langmuir,* 17, 4202, 2001.
23. Tamashiro, M.N. and Schiessel, H., *J. Chem. Phys.,* 119, 1855, 2003.
24. Grahame, D.C., *Chem. Rev.,* 41, 441, 1947.
25. Gouy, G., *J. Phys.,* 9, 457, 1910.
26. Chapman, D.L., *Phil. Mag.,* 25, 475, 1913.
27. Gouy, G., *Ann. Phys.* (N.Y.), 7, 129, 1917.

4 Membrane Equilibria for Interacting Macroions

Until we discovered the constancy of the surface potential from the uniaxial stress results, like most other people, I had been more interested in constant surface charge models. If you do not know how the valency of a macroion varies with the external conditions, it is reasonable to assume it to be constant unless given evidence to the contrary. Given the evidence that $\psi_0 \approx 70$ mV is roughly constant for the n-butylammonium vermiculite system, what other consequences follow from this? In particular, what happens if we apply the coulombic attraction theory with the constant surface potential boundary condition?

The situation illustrated in Figure 4.1a has by now become familiar. It depicts a gel composed of a parallel stack of plate macroions with a well-defined interplate spacing (in the colloidal range 10 to 100 nm) in equilibrium with a supernatant fluid. Let us think of the boundary of the gel as an effective membrane enclosing the macroions, transforming the picture into Figure 4.1b. This chapter is concerned with the calculation of the distribution of salt between the gel (I) and supernatant fluid (II) in the two-phase region of colloid stability.

Figure 4.1b is a typical illustration of the Donnan equilibrium [1]. A membrane impermeable to macroions (P^{n-}) but permeable to small ions (M^+, X^-) and solvent molecules (S) divides a solution into two regions. The situation is a common one in colloid science, and the fact that the equilibrium salt concentration in region II (the simple electrolyte solution), $[X^-]_{II}$, is greater than that in region I (the region occupied by the macroions), $[X^-]_I$, has been used in countless dialysis experiments. It is also well known [2] that equilibrium involves the establishment of not only a pressure difference but also an electrical potential difference across the membrane and that, in the simple case where the mobile ions behave as ideal solutes, the equilibrium condition is expressed as

$$\frac{[X^-]_{II}}{[X^-]_{I}} = \exp\left(\frac{e\varphi_{\text{Donnan}}}{k_B T}\right) = \exp(\Phi_{\text{Donnan}}) \tag{4.1}$$

where e is the electronic charge, k_B the Boltzmann constant, T the temperature, φ_{Donnan} the Donnan potential, or the difference in the average electric potential between the two regions, and Φ_{Donnan} a dimensionless Donnan potential, defined by

$$\Phi_{\text{Donnan}} = \frac{e\varphi_{\text{Donnan}}}{k_B T} \tag{4.2}$$

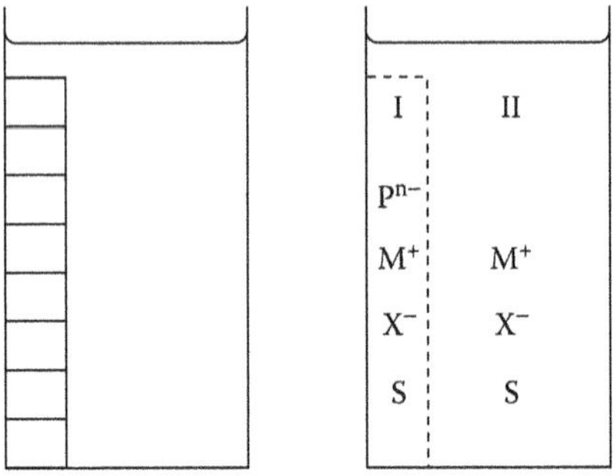

FIGURE 4.1 Schematic illustration of the phenomenon: (a) a swollen n-butylammonium vermiculite gel and (b) the components present in the two phases, the gel (I) and supernatant fluid (II). The dotted line represents an effective membrane enclosing the plate macroions P^{n-}. The symbols M^+, X^-, and S stand for univalent counterions (n-butylammonium ions), univalent co-ions (chloride ions) and solvent (water) molecules, respectively.

The case illustrated in Figure 4.1b is easily solved when we do not allow for interactions between the macroions. The solution given in Equation 4.1 follows simply from equating the chemical potentials of the small ions in regions I and II. What makes the Donnan equilibrium worth reconsidering is that the "membrane" in Figure 4.1a is not an externally imposed barrier; rather, it is created by the mutual interaction of the macroions themselves. There is no reason to believe that Equation 4.1, derived for the noninteracting case, should apply to the case where the volume of the macroionic phase is determined by the interactions. Indeed, we shall find that we need a new membrane equilibrium for dealing with the two-phase region of colloid stability [3, 4].

As elsewhere, we will confine our attention to a uni-univalent electrolyte solution for the sake of simplicity. The inverse Debye screening length κ is then given by

$$\left(\frac{\kappa^2}{8\pi\lambda_B}\right) = n \tag{4.3}$$

where n is the number density of the simple ions and λ_B is the Bjerrum length, defined by

$$\lambda_B = \frac{e^2}{4\pi\varepsilon_0\varepsilon_r k_B T} \tag{4.4}$$

with ε_0 being the permittivity of free space and ε_r the relative permittivity of the medium (dielectric constant). Such a definition of κ is only strictly valid in a *simple* electrolyte solution and is therefore, in the present case, strictly valid only in region II, the reservoir of macroion-free electrolyte solution.

We are going to make some brutal approximations to bring out the leading features of the behavior. The first one, which we shall reconsider later in the discussion, is that there is an infinite reservoir of salt solution outside the colloidal gel. In these circumstances, any salt that is fractionated out of the gel phase does not affect the salt concentration in the external fluid. This makes the equations much easier to handle when we are seeking the leading features, and the approximation can easily be corrected for afterward, as we shall see later in the chapter. Second, we are going to use the simplest possible form for the position of the Sogami minimum as determining the *d*-value in the gel phase. As we saw earlier, the full prediction for the position of the minimum x_{min} is given by Smalley [5] as

$$x_{min} = \frac{1}{\kappa}\left[4 + 2a\kappa \frac{\sinh(2a\kappa)}{1+\cosh(2a\kappa)}\right] \tag{4.5}$$

where $2a$ is the plate thickness. However, we will use the simplified prediction

$$\kappa x_{min} = 4 \tag{4.6}$$

As discussed previously, this approximation is certainly valid in dilute ($n < 10^{19}$ cm^{-3}) electrolyte solutions, and we use it for all salt concentrations as a first approximation. It is also straightforward to correct for the term containing a, as we shall see later in the chapter.

The third approximation is the least justifiable, but it is necessary if we are going to get a clear analytical prediction that we can understand. If we are to make use of the coulombic attraction theory prediction that the plates should be localized at a distance of four Debye screening lengths, we obviously have to put in a value for κ. We have already seen in Chapter 3 that the kappa value in the gel phase can be different than that calculated on the basis of the salt concentration in the external fluid, and there is no *a priori* reason why κ_{in} and κ_{ex} should be the same. Our third approximation is to calculate the κ value in the gel phase by assuming that neither the macroions nor the counterions contribute to the screening length. Equation 4.3 suffices to define κ in a simple ionic solution, but for macroionic solutions the situation is not so simple. To quote Schmitz [6], "There is some ambiguity in polyelectrolyte and colloidal systems as to whether or not the macroion contributes to the calculation of the screening length." The confusion surrounding the many possible methods for calculating κ has even led Yamanaka et al. [7] to raise the question of whether the Debye-Hückel screening length as defined by simple electrolyte theory has any meaning at all in polyion systems. Here we adopt the position used by Schmitz of treating the Debye-Hückel screening length as a scaling parameter for the purpose of characterizing the solution properties of charged particles that differ in ionic concentrations. We have already seen that κ_{in} is significantly lower than κ_{ex} in the n-butylammonium system when the salt concentration is very dilute. This situation would come about if κ_{in} were determined solely by a salt concentration lower than that in the external phase, as expected for a Donnan-type equilibrium. When we adopt the condition that the kappa to be inserted into Equation 4.6 can be calculated assuming that neither the macroions nor the counterions contribute to the screening length, "the Debye-Hückel screening length is calculated *solely* from the added electrolyte concentrations, and it follows that the macroion-macroion interaction potential is independent of the concentration of macroions" [6] (Schmitz's italics).

Our third approximation is a brutal one, but it is consistent with the prediction that the minimum in the pair potential always lies at $4/\kappa$ and should give reasonable results in the limit of dilute electrolyte solutions. With this simplification, we can obtain a clean analytical prediction for the salt-fractionation effect.

The distribution of small ions between regions I and II is described as the expulsion of a certain amount of ions of the same sign as the colloid. This ion exclusion effect is described in the manner of Klaarenbeek [8], whose result is encapsulated in a quantity g that expresses the ratio of the co-ion deficit to the total double-layer charge. Let us write

$$g = \frac{N_{\mathrm{de}}}{Z_0} \tag{4.7}$$

where, for a negatively charged wall, N_{de} is the deficit of negative ions per unit area and Z_0 is the total surface charge per unit area. The relationship between the surface charge, Z_0, and the dimensionless surface potential, Φ_{s}, is given by Sogami et al. [9] as

$$Z_0 = \left(\frac{\kappa}{2\pi\lambda_{\mathrm{B}}}\right)\sinh\left(\frac{\Phi_{\mathrm{s}}}{2}\right) \tag{4.8}$$

As g can also be calculated in terms of the surface potential (see below), N_{de} can be calculated as a function of Φ_{s}. When N_{de} is known, the average number density of the deficit, n_{de}, can be calculated from Equation 4.6; the position of the coulombic attraction theory minimum is connected to the average electrolyte concentration in the gel phase because it defines the average volume occupied per macroion. The number density of the deficit of negative ions in the gel phase is given by

$$n_{\mathrm{de}} = \frac{gZ_0}{l} \tag{4.9}$$

where l is the half separation of the plates (this definition is used because g is defined by an integral over a region of monotonically decaying electric potential; see below) given by

$$l = \frac{2}{\kappa} \tag{4.10}$$

Combining Equation 4.8 and Equation 4.10 gives

$$n_{\mathrm{de}} = \left(\frac{\kappa^2}{8\pi\lambda_{\mathrm{B}}}\right)2g\sinh\left(\frac{\Phi_{\mathrm{s}}}{2}\right) \tag{4.11}$$

The prefactor is none other than the number density of the simple ions in the supernatant fluid, n (see Equation 4.3). Equation 4.11 can therefore be rewritten as

$$n_{\mathrm{de}} = 2gn\sinh\left(\frac{\Phi_{\mathrm{s}}}{2}\right) \tag{4.12}$$

Because the number density of the deficit of negative ions in the gel phase is defined by

$$n_{de} = n - n_{gel} \tag{4.13}$$

where n_{gel} is the average number density of the negative ions in the fluid bounded by the macroions, the calculation yields n_{gel} in terms of the experimentally measurable and controllable quantity n as

$$n_{gel} = n\left[1 - 2g\sinh\left(\frac{\Phi_s}{2}\right)\right] \tag{4.14}$$

This in turn gives the ratio of these two quantities, the salt-fractionation factor s, as

$$s = \frac{n}{n_{gel}} = \frac{1}{1 - 2g\sinh\left(\frac{\Phi_s}{2}\right)} \tag{4.15}$$

and the result can be compared with Equation 4.1, the equation for the Donnan equilibrium. To complete this calculation, we need the quantity g.

For a single flat double layer, g is given by the integral [8]

$$g = \frac{\int_0^{\Phi_s} (1 - e^{-\Phi}) \frac{dx}{d\Phi} d\Phi}{\int_0^{\Phi_s} (e^{\Phi} - e^{-\Phi}) \frac{dx}{d\Phi} d\Phi} \tag{4.16}$$

where Φ_s, the dimensionless surface potential, has been expressed as a positive definite quantity (this definition was not used in reference [4]). This is a strictly accurate procedure for an isolated plate macroion, since the potential can be taken to decay to zero at an infinite distance from the plates. However, when the double layers overlap, the electric potential is a monotonically decaying function of distance in the half region from the plate surface to the midplane between two plates, and has some finite value, Φ_d, at the midplane. In this case, g should be calculated as

$$g = \frac{\int_{\Phi_d}^{\Phi_s} (1 - e^{-\Phi}) \frac{dx}{d\Phi} d\Phi}{\int_{\Phi_d}^{\Phi_s} (e^{\Phi} - e^{-\Phi}) \frac{dx}{d\Phi} d\Phi} \tag{4.17}$$

In general, the integrals in Equation 4.16 and Equation 4.17 cannot be solved analytically, so we adopt a fourth approximation in our general approach, namely a mathematical one for g. Before we do this, I wish to emphasize that the crucial new aspect of our approach is that the coulombic attraction theory permits a calculation of the salt-fractionation effect precisely because it defines the volume bounded by the macroions. Equation 4.7 and Equation 4.9 tell us that we can always calculate the total number of co-ions expelled in terms of an electrical integral for g, but that we can only use this to calculate the number density of co-ions expelled, and hence the salt-fractionation factor s, if we have a definite expression for the half separation of the plates l, which is taken to be $2/\kappa$. We then estimate g within this fixed box by two methods, using (a) Equation 4.16 and (b) Equation 4.17.

METHOD (A) NONOVERLAPPING DOUBLE LAYERS

Using Equation 4.16 to estimate g in region I at first appears to be a strange procedure because the case $\Phi_d = 0$ corresponds to no overlap of the double layers. The variation of electric potential with distance is then as sketched by the thin line in Figure 4.2. The thicker line in Figure 4.2 represents an exponentially decaying potential function (see method (b) below). Let us first pursue the mathematics of the $\Phi_d = 0$ case. The crude approximation

$$\int_0^{\Phi_s} f(\Phi)\frac{dx}{d\Phi}d\Phi = f\left(\frac{\Phi_s}{2}\right)\frac{dx}{d\Phi}\bigg|_{\Phi_s/2} \times \Phi_s \tag{4.18}$$

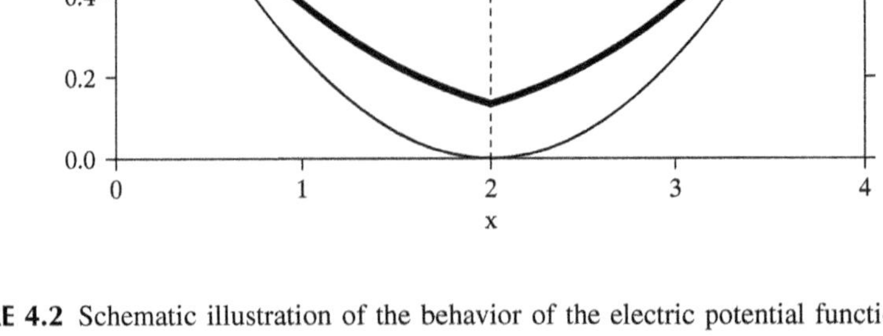

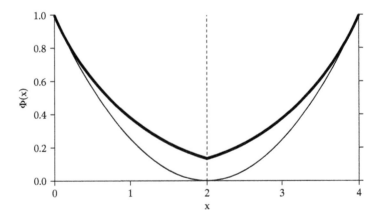

FIGURE 4.2 Schematic illustration of the behavior of the electric potential function in the two models. The thin line is a sketch of $\Phi(x)$ for the approximation $\Phi_d = 0$, and the thick line is a plot of an exponentially decaying function, for which $\Phi_d = 0.135\Phi_s$. The potential $\Phi(x)$ and the distance x are given in units of the dimensionless surface potential Φ_s and the Debye screening length $1/\kappa$, respectively. The dashed line at $x = 2/\kappa$ designates the midplane.

can be used provided that the integrand $f(\Phi)$ is a smoothly varying function. This is the fourth approximation in our approach, and the final one. Applying this approximation to both the numerator and denominator in Equation 4.16 gives

$$g = \frac{(1 - e^{\Phi_s/2}) \left.\dfrac{dx}{d\Phi}\right|_{\Phi_s/2} \times \Phi_s}{(e^{\Phi_s/2} - e^{-\Phi_s/2}) \left.\dfrac{dx}{d\Phi}\right|_{\Phi_s/2} \times \Phi_s} \tag{4.19}$$

and the cancellation of terms reproduces Klaarenbeek's result [8] as

$$g = \frac{(1 - e^{-\Phi_s/2})}{(e^{\Phi_s/2} - e^{-\Phi_s/2})} = \frac{(1 - e^{-\Phi_s/2})}{2\sinh\left(\frac{\Phi_s}{2}\right)} \tag{4.20}$$

Substituting Equation 4.20 into Equation 4.15, we obtain

$$s = \frac{n}{n_{\text{gel}}} = \exp\left(\frac{\Phi_2}{2}\right) \tag{4.21}$$

for the salt-fractionation factor s.

Comparing Equation 4.21 with Equation 4.1, we see that the use of Equation 4.20 leads to a mathematical formula for s similar to the formula for the Donnan equilibrium. Indeed, if we identify Φ_{Donnan} with $\Phi_s/2$, the expressions are identical. Because the average value of the electric potential function is equal to zero in region II and the approximation used in Equation 4.18 corresponds to taking the average potential in the gel to be $\Phi_s/2$, the relationship $\Phi_{\text{Donnan}} = \Phi_s/2$ is consistent with the definition of the Donnan potential in the case where the activity coefficients of the simple ions are taken to be unity. Because this assumption is implicit in the use of the Poisson-Boltzmann equation that underlies the preceding calculations, Equation 4.21 is the same as Equation 4.1.

Method (a), the use of the position of the coulombic attraction theory minimum with the $\Phi_d = 0$ value for g, leads to the same mathematical formula for s as that expressing the Donnan equilibrium. However, we cannot say that this constitutes a derivation of the Donnan equilibrium from the coulombic attraction theory because it does not correspond to a physical limit. If $\Phi_d = 0$ really were the case, there would be no reason for the macroions to remain at the minimum position of the interaction potential. Nevertheless, the identity of the two expressions is an interesting result. Because Equation 4.20 is derived in the case in which there is no double layer overlap and Equation 4.1 (the Donnan equilibrium) is likewise derived without reference to the overlap of the double layers, it is precisely in this limit that the calculation should reproduce the Donnan equilibrium. The fact that it does gives us some confidence that our approximations are not too drastic and should lead to physically significant results when applied to overlapping double layers.

METHOD (B) OVERLAPPING DOUBLE LAYERS

Of more interest from the physical point of view is the case where the double layers overlap, represented by a finite value of Φ_d. The exact mean field potential for plate macroions in region I is expressed in terms of elliptic integrals [9], but it varies approximately exponentially with distance from the surface, and we use the simple function $\Phi(x) = \Phi_s \exp(-\kappa x)$ to represent it, as shown by the thicker line in Figure 4.2. For $x = 2/\kappa$, the midplane potential is $e^{-2}\Phi_s = 0.135\Phi_s$, and we take this as the lower limit of the integral in Equation 4.17. Although the use of an exponentially decaying potential is itself a rough approximation to the electric potential function in region I (for example, it does not fulfill the symmetry requirement that the gradient of the potential goes to zero at the midplane), greater accuracy is not warranted in using the analogue of Equation 4.18 to solve for g. Using this approximation, g is recalculated as

$$g = \frac{(1-e^{-\Phi_s/2}) - (1-e^{-\Phi_d/2})}{(e^{\Phi_s/2} - e^{-\Phi_s/2}) - (e^{\Phi_d/2} - e^{-\Phi_d/2})} \tag{4.22}$$

for interacting double layers.

The values for g obtained from Equation 4.22 do not seem to be very different from those obtained from Equation 4.20, as shown in Table 4.1 and Figure 4.3a. It is easy to see that g must be equal to $^1/_2$ as Φ_s tends to zero by expanding the exponentials in the linear approximation (Debye limit). Naturally, Equation 4.15 gives us $s = 1$ in this limit, as an uncharged layer does not expel co-ions and salt is equally distributed between regions I and II. However, as shown in Table 4.2 and Figure 4.3b, the predicted salt-fractionation effect obtained by substituting Equation 4.22 into Equation 4.15 is markedly different from the Donnan equilibrium.

For low surface potentials ($\Phi_s < 1$), the salt-fractionation factors calculated (a) via the coulombic attraction theory and the electric integral solved by Equation 4.20 (i.e., via the Donnan equilibrium) and (b) via the coulombic attraction theory and the electric

TABLE 4.1
Some Illustrative Values of the Fractional Co-Ion Defect (g) as a Function of the Surface Potential (ψ_0)

ψ_0 (mV)	g ($\psi_d = 0$)	g ($\psi_d = 0.135\ \psi_0$)
0	0.50	0.50
25	0.38	0.35
50	0.27	0.24
70	0.20	0.17
75	0.18	0.16
100	0.12	0.09
200	0.02	0.01

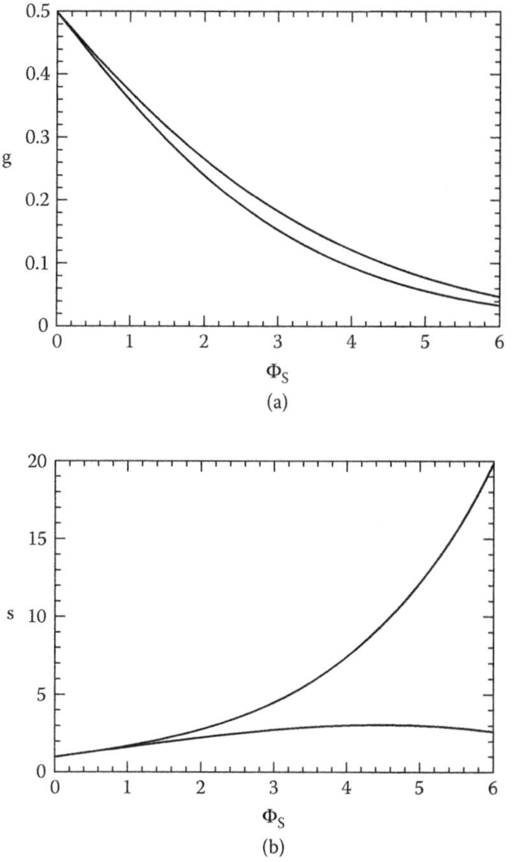

FIGURE 4.3 (a) The quantity g and (b) the salt-fractionation factor s as functions of the dimensionless surface potential Φ_s. In both (a) and (b), the upper and lower curves are obtained from the approximations $\Phi_d = 0$ and $\Phi_d = 0.135\Phi_s$, respectively.

TABLE 4.2
Some Illustrative Values of the Salt-Fractionation Factor (s) as a Function of the Surface Potential (ψ_0)

ψ_0 (mV)	s
0	1.0
25	1.6
50	2.3
70	2.8
75	2.9
100	3.1
200	2.3

integral solved by Equation 4.22 (i.e., for overlapping double layers) are practically indistinguishable, but for $\Phi_s > 2$ ($\phi_s > 50$ mV) the calculations diverge significantly, with the overlap of the double layers serving to suppress the salt-fractionation factor. A remarkable feature of the new membrane equilibrium for interacting macroions is that s has a turning point as a function of Φ_s. For surface potentials greater than 104 mV, the behavior of s is qualitatively different from that predicted by Donnan; s from method (a) always increase exponentially with Φ_s, whereas s from method (b) actually decreases as a function of Φ_s beyond this point. It is particularly noteworthy that s for interacting double layers is a very slowly varying function of Φ_s for $\Phi_s > 4$, taking the values 3.1 and 2.3 at $\Phi_s = 4$ and 8, respectively (see Table 4.2). At high surface potentials, the salt-fractionation effect is nearly constant as a function of surface potential. Indeed, s takes a value between 2.1 and 3.1 for any surface potential greater than 40 mV, up to unphysically high values.

We may have made some rough approximations, but we now have the leading features of the behavior that we sought, in terms of elementary analytic functions. The predictions for s can, of course, be tested. If ϕ_s is constant with respect to electrolyte concentration, we predict that s will also be constant with respect to c, which appears to be a novel result. For ϕ_s = const = 70 mV, as in the n-butylammonium vermiculite system, we predict $s = 2.8$. This second prediction is markedly different from the value $s = 4.0$ predicted from the Donnan equilibrium. In the next chapter we will describe our experimental tests of these two predictions.

Before we can make a meaningful comparison between theory and experiment, we need to relax the first of our approximations, namely that there is an infinite reservoir of salt solution outside the colloidal gel. In our early experiments on the n-butylammonium vermiculite system [10–12], described in Chapters 1 to 3, we used small crystals in large beakers of soaking solution, such that the volume fraction r of clay crystals in the condensed-matter system was less than 0.01. With $r < 0.01$, any salt fractionated out of the gel phase has a negligible effect on the salt concentration in the external fluid, which is mathematically but not experimentally convenient. We therefore have a strong practical reason for investigating the case $r \geq 0.01$. A yet stronger motivation for investigation of the case $r \geq 0.01$ is that it will enable us to construct a complete electrical theory of clay swelling [4].

The n-butylammonium vermiculite system is an example of a three-component system of a monodisperse colloid, electrolyte and solvent. There are four constituents in the macroionic solution — the negatively charged clay plates, n-butylammonium ions (counterions), chloride ions (co-ions), and water — but these may not vary independently because they are subject to the restriction that

$$[n-\text{Bu}^+]=[\text{plate}^-]+[\text{Cl}^-] \tag{4.23}$$

in an obvious notation. Hence the number of components is 4 − 1 = 3, and the composition of the system can be specified in terms of two concentration variables. In the following we shall refer to the solvent as water and to the electrolyte as salt.

Let us recall the schematic illustration of the raw phenomenon of the clay swelling in Figure 1.4. In the cases studied in Chapters 1 to 3, V^* was always much greater than V, the volume occupied by the macroions. We now define V_m to be the volume occupied by the macroions in the coagulated (crystalline) state, as in Figure 1.4a in the vermiculite system. This is an experimentally controlled variable. We define the sol concentration r by

$$r = \frac{V_m}{V^*} \tag{4.24}$$

where V^* is the total volume of the condensed-matter system. In the case of swelling illustrated by the transition from Figure 1.4a to Figure 1.4b, we have seen that V^* decreases by approximately 0.1% [10]. This is a very small fractional volume change compared with that observed in V, so in the following we ignore the electrical constriction of the solvent that accompanies swelling; that is, we take V^* = const. Although the phase boundary between the crystalline and gel states was investigated with respect to temperature and hydrostatic pressure [10, 11], we now restrict attention to T and P constant so that we can represent the phase behavior of the system on triangular graph paper.

For the n-butylammonium vermiculite system, the molecular weight of the salt is 109.5. The partial molal volume of the salt is nearly independent of the salt concentration, its average value being equal to 110 ml/mol [13]. The density of the salt is therefore approximately 1 g/cm^3, so volume fractions and mass fractions are identical for the simple electrolyte solution in this case. The n-butylammonium chloride salts out from simple electrolyte (macroion-free) solutions at about 4.5 M [14]. This value does not depend significantly on the presence of macroions in the solution and so is independent of the sol concentration r. This enables us to draw in the left-hand wedge in Figure 4.4a, which gives the phase diagram in mass fractions. For the density of the n-butylammonium vermiculite crystals, whose known mass used in the experiment determines V_m, we have taken the value of 1.86 g/cm^3 appropriate for the Eucatex samples studied in reference [14].

The left-hand wedge of Figure 4.4 represents a three-phase region of crystalline clay, solid salt and saturated salt solution, and this is labeled as region IV in the schematic Figure 4.4c. In Figure 4.4c, the r = 0.1, c = 0.01 M point has been placed at the center of the triangle and the scale has been distorted to show all four regions clearly on the same plot. The reason for this distortion is that if we plot the phase diagram in ordinary weight percentages, as shown in Figure 4.4a, the most interesting physics, that of gel formation, is confined to too small a region on the right-hand edge. We gain more insight into the phase behavior of the system if we rescale the molecular weight of the salt by a factor of 1000, as shown in Figure 4.4b. This has the effect of fanning out the plot around the c = 0.01 M line and shows the relevant phase boundaries clearly.

As the salt concentration is decreased below 4.5 M, the solid salt phase disappears, but the clay crystals do not swell until c is decreased below 0.2 M (at T = 4°C, P = 1 atm). To a first approximation, this value is also independent of r,

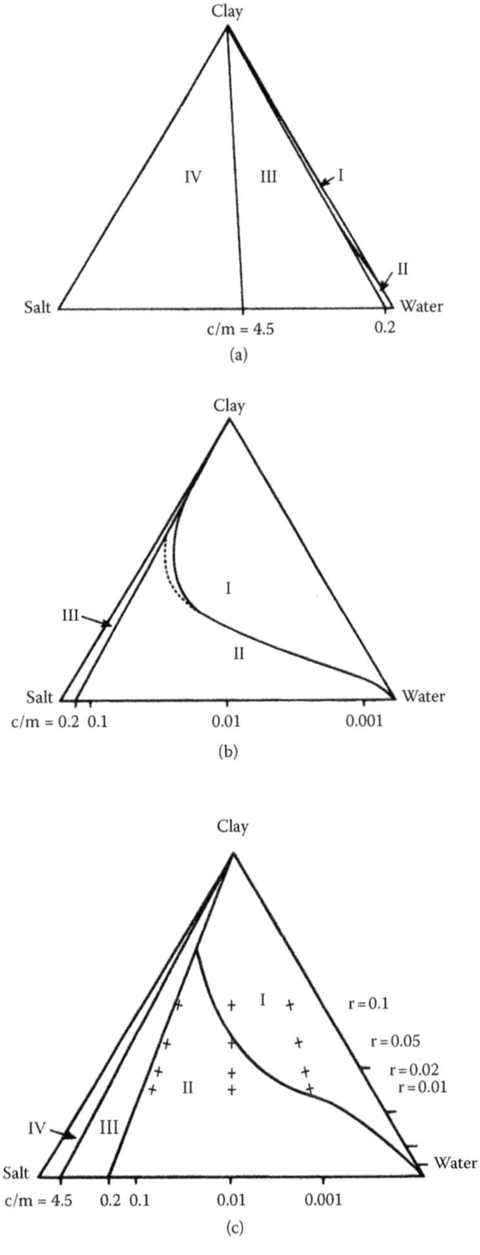

FIGURE 4.4 Phase diagram of the three-component system of clay (n-butylammonium vermiculite), salt (n-butylammonium chloride) and water at $T = 4°C$, $P = 1$ atm. (a) A straight mass-fraction plot. (b) The mass-fraction plot obtained when the molecular weight of the salt is rescaled by 1000. The curved phase boundary is calculated in the text, with the dotted line representing the reappraised phase boundary. (c) A schematic plot where the $r = 0.1$, $c = 0.01$ M point has been placed at the center of the triangle, and the scale has been distorted to show all four regions clearly. The labelling of the regons is explained in the text. The crosses indicate the points studied by Williams et al. [14].

which enables us to draw in the central wedge (region III) in Figure 4.4. This represents a two-phase region of salt solution and crystalline clay. In both regions III and IV, the macroions are in their primary minimum (crystalline, coagulated) state. To the right of region III, in electrolyte concentrations $c < 0.2$ M, the clay absorbs water macroscopically and swells into the secondary minimum (gel, flocculated) state, giving us the one-phase (I) and two-phase (II) regions of colloid stability. So far we have only looked at the two-phase region, where the gel exists in equilibrium with a supernatant fluid (dilute salt solution). The experiments described in Chapters 1 to 3 correspond to a cut parallel to the base of Figure 4.4c, closer to the base than the three points marked at $\{r = 0.01, c = 0.001 \text{ M}\}$, $\{r = 0.01, c = 0.01 \text{ M}\}$ and $\{r = 0.01, c = 0.1 \text{ M}\}$. For such a cut, the approximation that there is an infinite reservoir of salt solution is a reasonable one. However, as the amount of clay in the system is increased, and we head toward the apex of Figure 4.4c, we reach a point when the clay soaks up all of the available solution and we enter the one-phase region (I) under high-r, low-c conditions.

We now have two major aims: first, to calculate the position of the phase boundary between the one-phase and two-phase regions, already sketched in the figure; and second, to calculate the d-value between the layers as a function of r and c. I think that if we can do these two things, we can reasonably claim to have constructed a complete electrical theory of clay swelling.

We might guess from the outset that our first aim would be solved by the position of the coulombic attraction theory minimum given by Equation 4.6, where x_{min} is equal to 4 Debye lengths. However, this is the position of the minimum in the thermodynamic pair potential (Gibbs pair potential) between macroionic plates in the low r limit (the infinite reservoir limit). We now know that there will be a "feedback" effect on the swelling because the salt concentration in the finite reservoir will increase as the gel expels salt. We will therefore start the calculation in the low r limit, where we already have an analytical result. The calculation is much simpler if we adapt it straightaway to the n-butylammonium vermiculite system (in a way that is obvious to generalize) and get κ in lab units from

$$\kappa^2 = 0.107c \tag{4.25}$$

where κ is expressed in units of Å^{-1} and c is the salt concentration in moles per liter. A major advantage of our simple result that the salt-fractionation factor s is constant with respect to c (for a constant surface potential system) is that it enables us to solve the two problems mooted above analytically. We start by calculating the d-value in the two-phase region as an analytic function of r and c. This in turn enables us to calculate the phase boundary between the one-phase and two-phase regions of clay swelling.

The coulombic attraction theory prediction $x_{min} = 4/\kappa$ only yields an experimentally testable prediction of x_{min} vs. c if we know what to insert for c in Equation 4.25. In reference [5], before we obtained the uniaxial stress results, I had implicitly assumed

$$\kappa_{gel} = \kappa_{ex} = \kappa_b \tag{4.26}$$

where the symbols represent the inverse Debye screening lengths in the gel, supernatant fluid and globally, respectively. This gives an unambiguous experimental definition of κ because the supernatant fluid is a simple electrolyte solution, and its experimentally controlled, known concentration then determines κ_{ex} via Equation 4.25. However, Equation 4.26 does not take into account the new generalized Donnan equilibrium, which for the n-butylammonium vermiculite gels (with $\phi_s = \text{const} = 70$ mV) reduces to

$$\frac{c_{ex}}{c_{gel}} = s = 2.8 \tag{4.27}$$

To calculate κ in the gel phase, we continue to adopt the simple procedure used above in calculating s, that of assuming that neither the macroions nor the counterions contribute toward the screening length. Because κ is then determined solely by the electrolyte concentration, it seems logical to choose

$$\kappa_{gel}^2 = 0.107c_{gel} \tag{4.28}$$

in accordance with Equation 4.25. An immediate benefit of this choice is that it gives us a new prediction for the d-value as a function of c in the limit of low sol concentrations ($r < 0.01$). Within this limit, the volume of the whole condensed-matter system V^* is much greater than the volume occupied by the crystalline macroions, V_m, expressed as $V^* \gg V_m$, so the salt that is excluded from the gel phase when V_m expands to V (the volume occupied by the gel phase) has a negligible effect in the large volume $V^* - V$ that remains as the supernatant fluid. For the n-butylammonium vermiculite gels, we then have the relation $c_b = c_{ex} = 2.8c_{gel}$. Together with Equation 4.25 and Equation 4.28, this implies $\kappa_{ex} = 1.7\kappa_{gel}$, and substitution of κ_{gel} into Equation 4.6, where x_{min} is equal to 4 Debye lengths, gives the new prediction: $\kappa_{ex}x_{min} = 6.7$. Perhaps fortuitously, this gives quantitative agreement with the observed interlayer spacings in the low-r, low-c limit [11, 12]. For example, for $r < 0.01$, $c_{ex} = 0.001$ M ($\kappa = 0.01$ Å^{-1}), the prediction is $x_{min} = d_{gel} = 670$ Å, and the observations lie between 550 Å and 680 Å, as we have seen. In this sense, we could regard this method for calculating kappa as a way of paramaterizing the coulombic attraction theory to fit the data in the limit of low salt and sol concentrations. The uniaxial stress data at $r < 0.01$, $c = 0.001$ M give a value for κ_{gel} in agreement with the one calculated by this method. The agreement may again be fortuitous, but it shows that we have at least a consistent parameterization in the low r, low c limit. We must bear in mind that when we extend the approach to higher r values below, we can only expect qualitative accuracy from an approach that neglects the contribution of the counterions to the screening length. We should also note that the present approach of equating x_{min} with d_{gel}, the d-value in the gel phase, is consistent with the approach introduced in Chapter 2, has been used throughout this chapter, and will be used in the following discussion.

For the case $r > 0.01$, the equation $c_b = c_{ex}$ is no longer applicable because the salt excluded from the gel phase has a significant feedback effect on the concentration in the supernatant fluid. However, the problem is solved easily using Equation 4.6, Equation 4.27, and Equation 4.28 together with two simple conservation principles and the assumption that the swelling in clay minerals is perfectly homogeneous, so that we can relate the microscopic d-value directly to the volume of the macroion phase.

The equation of conservation of salt is

$$V_{\text{gel}}c_{\text{gel}} + V_{\text{ex}}c_{\text{ex}} = (V^* - V_{\text{m}})c_{\text{a}} \tag{4.29}$$

where c_{a} is the concentration of the added salt and where we have assumed that the macroions are initially salt-free. V_{gel} and V_{ex} are the volumes of the gel phase and supernatant fluid at equilibrium, and c_{gel} and c_{ex} are their respective (average) salt concentrations. The equation of the conservation of volume of the condensed-matter system (electrical constriction effects ignored) is

$$V_{\text{gel}} + V_{\text{ex}} = V^* \tag{4.30}$$

and the equation relating the volume of the gel phase to the d-value in the gel phase is

$$\frac{V_{\text{gel}}}{V_{\text{m}}} = \frac{d_{\text{gel}}}{2a} \tag{4.31}$$

where $2a$ is the c-axis repeat distance of the crystalline mineral, an experimentally determined quantity.

It is straightforward to show that the six equations (Equation 4.6 and Equation 4.27 to Equation 4.31) lead to a quadratic equation for d_{gel} in terms of r and c. For n-butylammonium vermiculite, $2a$ = 19.4 Å [11], and the result is conveniently expressed as

$$c(1-r)d_{\text{gel}}^2 + 14rd_{\text{gel}} - 420 = 0 \tag{4.32}$$

where c is the concentration of added salt in moles per liter, r is the sol concentration and d_{gel} is the d-value in angstroms, given by the positive (physical) root of the equation. Some illustrative values of d_{gel} as a function of r and c are given in Table 4.3.

As an example of a complete set of parameters specifying the solution at an $\{r,c\}$ point, we consider the addition of 99 cm^3 of an 0.001 M solution to 1 cm^3 of pure (salt-free) clay. In this case, V_{gel} = 30 cm^3, $c_{\text{gel}} = 4.4 \times 10^{-4}$ M, V_{ex} = 70 cm^3, and $c_{\text{ex}} = 1.2 \times 10^{-3}$ M. Note that c_{ex} is significantly higher than the value of the added electrolyte concentration, illustrating the feedback effect clearly. The result for the d-value therefore differs significantly from that obtained in the limit of infinite sol dilution ($r \to 0$, c = 0.001 M, d_{gel} = 670 Å); solution of Equation 4.32 gives d_{gel} = 580 Å, corresponding to an approximately 30-fold expansion of the crystal. Doubtless one of the reasons why we obtained a spread of results for the d-value at $c = 0.001$ M, $r < 0.01$ in the early experiments [11, 12] was that the sol concentration, although always small, was poorly controlled. This was corrected in the protocol for the experiments to be described in the next chapter.

The third column in Table 4.3 has been calculated as $d_{\text{max}} = 19.4/r$, which defines the maximum obtainable spacing when the clay has soaked up all the solvent. Both

TABLE 4.3
Sol Concentration Effect in n-Butylammonium Vermiculite Swelling in 0.001-M, 0.01 M and 0.1 M Soaking Solutions

	c = 0.001 M	
r	d_{gel} (Å)	d_{max} (Å)
→0	650	•
0.01	580	1940
0.02	530	970
0.05	390	390
	c = 0.01 M	
r	**d_{gel} (Å)**	**d_{max} (Å)**
→0	210	•
0.01	200	1940
0.02	190	970
0.05	180	390
0.10	150	190
0.15	130	130
	c = 0.1 M	
r	**d_{gel} (Å)**	**d_{max} (Å)**
→0	65	•
0.1	61	194
0.2	57	97
0.3	53	65
0.4	49	49

d_{gel} and d_{max} are monotonically decreasing functions of r, and the boundary with the one-phase region is defined by

$$d_{gel} = d_{max} = d^* = \frac{19.4}{r^*} \tag{4.33}$$

where r^* is the sol concentration at the phase boundary between the one-phase and two-phase regions of clay swelling. For the three example cases given in Table 4.3, namely c = 0.001 M, c = 0.01 M, and c = 0.1 M, r^* = 0.05, 0.15, and 0.39, respectively. The calculation can be repeated for any arbitrary value of c, which gives a complete calculation of the $\{r,c\}$ phase boundary. Some illustrative results are given in Table 4.4.

At the high salt concentration end of Table 4.4, for $c > 0.1$ M, we are really pushing mean field theory beyond its limits. This is shown by the third column of

TABLE 4.4
The (*r*,*c*) Phase Boundary in n-Butylammonium Vermiculite Swelling

c (M)	r^*	d^* (Å)	N_κ
10^{-5}	0.0050	3900	4.0
10^{-4}	0.016	1200	4.0
10^{-3}	**0.050**	**390**	**4.0**
0.003	**0.084**	**230**	**4.1**
0.01	**0.15**	**130**	**4.2**
0.03	**0.24**	**80**	**4.6**
0.1	**0.39**	**50**	**5.1**
0.2	0.50	39	5.7
0.3	0.57	34	6.1
0.4	0.62	31	6.5

Note: Boldface entries designate the best characterized experimental range.

the table, which gives the *d*-value at the phase boundary. For $c = 0.4$ M, $d_{gel} = 30$ Å, which is the approximate thickness of the clay plates with two adsorbed layers of n-butylammonium ions [14, 15]. This is the region of crystallization into the primary minimum, which occurs at $c = 0.2$ M at 4°C, where the phase diagram in Figure 4.4b has been plotted. The predicted increase in swelling between 0.4 and 0.2 M soaking solutions is 8 Å. This contains an insufficient number of layers of water molecules (about three) for electrical theory, treating the solvent as a dielectric medium, to be appropriate. The results for $c > 0.1$ M have therefore to be regarded with caution: $c = 0.1$ M should be considered the upper limit of reliability in Table 4.4, and the results have been given in plain type rather than boldface type to reflect this. The reason for the lower limit of reliability in Table 4.4 is completely different. In this region, the theory should be most applicable, but the range of salt concentrations below 10^{-3} M is difficult to control experimentally because of the effect of salt leaching out of the crystals, even after they have been subjected to a lengthy washing procedure [14].

It is apparent from columns 1 and 3 of Table 4.4 that d^* varies approximately as the inverse square root of *c*. This is made explicitly clear in the fourth column of the table, which gives N_κ, the number of Debye screening lengths between the plates at the phase boundary. N_κ has been calculated as $N_\kappa = \kappa d^*$, where $\kappa^2 = 0.107c$, with *c* being the global salt concentration. This is an easily controlled experimental variable, and we can use it because all the salt is trapped between the plates in the one-phase region of clay swelling. In this case, $c_b = c_{ex} = c_{gel}$, and we can use $\kappa^2 = 0.107c_b$ to define κ unambiguously.

The equilibrium separation of the particles in the two-phase region is roughly inversely proportional to the concentration of the salt solution, and so is the phase

TABLE 4.5
Reappraised (*r*,*c*) Phase Boundary

c (M)	4/κ (Å)	f (a,κ)(Å)	x_{min}(Å)	N_κ
0.001	400	2	402	4.0
0.003	231	3	234	4.1
0.01	126	6	132	4.2
0.03	73	9	82	4.5
0.1	40	15	55	5.5

boundary. Equation 4.6, which contains this dependence, is, as we have seen, only the small κ approximation to the equilibrium separation of plate macroions, the full equation [5] being

$$x_{\min} = \frac{1}{\kappa}\left\{4 + 2a\kappa\frac{\sinh(2a\kappa)}{1+\cosh(2a\kappa)}\right\} \tag{4.34}$$

The two terms in Equation 4.34 are given separately in the second and third columns of Table 4.5, where the second term has been labeled $f(a,\kappa)$.

In column 4 of Table 4.5 we have given the predictions for $x_{\min}$ as a function of c, and in the fifth column we have expressed this as the appropriate number of Debye screening lengths, N_κ. The right-hand columns of Table 4.5 and the boldface part of Table 4.4 show a remarkable agreement. This is no accident. The essential feature of the function given by Equation 4.34 for $x_{\min}$ is that, for separations greater than about $4/\kappa$, the plates attract each other, so that the clay does not soak up any more solvent beyond this point, and the two-phase region begins. This suggests an easy method for calculating directly the position of the phase boundary by using $x_{\min} = d^*$. The reappraised phase boundary obtained by this method has been plotted as the dotted line in Figure 4.4b. It only deviates from the solid curve, that obtained by approaching the phase boundary from the two-phase region, at the high-salt-concentration end. Around $c = 0.1$ M the salt-fractionation effect deviates from the calculated value of 2.8 because this has been derived from the approximate Equation 4.6, so we should not expect quantitative agreement here. It has already been remarked that this salt concentration marks the upper limit of applicability of mean field theory.

Of course, we could have guessed from the outset that the position of the secondary minimum in the coulombic attraction theory would define the phase boundary, but then we would have missed out on some interesting results in the two-phase region. Although some of the approximations we have made in Chapter 4 were brutal, they have enabled us to bring out the leading features of applying the coulombic attraction theory to a constant surface potential system [4, 16]. These are (1) that the ratio s of the salt concentration in the supernatant fluid to the average salt concentration in the gel phase is constant, (2) that for a surface potential of 70 mV, s is equal to 2.8, and (3) that for such a constant surface potential, the interlayer d-values in the gels are as given by Equation 4.32. These are bold predictions, way beyond the scope of DLVO theory and likewise nowhere to be found in the work of the many theoreticians who have used sophisticated computer programs to study the interactions of plate macroions. We tested these predictions on the n-butylammonium vermiculite gels. As you can imagine, I would not be writing this book if we had not obtained excellent agreement between theory and experiment. The next chapter describes the experimental work in detail.

REFERENCES

1. Donnan, F.G., *Chem. Rev.,* 1, 73, 1924.
2. Everett, D.H., *Basic Principles of Colloid Science,* Royal Society of Chemistry, London, 1988.
3. Smalley, M.V., *Prog. Colloid Polym. Sci.,* 97, 59, 1994.
4. Smalley, M.V., *Langmuir,* 10, 2884, 1994.
5. Smalley, M.V., *Mol. Phys.,* 71, 1251, 1990.
6. Schmitz, K.S., *Macroions in Solution and Colloidal Suspension,* VCH, New York, 1993.
7. Yamanaka, J., Matsuoka, H., Kitano, H., Ise, N., Yamaguchi, T., Saeki, S., and Tsubokawa, M., *Langmuir,* 7, 1928, 1991.
8. Klaarenbeek, F.W., *Over Donnan-evenwichten bij solen van arabische gom,* Ph.D. thesis, Utrecht, 1946.
9. Sogami, I.S., Shinohara, T., and Smalley, M.V., *Mol. Phys.,* 76, 1, 1992.
10. Smalley, M.V., Thomas, R.K., Braganza, L.F., and Matsuo, T., *Clays Clay Min.,* 37, 474, 1989.
11. Braganza, L.F., Crawford, R.J., Smalley, M.V., and Thomas, R.K., *Clays Clay Min.,* 38, 90, 1990.
12. Crawford, R.J., Smalley, M.V., and Thomas, R.K., *Adv. Colloid Interface Sci.,* 34, 537, 1991.
13. Desnoyers, J.E. and Arel, M., *Can. J. Chem.,* 45, 359, 1967.
14. Williams, G.D., Moody, K.R., Smalley, M.V., and King, S.M., *Clays Clay Min.,* 42, 614, 1994.
15. Rausell-Colom, J.A., *Trans. Faraday Soc.,* 60, 190, 1964.
16. Smalley, M.V., Schärtl, W., and Hashimoto, T., *Langmuir,* 12, 1331, 1996.

COLOR FIGURE 5.4 Photograph of a typical sample after swelling for two weeks. The height of the sample jar is approximately 5 cm.

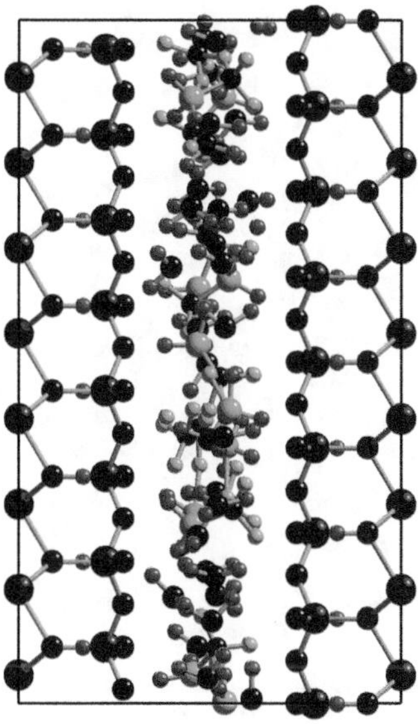

COLOR FIGURE 9.4 Molecular graphics snapshot of hydrated methylammonium vermiculite, with color coding to show the species that we target by using isotopic substitution in conjunction with neutron diffraction. Red = hydrogen; blue = nitrogen; green = methyl; black = unsubstituted species (oxygen, magnesium, silicon and carbon). In the system illustrated, the water content is 2.5 molecules per methylammonium counterion, and the clay layer spacing along the c*-axis is 12.3 Å. During the neutron diffraction experiments, the samples were aligned so that the clay layers were perpendicular to the scattering vector Q.

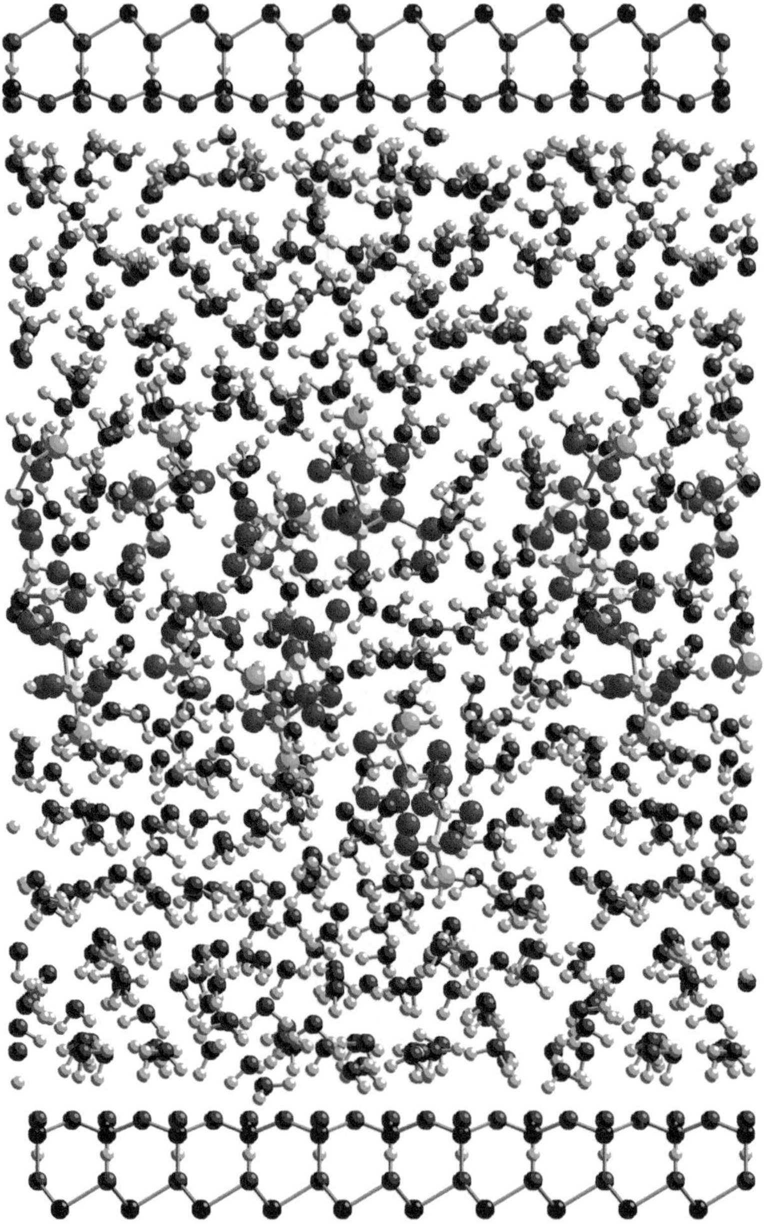

COLOR FIGURE 9.10 Molecular graphics snapshot of a d = 43.6 Å propylammonium vermiculite gel, illustrating how the majority of the counterions are separated from the charged clay surfaces by two layers of partially oriented water molecules. The atoms are color-coded as follows: labeled hydrogen/deuterium (red); oxygen (blue); hydrogen (white); carbon (yellow); nitrogen (green); and silicon, magnesium, aluminum (black).

5 The Sol Concentration Effect in Clay Swelling

In the early 1990s, the instruments at the Institut Laue-Langevin (ILL), Grenoble, were out of action due to refurbishment of the reactor. We therefore moved our SANS (small-angle neutron scattering) experiments to the LOQ (low Q) instrument at ISIS, Didcot, U.K. The standard of the canteen lunches went down when our neutron-scattering experiments moved from Grenoble to Didcot, but the LOQ instrument and the new, bigger cells turned out to be perfect for studying the sol-concentration effect in n-butylammonium vermiculite swelling [1].

We studied the 12 $\{r,c\}$ points marked by crosses in the triangular graph of the three-component system shown in Figure 4.4c. I had gotten used to seeing the semitransparent yellow gels evident at $r = 0.01$, and it was quite a shock to see the heavy, blackish gels formed at $r = 0.1$. At $r = 0.1$, with the clay occupying 10% of the condensed-matter system, it is obvious that the maximum expansion of the dark, translucent crystals is tenfold; there is not any more water to soak up. As we saw in Chapter 4, at $c = 0.001$ M (we will shortly see why this is the lowest concentration that can be studied in practice), the phase boundary lies at around a 20-fold expansion in a $c = 0.001$-M solution according to the coulombic attraction theory, so the $r = 0.1$, $c = 0.001$-M point lies inside the one-phase region. This was the case experimentally. Under these conditions the clay had soaked up all the available solution and had formed a single pastelike mass, resembling a little clod of earth. For unoriented materials, this is the region of pastes rather than suspensions. The $r = 0.1$, $c = 0.001$-M sample was the only one studied in the one-phase region because of our leading interest in the salt-fractionation effect. The phase boundary at $c = 0.001$ M should be at around $r = 0.05$ according to the theory of Chapter 4, and this was roughly the case. Near the boundary, it is not easy to be sure if there is excess water clinging to the sides of the gels. There seemed to be a gradual variation of properties between the $r = 0.01$ and $r = 0.1$ samples at the two intermediate sol concentrations studied, namely $r = 0.02$ and $r = 0.05$.

The neutron-diffraction experiments were done using the LOQ small-angle scattering diffractometer at the ISIS spallation neutron source, described in reference [2]. The LOQ instrument was chosen because it is designed for looking at spacings between 30 and 900 Å simultaneously and so was ideal for studying the gels. The samples were prepared 2 weeks in advance to ensure fully homogeneous swelling. The crystals, in their fully hydrated $d = 19.4$ Å crystalline phase, were washed and dried to remove excess water. After drying, the crystals were cut to dimensions of approximately $6 \times 6 \times 0.5$ mm. These were individually weighed, and the volume of the crystal in its fully hydrated state was calculated using the density. Solutions of n-butylammonium chloride in D_2O were made up at concentrations of 0.1, 0.01 and 0.001 M and added to the crystals to produce the appropriate sol concentrations. The cells were then sealed with Parafilm to (a) prevent exchange of H_2O in the atmosphere with D_2O in the solution and (b) prevent D_2O from evaporating, which would cause a decrease in the volume of solution and an increase in the sol concentration *r*.

Quartz sample cells of dimensions $1 \times 1 \times 5$ cm were used, quartz being practically transparent to neutrons at the wavelengths utilized on LOQ. Due to the fragile nature of the swollen gels, particularly those soaked in the more dilute solutions (with respect to both *r* and *c*) where the extent of swelling is greatest, the vermiculite crystals were placed directly into the quartz cells after weighing and left to swell *in situ* to minimize the amount of handling required when swollen. It was necessary to swell the crystals in D_2O rather than H_2O solutions because of the large, incoherent, neutron-scattering cross section of hydrogen that would otherwise have obscured the scattering of interest. The small-angle scattering from D_2O is of low intensity and completely unstructured over the Q range used here.

ISIS is a spallation, or pulsed, source and so time-of-flight instrumentation is used to measure the neutron wavelengths. This also means that the instrument can employ a "fixed" (or constant θ) scattering geometry. This in turn has the advantage that in a single run/measurement, data can be collected over a wide range of Q values, where Q, the modulus of the scattering vector, is given by

$$Q = \frac{4\pi \sin\theta}{\lambda} \tag{5.1}$$

where λ is the neutron wavelength and 2θ is the scattering angle. On LOQ a wavelength-limiting chopper operating at 25 Hz provides an incident neutron beam with wavelengths between 2 and 10 Å. Typically the Q-range of LOQ is 0.007 to 0.2 Å^{-1}.

The geometry of the neutron-diffraction experiments is shown in Figure 5.1a. Due to the parallel plate geometry of the gel samples under investigation, the incident beam was collimated by passage through a 2×8-mm slit placed immediately before the samples. This produced a rectangular beam of neutrons parallel to the clay plates rather than the 8-mm-diameter circular collimation usually employed. The samples were mounted on a temperature-controlled 20-position sample changer under the control of the instrument computer. Neutrons scattered by the gel samples were recorded on a large two-dimensional "area" detector, the active area of which was 64×64 cm, situated approximately 4.5 m behind the samples. This was a ^3He gas detector, software coded as 64×64 pixels $\times$ 102 time channels. A typical scattering pattern from a gel sample (over all wavelengths) is shown in Figure 5.1b.

Figure 5.1b represents the corrected data from the detector, with intensity contours plotted as a function of the scattering vector perpendicular (Q_z) and parallel (Q_y) to the layers. As shown in Figure 5.1a, the samples were mounted with the clay layers horizontal to the ground, and the scattering pattern in Figure 5.1b consists of two lobes of intensity above and below the plane of the layers in the gel, defined as the *xy*-plane in Figure 5.1a. If the layers were perfectly parallel, then coherent

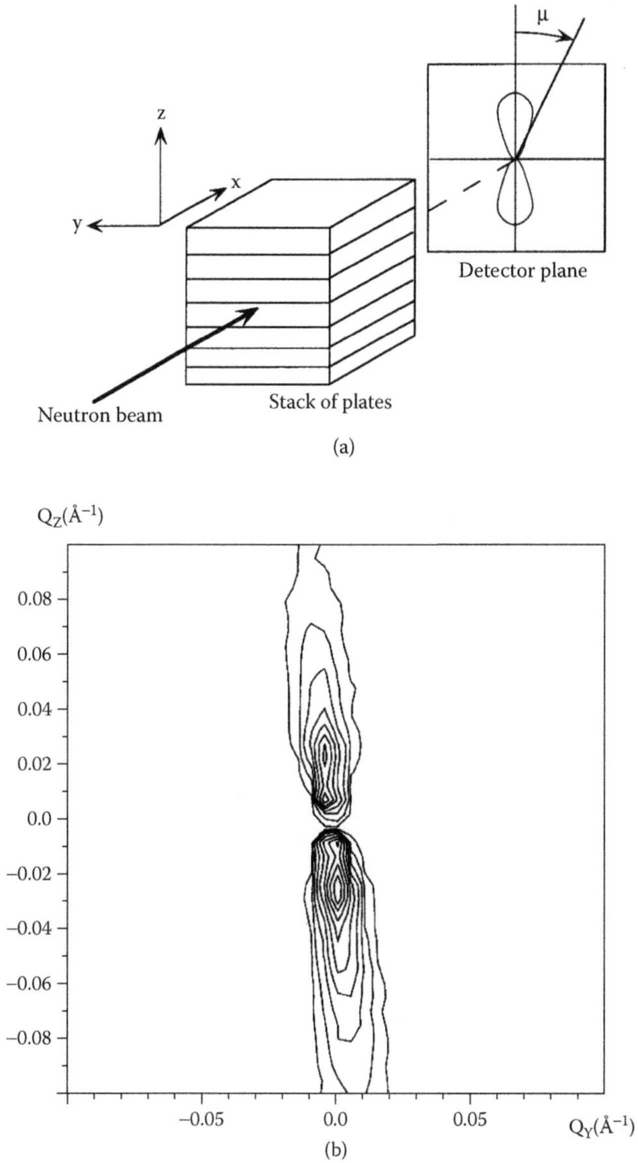

FIGURE 5.1 (a) The geometry of the LOQ scattering experiments. (b) A contour plot of a typical neutron-scattering pattern. The momentum transfers (Å^{-1}) perpendicular and parallel to the silicate layers lie along the vertical and horizontal axes, respectively.

scattering would only occur along the z-axis perpendicular to the layers, but in reality the gels have mosaic spreads of about 5 to 10°, as we saw earlier, so scattering occurs to either side of this axis such that almost all of the scattering falls within two cones of 60° azimuthal width above and below the horizontal plane, corresponding to $\mu = 30°$ in Figure 5.1a.

The slight tilting of the line of most intense scattering from the z-axis in Figure 5.1b and the asymmetries in the cones of scattering in the $+z$ and $-z$ directions are due to the average orientation of the stack of layers lying slightly out of the xy-plane in the sense of rotations about the x-axis and y-axis, respectively. Such tilts merely reflect the fact that the clay does not swell exactly perpendicular to the ground. The existence of pronounced intensity maxima within the lobes of scattering is due to the interference effect between the vermiculite layers, which turned out to have well defined d-values for all the $\{r,c\}$ points studied. The intensity $I(Q)$ was obtained as a function of Q by radially summing the intensity at constant Q around the detector while limiting the azimuthal range to between ±30° of the line joining the intensity maxima in the upper and lower lobes, which was normally close to the z-axis. This gave a sharper diffraction pattern compared with radial summations over all μ. All of the sample runs were corrected for the background scattering using a quartz cell containing D_2O, with the appropriate transmission corrections made using the straight-through beam. In this way the scattering patterns analyzed arose purely from the gels.

The object of the experiments was to find not only the c-axis spacing d in the gel but also the phase-transition temperature T_c for 16 sets of $\{r,c\}$ conditions. These were $r = 0.01$, 0.02, 0.05 and 0.10 for each of $c = 0$, 10^{-3}, 10^{-2} and 10^{-1} M, although the actual salt concentrations were greater due to salt trapped inside the crystals that is released when they swell. From the point of view of controlling the salt concentration at finite ($r \geq 0.01$) sol concentrations and measuring the salt-fractionation effect, an unfortunate experimental feature of the vermiculite samples is that they inevitably contain a certain amount of salt in addition to any added in the aqueous solution. This arises because the samples are prepared by soaking in molar solutions. The samples were washed before use, normally 20 washes in demineralized water at 80°C (well above the swelling transition temperature $T_c = 40$°C previously observed in the low r, low c limit), but there was always some salt left in the crystals. This creates an artifact in the effect of the sol concentration because the salt trapped in the crystal is diluted to a varying extent depending on the volume fraction. Our experiments to correct for the trapped-salt effect are described in detail below, together with our measurements of the salt-fractionation effect. Here we simply note that the real (global) salt concentrations at the 16 points studied are as given in Table 5.1. It is clear from the table that we cannot go below 10^{-3} M in practice.

TABLE 5.1
Real Salt Concentrations (mM) in Neutron-Diffraction Experiments and Laboratory T_c Experiments

r	c = 0 M	c = 10^{-3} M	c = 10^{-2} M	c = 0.1 M
0.01	1 ± 0.5	2 ± 0.5	11 ± 0.5	101 ± 0.5
0.02	2 ± 1.0	3 ± 1.0	12 ± 1.0	102 ± 1.0
0.05	5 ± 2.5	6 ± 2.5	15 ± 2.5	105 ± 2.5
0.10	10 ± 5.0	11 ± 5.0	20 ± 5.0	110 ± 5.0

Initially, the sample changer was maintained at 2°C. For each $\{r,c\}$ point four samples were studied; this was necessary due to sample-to-sample variation and to ensure at least one of each set gave a clear gel peak. It was found that good statistics could be obtained in about 15 minutes. The sample giving the clearest scattering pattern was selected from each set of four, and those 16 were run at higher temperatures, increasing the temperature in 2°C steps. These temperature scans were only run for about 3 minutes because beam time was limited and because the collapse of a gel could be seen by the disappearance of the peak, thus eliminating the need for the very good statistics of the earlier sets of data. Samples of the diffraction patterns obtained are shown in Figure 5.2 to illustrate various features.

Figure 5.2 shows two sets of results for four "identical" samples at 2°C, (a) at $r = 0.05$, $c = 0$ and (b) at $r = 0.1$, $c = 0.001$ M. It is clear that intense peaks are

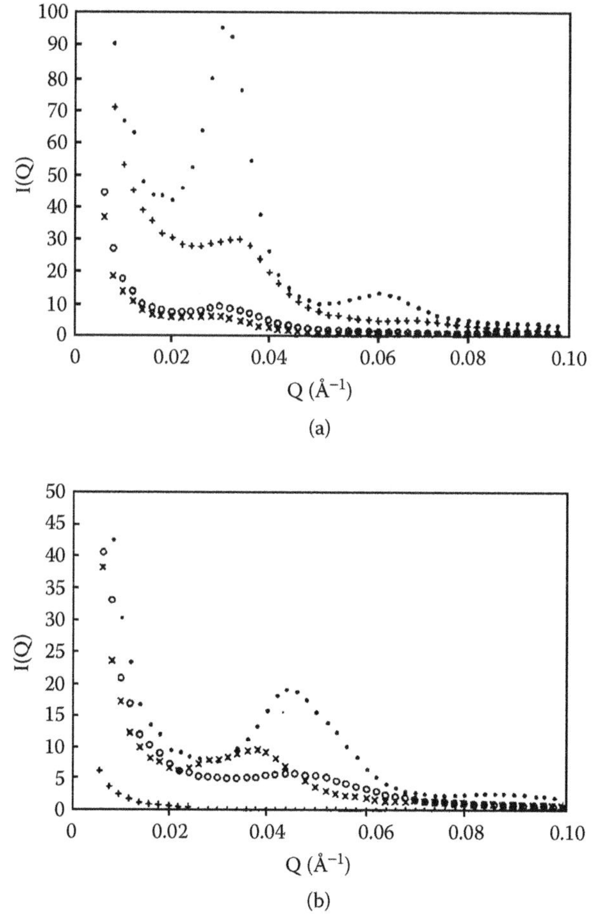

FIGURE 5.2 $I(Q)$ (arbitrary units) vs. Q patterns obtained at 2°C for four different samples. The top panel shows the four traces obtained from the four different samples prepared at $r = 0.05$, $c = 0$. The lower panel shows the four traces obtained at $r = 0.1$, $c = 0.001$ M.

observed for samples at higher sol concentrations, as with the case $r < 0.01$ studied previously. Figure 5.2a shows a set for which there is little sample-to-sample variability. This was a typical set. The four plots show peaks of different intensities, but this is due merely to physical differences in the size of the gels; the wider the section of gel in the neutron beam, the more intense is the scattering. One of the plots has a distinct second-order peak, which was fairly rare. Figure 5.2b shows the worst set of data in terms of sample-to-sample variability. In this case there were two plots giving clear peaks with different Q values as well as one that gave no peak at all. Although the vermiculite gels form a useful model of a one-dimensional colloid, it is worth remembering that the gels come from crystals mined out of the ground and are not some ideal desktop model. Individual samples can vary greatly, in part due to variation of the trapped salt concentration quantified below. Gels can also be incompletely homogeneous, particularly near the boundary of the one-phase and two-phase regions. In general, experimental results across the whole three-component phase diagram can be no more accurate than within about 10% due purely to sample variation. Accordingly, in no case will our results be given to more than two significant figures. The d-values were obtained by applying the simple equation

$$d = \frac{2\pi}{Q_{max}} \tag{5.2}$$

to the Q-value at the maximum of the first-order diffraction effect, Q_{max}, and averaging over the samples that gave clear peaks. The results are given in Table 5.2. For the example traces shown in Figure 5.3, the d-value was equal to 210 Å in case (a) and 150 Å in case (b). In an ideal world, we might have analyzed the neutron-diffraction patterns quantitatively using a one-dimensional paracrystalline lattice model [3], but whenever we saw higher order peaks they were linear in Q, and the sharpness of the diffraction effect permits the immediate approximate calculation.

Having determined d as a function of r and c, we proceeded to do the same for T_c. Although we have no theoretical prediction for this quantity, an investigation of

TABLE 5.2
Interlayer *d*-Values and Phase Transition Temperatures T_c at 16 {*r*,*c*} Points Obtained by Neutron Diffraction Experiments

r	c = 0 M	c = 10^{-3} M	c = 10^{-2} M	c = 0.1 M
0.01	d = 520 Å $T_c > 34$°C	d = 480 Å $T_c > 34$°C	d = 310 Å $T_c = 27 \pm 1$°C	d = 120 Å $T_c = 13 \pm 1$°C
0.02	d = 450 Å $T_c = 34 \pm 1$°C	d = 490 Å $T_c > 34$°C	d = 305 Å $T_c = 27 \pm 1$°C	d = 120 Å $T_c = 12 \pm 1$°C
0.05	d = 210 Å $T_c = 20 \pm 1$°C	d = 210 Å $T_c = 23 \pm 1$°C	d = 200 Å $T_c = 22 \pm 1$°C	d = 110 Å $T_c = 9 \pm 1$°C
0.10	d = 135 Å $T_c = 20 \pm 1$°C	d = 150 Å $T_c = 22 \pm 1$°C	d = 130 Å $T_c = 19 \pm 1$°C	d = 95 Å $T_c = 9 \pm 1$°C

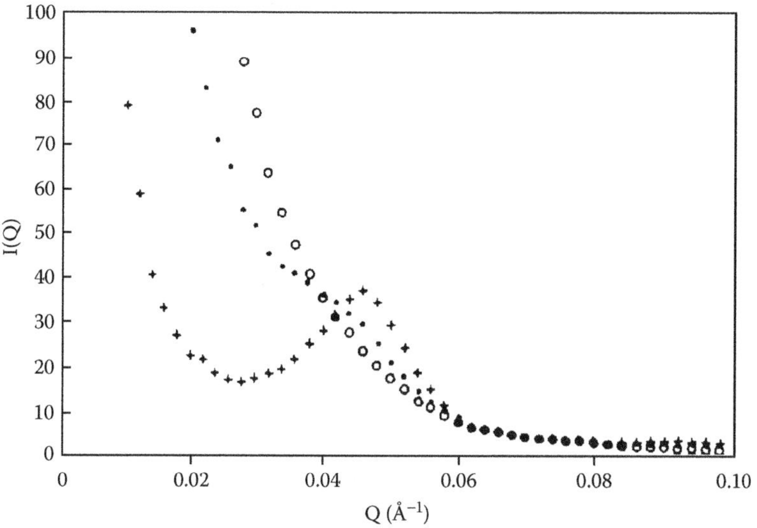

FIGURE 5.3 $I(Q)$ vs. Q plots obtained for one sample at $r = 0.1$, $c = 0.01$ M. The crosses (+), stars (*) and circles (o) show the scans obtained at 18, 20 and 22°C, respectively. In this case $T_c = 19 \pm 1$°C.

the $\{r,c,T_c\}$ phase boundary (at $P = 1$ atm) seemed worthwhile. Figure 5.3 shows the runs for a single sample at $r = 0.1$, $c = 0.01$ M, at 18, 20 and 22°C. The traces clearly show the disappearance of the peak as the gel collapses. In this case $T_c = 19 \pm 1$°C. The T_c values obtained in this way at the 16 $\{r,c\}$ points studied are given together with the corresponding d-values in the gel phase in Table 5.2.

Neutron scattering experiments provide the sharpest information on d and T_c because they probe the microscopic structure of the gels. However, the samples used in the neutron experiments were too small to provide a reliable estimate of s. The salt-fractionation factor was therefore investigated by simple volumetric and gravimetric experiments on larger samples, as shown in Figure 5.4. Prior to performing an experiment swelling the clay, the crystals were first washed thoroughly to remove any molar solution that might be trapped in surface imperfections. This was achieved by rinsing the crystals with 500 cm^3 of distilled water at 60 to 80°C, as described previously. Even after this thorough washing procedure, it was discovered that an unknown amount of salt was always present in the original vermiculite crystal. After over a year of soaking in hot molar chloride solutions to obtain the pure n-butylammonium form of the vermiculite, and after storage in molar n-butylammonium chloride solutions, it is not surprising that the crystals, which contain cracks and pores, contain a substantial amount of n-butylammonium chloride. This salt leaches out into the solution upon swelling, and so the total amount of salt present is unknown until the final concentrations in both the supernatant fluid and the gel phase and the volumes occupied by both phases have been determined.

The average density of the vermiculite crystals was found to be 1.86 ± 0.01 g cm^{-3}, and the volume of vermiculite used in the experiments was determined by weighing

FIGURE 5.4 Photograph of a typical sample after swelling for two weeks. The height of the sample jar is approximately 5 cm. See color insert following page 76.

the crystals and using this density value. In a typical experiment, approximately 0.5 cm^3 of washed and dried vermiculite crystals were accurately weighed out and their volume calculated from the density. The crystals were placed in a sealable bottle and distilled water was added until the ratio of water to clay in the bottle was 50:1, that is, $50 \times V_m$ of water was added, where V_m is the volume of the macroions in their crystalline state. The bottle was then sealed to prevent evaporation, which would change the sol concentration, and allowed to stand for two weeks to enable the crystals to swell freely and reach equilibrium. After two weeks had passed the appearance of the samples was as shown in Figure 5.4.

After equilibrium had been reached, it was straightforward to pipette off the supernatant fluid, and its volume and concentration were easily measured, the latter by the Volhardt chloride ion titration described below. To measure the internal volume and concentration, however, it was first necessary to remove the solution from inside the gel. This proved to be more difficult. It was thought at first that this could be done by putting physical pressure on the gel. Various methods were tried but failed owing to the fragility of the gel stacks, which were easily mashed into a homogeneous paste. Instead, it was decided to collapse the gels chemically by adding a salt containing a cation that inhibits swelling. Sodium, potassium and multivalent metal cations inhibit swelling. The unhydrated K ion fits in the cavities on the plate surface and so binds the plates closely together, the K Eucatex vermiculite having a spacing of only 10.2 Å [4]. Potassium bicarbonate was chosen, as $KHCO_3$ is rapidly dissolved on the damp surface of the gel. This is important, as it cannot be added as a solution, since this would alter the chloride concentration. This method was very effective. K ions diffuse rapidly into the gel, causing it to release the interlayer fluid as a clear solution.

A small amount of $KHCO_3$ crystals were sprinkled over the gel, which was then left for 24 hours. The solution released from inside the gel was then removed by pipette and its volume measured. Samples of the solutions from both inside and outside the gel were then titrated for Cl^- ion concentration using the Volhardt titration. In the Volhardt titration for chloride ions, an excess of $AgNO_3$ is added to the sample, precipitating out the Cl^- as AgCl. The remaining Ag^+ ions are then back titrated using KSCN and a concentrated $HNO_3/Fe(NO_3)_3$ indicator. At the end point, when the remaining Ag has precipitated out as AgSCN, a blood-red color is seen from the FeSCN complex formed in solution. The original chloride concentration can then be calculated. The Volhardt titration is ideal for chloride solutions of concentrations down to about 0.005 M, but below this the end point becomes increasingly harder to see accurately, so that the accuracy is slightly lower in the more dilute salt solutions studied. Large sets of samples (between 16 and 32) were used for each $\{r,c\}$ point in an effort to obtain statistically significant average results, as the amount of trapped salt turned out to be a variable with a wide distribution.

In the case of no added salt, the total number of moles of salt n_t trapped inside a crystal is given by

$$n_t = c_{ex}V_{ex} + c_{gel}V_{gel} \tag{5.3}$$

where c_{ex} and c_{gel} are the salt concentrations in the supernatant fluid and the gel phase and V_{ex} and V_{gel} are their respective volumes. Because V_m, the volume of the crystals, is known, the concentration of trapped salt c_t was determined as

$$c_t = \frac{n_t}{V_m} \tag{5.4}$$

Let us consider only the data obtained after 20 washes, mimicking the preparation method used in the neutron scattering measurements. The most reliable data for the trapped salt concentration are obtained when there is no added salt. As shown in Table 5.3, three data sets, each with 32 samples ($N_s = 32$), were obtained with no added salt after 20 washes; these gave $c_t = 0.04$, 0.12 and 0.09 M for the $r = 0.01$, 0.02 and 0.05 batches, respectively. There was no study of $r = 0.1$ samples in this case because results for the salt fractionation effect cannot be obtained in the one-phase region and are difficult to obtain near its boundary because of the small volumes of the supernatant fluids. For $c = 0.1$ M, the added salt concentration is too high for reliable measurements of c_t, but we can also take the data sets for samples after 20 washes for $c = 0.03$ M. If we omit the anomalous $r = 0.01$ result (see below), the two remaining data sets, each with 16 samples ($N_s = 16$), both gave $c_t = 0.15$ M, so the value averaged over 128 samples studied in the laboratory was $c_t = 0.10$ M. The spread of the results suggests that c_t varies between sets of crystals (in each case, all of the crystals used in an experiment were washed together) and that an appropriate range to take for c_t is 0.10 ± 0.05 M. We therefore conclude that

$$c_t = 0.10 \pm 0.05\text{M} \tag{5.5}$$

for the salt concentration trapped inside the crystals before swelling. This 0.1 M "accidental" salt concentration inside the crystals contributes a global salt

TABLE 5.3
Twelve Sets of Results Obtained for the Salt Fractionation Effect

r	c (M)	W	N_s	c_{ex} (mM)	c_{gel} (mM)	s
0.01	0	20	32	1.4 ± 0.4	0.8 ± 0.5	1.6 ± 1.5
0.01	0	0	16	3.3 ± 0.4	1.2 ± 0.3	2.7 ± 1.0
0.01	0.03	20	16	30	5	6.0
0.01	0.10	20	16	105 ± 1.8	39 ± 7.7	2.7 ± 0.6
0.02	0	20	32	1.1 ± 0.5	0.5 ± 0.3	2.2 ± 2.0
0.02	0	0	20	13 ± 2.2	8.0 ± 1.9	1.6 ± 0.7
0.02	0.03	20	16	32 ± 2.0	8.6 ± 1.1	3.7 ± 0.7
0.02	0.10	20	16	110 ± 2.5	40 ± 2.4	2.7 ± 0.2
0.05	0	20	32	11 ± 2.6	4.7 ± 1.7	2.3 ± 1.4
0.05	0	0	16	37 ± 2.4	15 ± 1.7	2.5 ± 0.4
0.05	0.03	20	16	40 ± 4.0	14 ± 1.5	2.9 ± 0.6
0.05	0.10	20	16	120 ± 2.7	38 ± 4.3	3.2 ± 0.4

Note: r is the sol concentration, c is the salt concentration of the solution added to the crystals, W is the number of washes that the crystals were given and N_s is the number of samples used under each set of conditions. Measured salt concentrations in the supernatant fluid c_{ex} and the gel phase c_{gel} and the salt fractionation factor $s=c_{ex}/c_{gel}$, are given in the form $\alpha \pm \sigma$, where α is the average result and σ its standard deviation. The exception is the $r = 0.01$, $c = 0.03$ M result, for which only the average values are given.

background of 10^{-2} M at $r = 0.1$ and 10^{-3} M at $r = 0.01$. The wide range of this random variable prevents us from making accurate measurements below 10^{-3} M, but this still allows us two orders of magnitude in c in which to study the salt fractionation effect.

All of the 128 measurements described previously were obtained from pairs of titrations on the supernatant fluids and the fluids obtained from the gels. In each case, the salt fractionation factor s was determined from the ratio c_{ex}/c_{gel} of the two measurements. After determining the background electrolyte concentration from Equation 5.5, further measurements were carried out with added salt at $c = 0.03$ M and $c = 0.1$ M. The 12 sets of results obtained for the salt fractionation factor are given in Table 5.3, where W denotes the number of washes; a few unwashed samples were studied for comparison. The measured salt concentrations in the supernatant fluid c_{ex} and the gel phase c_{gel} have been given in the form $\alpha \pm \sigma$ in Table 5.3, where α is the average result and σ its standard deviation. The exception is the $r = 0.01$, $c = 0.03$ M result, for which only the average values are given. This result was not regarded as significant because the overall average of s over the 244 pairs of titrations performed was equal to 2.7, so that the result in this case lay more than three standard deviations from the mean. After this abnormal result had been discarded, the average over the remaining 11 sets of results was 2.6.

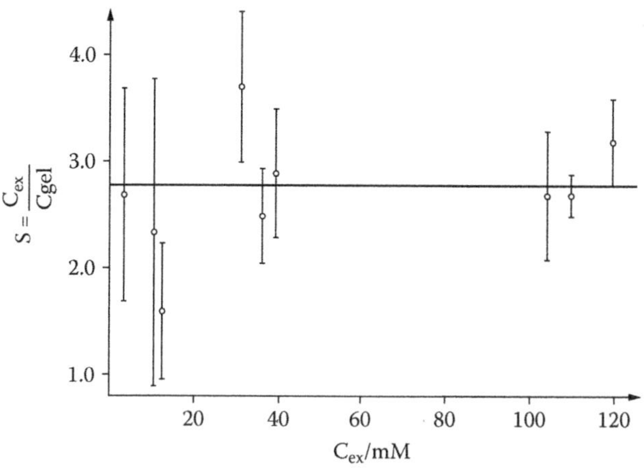

FIGURE 5.5 The salt fractionation effect in n-butylammonium vermiculite swelling. The salt fractionation factor $s = c_{ex}/c_{gel}$ is plotted as a function of c_{ex}, the salt concentration in the supernatant fluid. The solid line shows the coulombic attraction theory prediction, $s = 2.8$.

In each case s was determined both as the average over the N_s individual determinations of c_{ex}/c_{gel} and as the ratio $\langle c_{ex}\rangle/\langle c_{gel}\rangle$, and these two results were found to be equal to two significant figures. The range of s expressed after the ± sign in the final column of Table 5.3 was therefore obtained by expressing the standard deviations in c_{ex} and c_{gel} as percentages of their means, adding these two figures to give a compound percentage error and multiplying this number by the average value of s. These numbers have been plotted as the error bars in Figure 5.5, which shows $s = c_{ex}/c_{gel}$ as a function of the equilibrium salt concentration in the supernatant fluid, c_{ex}. The most obvious feature of Table 5.3 and Figure 5.5 is that there seems to be no systematic variation of s with respect to c_{ex}, in agreement with the coulombic attraction theory prediction. The quantitative prediction, $s = 2.8$, which is also plotted in Figure 5.5, fits most of the data, there being only two points whose error bars lie outside the predicted value of 2.8, one above and one below. It should be noted, however, that the percentage compound error in s is very large at the lower salt concentrations because of the lesser accuracy of the Volhardt titration, and so we would not be sensitive to deviations from the prediction for $c_{ex} < 2$ mM. The two results in this range have been omitted from Figure 5.5 because the percentage error in these cases approaches 100%. In the range 3 mM $< c_{ex} <$ 120 mM, we conclude that s is constant and equal to 2.6 ± 0.4.

We used the measurements of the amount of water absorbed in the laboratory experiments to give an independent check of the d-values obtained in the neutron experiments; we did this by assuming that the swelling is perfectly homogeneous. In these circumstances, for each volume of water that a unit volume of clay absorbs, the interlayer spacing must have increased by 19.4 Å. The conditions used are listed in Table 5.3 and the results are given in Table 5.4. There is a systematic variation

TABLE 5.4
Interlayer Spacing *d* as a Function of Salt Concentration *c* from Laboratory Experiments

N_s	c (M)	d (Å)	CAT ($dc^{0.5}$/ÅM$^{0.5}$)	DLVO ($d^4\exp(-dc^{0.5})$)
32	0.0010	620	20	3.0×10^{-2}
32	0.0014	600	22	3.6×10^{1}
16	0.0033	540	31	2.9×10^{-3}
32	0.011	235	24	0.12
20	0.013	260	30	4.2×10^{-4}
16	0.030	300	52	2.1×10^{-13}
16	0.032	290	52	1.8×10^{-13}
16	0.037	240	46	3.5×10^{-11}
16	0.040	235	47	1.2×10^{-11}
16	0.105	220	71	3.4×10^{-22}
16	0.110	210	69	2.1×10^{-21}
16	0.120	195	68	4.2×10^{-21}

Note: The quantities calculated in columns 4 and 5 are predicted to be constant by the coulombic attraction theory (CAT) and DLVO theory, respectively.

between the two sets of results, with those from the neutron experiments (Table 5.2) giving lower layer spacings than the laboratory experiments. This could have been because the samples used in the neutron experiments contained a higher than average trapped-salt concentration, or it could possibly be due to the slightly different nature of the samples used in the two types of experiments. In the case of many crystals soaked in a single jar, there may be absorption of water into intergel regions as well as the interlayers within the gel, leading to an apparently greater *d*-value. There may also have been a weak isotope effect due to the use of D_2O instead of H_2O to prepare the neutron samples, though this was not noted at $r < 0.01$ in the experiments described in Chapter 1. In any event, the difference is small, and so both sets of results are discussed together below.

Our final laboratory experiments on the sol concentration effect investigated T_c, the temperature above which the gels become unstable, for gels prepared at each of the 16 $\{r,c\}$ points studied in the neutron experiments. D_2O was used in these experiments so that the results could be directly compared with those of the neutron experiments without having to consider any possible isotope effect. The phase transition was approached from either side by carrying out two types of experiments. First, a set of 16 samples were placed in a water bath at 55°C immediately after preparation so that they would not swell. The temperature of the water bath was then reduced by 1°C at 24-hour intervals, and the temperature at which the first signs of swelling occurred was noted, as judged by traveling microscope experiments. In the second method, two sets of samples were placed in a water bath at 4°C and left for two weeks to swell to equilibrium. The temperature of the water bath was then increased by 1°C at 24-hour intervals, and

TABLE 5.5
Phase Transition Temperatures (°C) Obtained by Three Methods

(a) Neutron Diffraction

r	c = 0 M	c = 10^{-3} M	c = 10^{-2} M	c = 0.1 M
0.01	>34	>34	27 ± 1	13 ± 1
0.02	34 ± 1	> 34	27 ± 1	12 ± 1
0.05	20 ± 1	23 ± 1	22 ± 1	9 ± 1
0.10	20 ± 2	22 ± 3	19 ± 1	9 ± 1

(b) Laboratory Cooling Experiments

r	c = 0 M	c = 10^{-3} M	c = 10^{-2} M	c = 0.1 M
0.01	37 ± 1	36 ± 2	26 ± 1	13 ± 1
0.02	33 ± 1	34 ± 1	27 ± 1	12 ± 1
0.05	31 ± 1	31 ± 1	25 ± 1	12 ± 1
0.10	27 ± 1	26 ± 1	22 ± 1	11 ± 1

(c) Laboratory Heating Experiments

r	c = 0 M	c = 10^{-3} M	c = 10^{-2} M	c = 0.1 M
0.01	>39	>39	28 ± 2	14 ± 1
0.02	36 ± 4	37 ± 3	28 ± 2	15 ± 1
0.05	34 ± 2	32 ± 2	28 ± 1	12 ± 1
0.10	26 ± 2	26 ± 2	26 ± 2	13 ± 1

the temperature at which the gels collapsed was noted. The results from the two sets of experiments are listed in the above order in Table 5.5b and Table 5.5c, together with the results from the neutron scattering experiments in Table 5.5a.

The three sets of results are in good agreement, the variations representing an error of only a few percent within an absolute temperature of approximately 300 K. Of the three sets, the neutron data are probably the most precise, since the collapse of the gel was clearly evident by the disappearance of the diffraction peak within the range of 2 K, as shown in Figure 5.3. In the first set of laboratory experiments, the onset of swelling with increasing temperature was fairly clear, but the second method was less accurate, as it is much harder to see the exact temperature at which collapse occurs.

When a gel collapses, it does not regain its original crystalline appearance; groups of plates conglomerate, but the structure as a whole breaks up and does not decrease significantly in volume. The most reliable indication of collapse is the appearance of the metallic sheen of the crystal, but this too can be hard to spot, and so the errors are larger in Table 5.5c. This method, however, was the only one for which more than one sample under each set of $\{r,c\}$ conditions was used, and so the errors in Table 5.5c do contain an indication of the level of sample-to-sample variation. In any case, this is the most likely source of variation in the results, rather than experimental error, because of the wide variation in trapped salt concentration. It can be seen from Table 5.1 that, for low c and high r conditions, this error is greater than that in the values obtained for T_c, and so we will use an average over all the results. The largest inconsistency is that the neutron experiments give consistently lower results than the two laboratory methods. This suggests that the batch of crystals used for the neutron experiments had a higher than average concentration of trapped salt, a conclusion that is consistent with the results obtained for the d-values. The results for the latter two methods are very consistent. The fact that they approach the phase transition from opposite sides shows that there are no hysteresis loops.

The way we have presented the data in Table 5.2 and Table 5.5 corresponds to the $\{r,c\}$ map of Figure 4.4. It turns out that the sol concentration variable r is a weak variable in the two-phase region compared with the electrolyte concentration c. The fact that c is the dominant variable has been taken into account in the way that we presented the data for the d-values obtained from the laboratory experiments in Table 5.4. Here we have recognized that the leading feature of increasing r is mainly to increase the salt concentration in the supernatant fluid slightly, thus suppressing the swelling in the same way as an increase in the salt concentration at infinite sol dilution. All $\{r,c\}$ points can then be expressed in terms of the single variable c for small r. This has a global averaging effect on the data, giving us much better statistics for d and T_c as functions of c. In the former case, the data from the neutron experiments have been presented in the approximate d vs. c way in Table 5.6. In the latter case, the averaged data from Table 5.5 have been plotted in the approximate T_c vs. c way in Figure 5.6. Such averaging is possible only because of the constancy of the salt fractionation effect.

The T_c values in Figure 5.6 have been taken as the average over Table 5.5a, Table 5.5b and Table 5.5c, and the c values have been taken from Table 5.1 using the average value for the trapped salt concentration. The gradient of the graph is $-0.077\ K^{-1}$, corresponding to a decrease of 13 K per log unit. This value is in excellent agreement with that obtained previously [5] at $r < 0.01$, confirming the consistency of the mapping of the $\{r,c\}$ variation. In electrical terms it corresponds to a decrease of only 1 mV per decade of salt concentration, in keeping with our general finding of the constancy of the surface potential and in contrast to the 58 mV per decade variation predicted by the Nernst equation. Understanding the non-Nernstian behavior of these clay systems is a question of great interest in colloid science, as we saw in Chapter 4 and will revisit in Chapter 7.

The d vs. c mapping enables us to make a comparison between the coulombic attraction theory and DLVO (Derjaguin-Landau-Verwey-Overbeek) theory over the whole two-phase region. As we saw in Chapter 2, the leading predictions of the two theories are as follows.

DLVO:

$$d^4 \exp(-\kappa d) = \text{const} \tag{5.6}$$

CAT:

$$d = \frac{4}{\kappa} \tag{5.7}$$

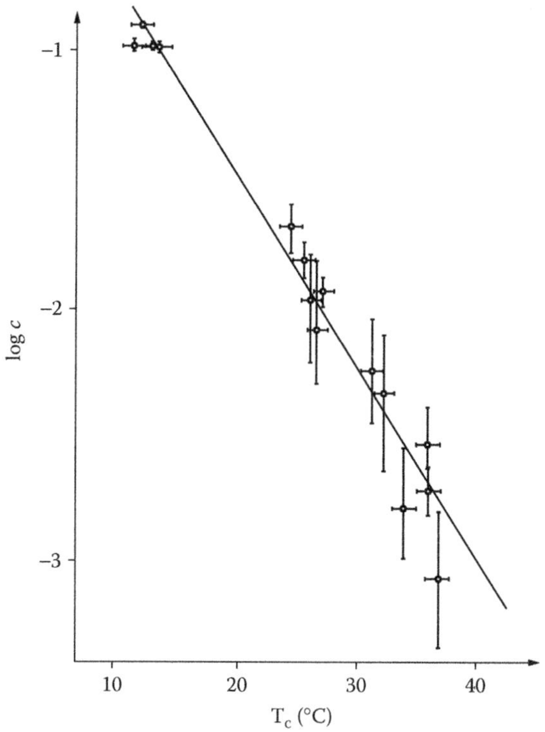

FIGURE 5.6 Logarithmic plot of the salt concentration c against the phase-transition temperature T_c. The gradient of the plot is equal to $-0.077\ K^{-1}$.

with

$$\kappa^2 = 0.1c \tag{5.8}$$

Combining Equation 5.8 with Equation 5.6 and Equation 5.7, the predictions become

DLVO:

$$d^4 \exp(-dc^{1/2}) = \text{const} \tag{5.9}$$

CAT:

$$d = 13c^{-1/2} \tag{5.10}$$

where d is in Å and c is in moles per liter. These predictions have been tested in the two right-hand columns of Table 5.4 and Table 5.6. It is very clear that the prediction of DLVO theory does not fit the experimental data. In Table 5.6 the calculated values of $d^4\exp(-dc^{0.5})$, predicted by DLVO theory to be a constant, vary by 11 orders of magnitude. In Table 5.4 there is an even greater range, the same quantity varying by 23 orders of magnitude. The coulombic attraction theory fits

TABLE 5.6
Interlayer Spacing *d* as a Function of Salt Concentration *c* from Neutron Diffraction Measurements in the Sol Concentration Range *r* between 0.01 and 0.1

c (M)	*d* (Å)	$dc^{0.5}$ (Å$M^{0.5}$)	d^4 ($\exp(-dc^{0.5})$)
0.0012	520	18	3.0×10^{-2}
0.0023	480	23	5.4
0.0027	450	23	4.2
0.0042	490	32	7.3×10^{-4}
0.0071	210	18	3.0×10^{1}
0.0085	210	19	1.1×10^{1}
0.012	310	34	1.6×10^{-5}
0.015	305	37	7.4×10^{-7}
0.016	135	17	1.4×10^{1}
0.018	150	20	1.7
0.021	200	29	4.1×10^{-4}
0.027	130	21	0.22
0.10	120	38	6.5×10^{-9}
0.11	120	40	8.8×10^{-10}
0.12	110	38	4.6×10^{-9}
0.13	95	34	1.4×10^{-7}

Note: The quantities calculated in columns 3 and 4 are predicted to be constant by the CAT and DLVO theory respectively.

the experimental results much more closely; the calculated values of $dc^{0.5}$ vary only by a factor of 2 for the neutron data and by a factor of 3 for the laboratory experiments. Although this may seem a poor approximation to a constant, the full theoretical prediction varies from about $7/\kappa$ in a 0.1 M solution to $4/\kappa$ in dilute salt solutions, and it must be remembered that there is a large degree of variation in the clays themselves, particularly with the amount of trapped salt. We are therefore looking for general rather than precise agreement. On this basis, looking at the general behavior of *d* over the whole two-phase region, the coulombic attraction theory is consistent with the experimental results; the DLVO theory is not.

Our experimental conclusion — that *s* is constant and equal to 2.6 ± 0.4 in the range 3 mM < c_{ex} < 120 mM — accords well with the prediction from the new generalized Donnan equilibrium made in Chapter 4. We recall that the coulombic attraction theory, with the constant surface potential boundary condition $\phi_s = 70$ mV, predicts that *s* is constant and equal to 2.8. A factor of ×40 in *c* provides a severe test of the prediction and it passes, although the quantitative agreement between the theoretical and experimental values of *s* in this case should be treated with caution because of the severity of the approximations used in deriving the theoretical result. The pure Donnan prediction that $s = 4.0$ for $\phi_s = 70$ mV is definitely invalidated by the data. This comparison calls for a basic reworking of membrane equilibria for interacting macroions.

As far as I am aware, independent experimental evidence for the values of the surface potential and salt fractionation factor have not been obtained for any system other than the n-butylammonium vermiculite gels. For this isolated system, the predicted values of *s* from the Donnan equilibrium and the new equilibrium based on the coulombic attraction theory, namely 4.0 and 2.8, respectively, are definitely distinguished by the experimental results. It would be highly desirable to obtain further tests of our prediction for *s* in systems of interacting plate macroions, both in clay science and lamellar surfactant phases.

Sometimes, establishing a new theory can be a bit of a slog, and wild speculation gives way to cautious progress. The salt fractionation effect has been presented here in terms of painstaking experimental results and cold theoretical comparisons. Let us now push the boat out. It often seems that mean field theory gives results that are accurate beyond the range for which they were calculated. Certainly, I would not push our theoretical basis above 0.1 M in view of the approximations underlying the theory, but what if the prediction for *s* in Chapter 4 actually worked up to, say, half molar? The near constancy of *s* for $\phi_s > 40$ mV then has an intriguing consequence if we generalize the result to any system of interacting macroions in solution, including biological systems. It is well known that the average salinity of cells is approximately 0.2 M, whereas the average salinity of seawater is about 0.5 M. A cell can be viewed as a system of interacting macroions, and the origins of life are probably to be found in aggregates of charged macroions in solution. The fact that biological cells have a salt concentration two to three times lower than the average salinity of the earth's most abundant electrolyte solution would follow naturally from the fact that s lies between 2 and 3 for all systems with $\phi_c > 40$ mV.

REFERENCES

1. Williams, G.D., Moody, K.R., Smalley M.V., and King, S.M., *Clays Clay Min.,* 42, 614, 1994.
2. ISIS User Guide, Boland, B.C. and Whapham, S., Eds., SERC, Report RAL 92-041, Rutherford Appleton Laboratory, 1992.
3. Hashimoto, T., Todo, A., and Kawai, H., *Polymer J.,* 10, 521, 1978.
4. Humes, R.P., Interparticle Forces in Clay Minerals, D.Phil. thesis, Oxford University, Oxford, U.K., 1985.
5. Braganza, L.F., Crawford, R.J., Smalley, M.V., and Thomas, R.K., *Clays Clay Min.,* 38, 90, 1990.

6 The Exact Mean Field Theory Solution for Plate Macroions

The preceding considerations show that the coulombic attraction theory is well adapted to explain the existence, extent and properties of the two-phase region of colloid stability. Such considerations also show that the n-butylammonium vermiculite system is governed by electrical forces. It is highly unlikely that we could accurately predict both the *d*-values and salt-fractionation effect in the two-phase region if any of the other forces that are commonly introduced into colloid theory, such as hydrophobic forces or van der Waals forces, played any significant role. The reason why we see the electrical phenomena so clearly in this model system is because the n-butylammonium ion approximates, as closely as any ion does, to ideal behavior. The enthalpy of solution of simple n-butylammonium salts is nearly equal to zero [1], and their partial molar volumes are nearly independent of concentration [2], implying that there are no special ion–solvent effects between n-butylammonium ions and water. As we have seen, the density of the n-butylammonium chloride is equal to 1.00 g cm^{-3}; this simple salt is as ideal as possible.

Before setting out on the exact mean field theory solution to the one-dimensional colloid problem, I wish to emphasize that the existence of the reversible phase transition in the n-butylammonium vermiculite system provides decisive evidence in favor of our model. The calculations presented in this chapter are deeply rooted in their agreement with the experimental facts on the best-studied system of plate macroions, the n-butylammonium vermiculite system [3]. We now proceed to construct the exact mean field theory solution to the problem in terms of adiabatic pair potentials of both the Helmholtz and Gibbs free energies. It is the one-dimensional nature of the problem that renders the exact solution possible.

In the Debye-Hückel theory of simple electrolytes [4], the Helmholtz free energy of the system is derived through the charging-up procedure by linearizing the Poisson-Boltzmann (PB) equation. The linear approximation is allowed in the case of simple electrolyte solutions because the strength of the mean electric field in the solution is weakened by the self-consistent cancellation of the contributions from the cations and anions, which form surrounding atmospheres with each other. Application of such a mean field description to macroionic solutions leads to a Helmholtz pair potential of the screened-coulomb type for macroions, alike in DLVO (Derjaguin-Landau-Verwey-Overbeek) theory [5, 6], Sogami-Ise (SI) theory [7, 8] and my own application of Sogami-Ise theory to plate macroions [9]. Maybe you find it surprising, then, that the Gibbs free energy has a long-range attractive branch in references [7–9]?

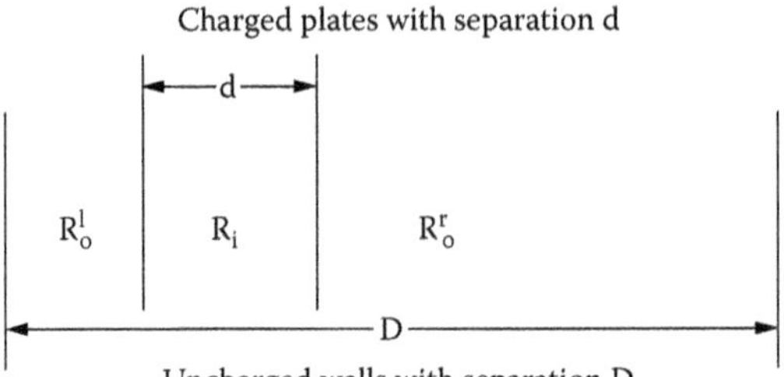

FIGURE 6.1 The geometry of the problem. Regions R_i, R_0^l and R_0^r are assumed to be in thermodynamic equilibrium.

Many colloid scientists did, and there has been a long argument about this difference, to be discussed in the following chapter. These debating issues notwithstanding, it remains an interesting question as to whether or not the Helmholtz free energy itself leads to a long-range attraction when the problem is solved without resort to the linearization approximation. The investigation of this question will form the first part of this chapter. In the second part we will go on to calculate the Gibbs free energy without resort to the linearization approximation, using an integral representation of the Gibbs free energy that we derived recently [10].

Let us consider $N + 1$ infinite plates with uniform surface charges and infinitesimal thickness immersed in a 1-1 electrolyte solution parallel to flat walls of a container of finite width. The obviously interesting two-plate problem ($N = 1$), which will form the bulk of the discussion, is illustrated in Figure 6.1. The solution is divided into regions R_i confined by the ith and $(i + 1)$th plates with separation d_i ($i = 1, 2, \ldots, N$), region R_0^l with separation D_l between the left container wall and the first plate, and region R_0^r with separation D_r between the $(N + 1)$th plate and the right container wall. The distance D between the left and right container walls is given by

$$D = D_l + D_r + \sum_{i=1}^{N} d_i \tag{6.1}$$

As a basic postulate, we assume that thermal equilibrium is achieved throughout all regions of the solution and, accordingly, the number density of the small ions with valency z_j ($z_\pm = \pm 1$), which are described as point particles, is determined by the common Boltzmann distribution

$$n_j(x) = n_0 \exp\{-z_j \Phi(x)\} = n_0 \exp\{-z_j \beta e \varphi(x)\} \tag{6.2}$$

for all the regions R_i, R_0^l and R_0^r. Here x is the space coordinate taken normally to the plate surfaces with its origin at the midpoint between the container walls, and $\Phi(x) = \beta e \varphi(x)$ is the ratio of the mean electric potential $\varphi(x)$ and the thermal energy $k_B T = 1/\beta$. The normalization constant n_0 takes a common value throughout all regions. The solvent is treated as a continuous medium with uniform dielectric constant ε.

In the mean field description, the potential $\Phi(x)$ is assumed to obey the PB equation

$$\frac{d^2\Phi(x)}{dx^2} = \kappa^2 \sinh\Phi(x) \tag{6.3}$$

where κ is Debye's screening parameter defined by

$$\kappa = (8\pi\lambda_B n_0)^{1/2} \tag{6.4}$$

with the Bjerrum length

$$\lambda_B = \frac{e^2}{\varepsilon k_B T} \tag{6.5}$$

The geometrical symmetry of the plate configuration requires the derivative of the potential to vanish at the midpoint $x = x_i^0$ of the region R_i, i.e.,

$$\left.\frac{d\Phi(x)}{dx}\right|_{x=x_i^0} = 0 \tag{6.6}$$

The surfaces of all the plates are considered naturally to form equipotential surfaces as

$$\Phi(x_i \pm 0) = \Phi_S \tag{6.7}$$

where x_i is the coordinate of the ith plate. The left and right outer surfaces of the plate system have the uniform surface charge densities Z_0^l and Z_0^r (nm^{-2}), i.e.,

$$\begin{aligned} -\left.\frac{d\Phi(x)}{dx}\right|_{x=x_l-0} &= -4\pi Z_0^l\lambda_B \\ -\left.\frac{d\Phi(x)}{dx}\right|_{x=x_{N+1}+0} &= 4\pi Z_0^r\lambda_B \end{aligned} \tag{6.8}$$

All other inner surfaces have uniform surface charge densities Z_i as

$$\begin{aligned} -\left.\frac{d\Phi(x)}{dx}\right|_{x=x_{i+1}-0} &= -4\pi Z_i\lambda_B \\ -\left.\frac{d\Phi(x)}{dx}\right|_{x=x_i+0} &= 4\pi Z_i\lambda_B \end{aligned} \tag{6.9}$$

In consideration of the fact that the surface charge of clay plates is negative, we assume Z_i, Z_0^{l}, $Z_0^{\mathrm{r}} < 0$. Here we will impose the Neumann-type boundary condition that the container surfaces have no charge. This boundary condition is expressed by

$$\left.\frac{d\Phi(x)}{dx}\right|_{x=x_c^{\mathrm{l}}+0} = \left.\frac{d\Phi(x)}{dx}\right|_{x=x_c^{\mathrm{r}}-0} = 0 \tag{6.10}$$

where x_c^{l} and x_c^{r} are the coordinates of the left and right inner surfaces of the container. This condition (Equation 6.10) can be changed to take into account situations where the container surfaces are not neutral.

To derive the mean potential, one more boundary condition must be imposed in addition to the conditions described in Equation 6.6 to Equation 6.10. There are two choices for the remaining boundary condition. One is a model with the Dirichlet boundary condition in which the value of the surface potential Φ_S is specified (Dirichlet model). The other is a model with the Neumann boundary condition in which the values of the surface charge densities Z_0^{l} and Z_0^{r} are given (Neumann model).

We introduce the general region R to be later attributed to each region R_i, R_0^{l} and R_0^{r} by parameter replacement. It is sufficient to derive the free energy in the general region R by solving the PB equation under appropriate boundary conditions and then to apply the result to all the regions to obtain the free energy of the system. The coordinate origin $x = 0$ is taken at the point where the potential Φ has its extremum:

$$\left.\frac{d\Phi(x)}{dx}\right|_{x=0} = 0 \tag{6.11}$$

i.e., x_i^0 in the inner region R_i and x_c^{l} (x_c^{r}) in the outer region R_0^{l} (R_0^{r}). The first integral of the PB equation in R is found to be

$$\left[\frac{d\Phi(x)}{dx}\right]^2 - 2\kappa^2\cosh\Phi(x) = -2\kappa^2\cosh\Phi(0) \tag{6.12}$$

It is convenient to introduce the auxiliary quantities

$$g(x) = k^{-1/2}\exp\left(\frac{1}{2}\Phi(x)\right)$$

$$k = \exp(\Phi(0)) \tag{6.13}$$

For a system with negative surface charge, $\Phi(x) \le \Phi(0) < 0$. Therefore, the values of k and $g(x)$ are restricted, respectively, to the intervals

$$0 < k < 1$$

$$0 < g(x) \le g(0) = 1 \tag{6.14}$$

Consequently, Equation 6.12 is converted into an elliptic integral as

$$x = \pm \frac{2k^{1/2}}{\kappa} \int_1^{g(x)} \frac{dt}{[(1-t^2)(1-k^2t^2)]^{1/2}}$$

or

$$x = \pm \frac{2k^{1/2}}{\kappa} (F(\sin^{-1} g(x), k) - K) \tag{6.15}$$

where the −(+) sign on the right-hand side is for $x > 0$ ($x < 0$) and F is the elliptic integral with modulus k of the first kind in Legendre's normal form [11, 12] defined by

$$F(\varphi, k) = \int_0^{\varphi} \frac{d\theta}{(1-k^2 \sin^2 \theta)^{1/2}} \tag{6.16}$$

and $K = F(\pi/2, k)$ is the perfect elliptic integral of the first kind. Hence the potential $\Phi(x)$ is expressed in terms of Jacobi's elliptic function sn with modulus k [11, 12] as follows:

$$\Phi(x) = 2 \ln \left[\mathrm{sn} \left(-\frac{\kappa}{2k^{1/2}} |x| + K(k), k \right) \right] + \Phi(0) \tag{6.17}$$

Consider that the region R is sandwiched by plates with the surface charge densities $Z_L \le 0$ at the position $x = -l_L$ and $Z_R \le 0$ at $x = l_R$. The boundary conditions are

$$-\left. \frac{d\Phi(x)}{dx} \right|_{x=-l_L+0} = 4\pi Z_L \lambda_B$$

$$-\left. \frac{d\Phi(x)}{dx} \right|_{x=l_R-0} = -4\pi Z_R \lambda_B \tag{6.18}$$

and the interplate distance is $d = l_L + l_R$. Substituting Equation 6.17 into Equation 6.18, we obtain

$$4\pi |Z_L| \lambda_B = \frac{\kappa}{k^{1/2}} \times \frac{\mathrm{cn}(-(\kappa/2k^{1/2})l_L + K(k), k)\mathrm{dn}(-(\kappa/2k^{1/2})l_L + K(k), k)}{\mathrm{sn}(-(\kappa/2k^{1/2})l_L + K(k), k)} \tag{6.19}$$

and

$$4\pi |Z_R| \lambda_B = \frac{\kappa}{k^{1/2}} \times \frac{\mathrm{cn}(-(\kappa/2k^{1/2})l_R + K(k), k)\mathrm{dn}(-(\kappa/2k^{1/2})l_R + K(k), k)}{\mathrm{sn}(-(\kappa/2k^{1/2})l_R + K(k), k)} \quad (6.20)$$

By giving the values of Z_L (or $\Phi(-l_L)$), Z_R (or $\Phi(l_R)$) and d, the values of k, l_L and l_R are determined.

Noting that the potential $\Phi(x)$ is symmetric with respect to the origin, the Boltzmann distribution (Equation 6.2) is integrated to give the numbers of small ions in the region R as

$$\begin{aligned} N_+ &= \int_{-l_L}^{l_R} n_0 e^{-\Phi(x)} dx \\ &= \left\{ |Z_L| + \frac{n_0 l_L}{k} + \frac{2n_0}{\kappa k^{1/2}} \left[E\left(-\frac{\kappa}{2k^{1/2}} l_L + K(k) \right) - E(K(k)) \right] \right\} + \{L \to R\} \end{aligned} \quad (6.21)$$

and

$$\begin{aligned} N_- &= \int_{-l_L}^{l_R} n_0 e^{\Phi(x)} dx \\ &= \left\{ \frac{n_0 l_L}{k} + \frac{2n_0}{\kappa k^{1/2}} \left[E\left(-\frac{\kappa}{2k^{1/2}} l_L + K(k) \right) - E(K(k)) \right] \right\} + \{L \to R\} \end{aligned} \quad (6.22)$$

where

$$E(K(k)) = E(\pi/2, k)$$

$$E\left(-\frac{\kappa}{2k^{1/2}} l + K(k) \right) = E(\theta, k) \quad (6.23)$$

in which $E(\varphi,k)$ is the Legendre elliptic integral of the second kind [11, 12]

$$E(\varphi, k) = \int_0^{\varphi} (1 - k^2 \sin^2 \phi)^{1/2} d\phi \quad (6.24)$$

and the value of θ is fixed by

$$F(\theta, k) = -\frac{\kappa}{2k^{1/2}} l + K(k) = -\frac{\kappa}{4k^{1/2}} d + F(\pi/2, k) \quad (6.25)$$

and

$$E(u) = E(\theta, k)$$

$$u = \int_0^\theta \frac{d\phi}{(1-k^2\sin^2\phi)^{1/2}} = F(\theta, k) \tag{6.26}$$

The difference between Equation 6.21 and Equation 6.22 confirms the condition for charge neutrality as

$$N_+ - N_- = |Z_L| + |Z_R| \tag{6.27}$$

The Helmholtz free energy F consists of an electric part F^{el} and an osmotic part F^{osm}. In Appendix A of reference [13], the electric part F^{el} is represented generally as a functional of the mean electric potential $\Phi(x)$ as follows:

$$\beta F^{el} = Z_L \Phi(x)\Big|_{x=-l_L+0} + Z_R \Phi(x)\Big|_{x=l_R-0} - \frac{1}{8\pi\lambda_B}\int_{l_L}^{l_R}\left[\frac{d\Phi(x)}{dx}\right]^2 dx$$
$$-N_+ \ln\left[\frac{1}{d}\int_{l_L}^{l_R} e^{-\Phi(x)}dx\right] - N_- \ln\left[\frac{1}{d}\int_{l_L}^{l_R} e^{\Phi(x)}dx\right] \tag{6.28}$$

Substitution of the $\Phi(x)$ in Equation 6.17 into this formula yields the electric part of the Helmholtz free energy of the region R as

$$\beta F^{el} = -\left\{2|Z_L| \ln\left[k^{1/2} sn\left(-\frac{\kappa}{2k^{1/2}} l_L + K(k), k\right)\right]\right\}$$
$$-\{L \to R\} - N_+\left[1 + \ln\left(\frac{N_+}{n_0 d}\right)\right] - N_-\left[1 + \ln\left(\frac{N_-}{n_0 d}\right)\right] + n_0 d\frac{1+k^2}{k} \tag{6.29}$$

where $N_\pm$ are the numbers of small ions obtained from Equation 6.21 and Equation 6.22. Using the formula for an ideal gas [14], the osmotic part F^{osm} of the Helmholtz free energy is found to be

$$\beta F^{osm} = -N_+\left\{1 + \ln\left[\frac{d}{N_+}\left(\frac{m_+}{m_p}\right)^{3/2}\lambda_S^{-3}\right]\right\} - N_-\left\{1 + \ln\left[\frac{d}{N_-}\left(\frac{m_-}{m_p}\right)^{3/2}\lambda_S^{-3}\right]\right\} \tag{6.30}$$

Here m_+ and m_- are the masses of the counterion and co-ion, m_p is the proton mass and

$$\lambda_S \equiv \left(2\pi\hbar^2\beta m_p^{-1}\right)^{1/2} \tag{6.31}$$

is the unit of length characterizing the size of the quantum cells in phase space [14]. In the osmotic part of the Helmholtz free energy (Equation 6.30), neither the effect of the solvent nor the contribution of the nonelectric charge limit of the charged plates are included. These effects are regarded as a common background contribution working uniformly for all regions.

Using the formulae in Equation 6.29 and Equation 6.30, the Helmholtz free energy is obtained for each region under the imposed boundary conditions. For the region R_i, the electric part of the Helmholtz free energy is

$$\begin{aligned} \mathrm{BF}_i^{\mathrm{el}} &= -4\,|Z_i|\ln\left[k_i^{1/2}\,\mathrm{sn}\left(-\frac{\kappa^{\mathrm{c}}}{4k_i^{1/2}}di + k(k_i), k_i\right)\right] \\ &\quad - N_{i+}\left[1+\ln\left(\frac{N_{i+}}{n_0 d_i}\right)\right] - N_{i-}\left[1+\ln\left(\frac{N_{i-}}{n_0 d_i}\right)\right] + n_0 d_i \frac{1+k_i^2}{k_i} \end{aligned} \tag{6.32}$$

and the osmotic part is

$$\beta F_i^{\mathrm{osm}} = -N_{i+}\left\{1+\ln\left[\frac{d_i}{N_{i+}}\left(\frac{m_+}{m_{\mathrm{p}}}\right)^{3/2}\lambda_{\mathrm{S}}^{-3}\right]\right\} - N_{i-}\left\{1+\ln\left[\frac{d_i}{N_{i-}}\right]\left(\frac{m_-}{m_{\mathrm{p}}}\right)^{3/2}\lambda_{\mathrm{S}}^{-3}\right\} \tag{6.33}$$

For the region $R_0^{\mathrm{l}} \cup R_0^{\mathrm{r}}$, the electric part of the Helmholtz free energy is

$$\begin{aligned} \beta F_{\mathrm{o}}^{\mathrm{el}} &= -\left\{2\,|Z_{\mathrm{o}}^{\mathrm{l}}|\ln\left[\left(k_{\mathrm{o}}^{\mathrm{l}}\right)^{1/2} sn\left(-\frac{\kappa}{4\left(k_{\mathrm{o}}^{\mathrm{l}}\right)^{1/2}}D_1 + K\left(k_{\mathrm{o}}^{\mathrm{l}}\right), k_{\mathrm{o}}^{\mathrm{l}}\right)\right] + N_{\mathrm{o}+}^{l}\left[1+\ln\left(\frac{N_{\mathrm{o}+}^{\mathrm{l}}}{n_{\mathrm{o}} D_1}\right)\right]\right. \\ &\quad \left. + N_{\mathrm{o}-}^{\mathrm{l}}\left[1+\ln\left(\frac{N_{\mathrm{o}-}^{\mathrm{e}}}{n_{\mathrm{o}} D_1}\right)\right] - n_{\mathrm{o}} D_l \frac{1+\left(k_{\mathrm{o}}^{\mathrm{l}}\right)^2}{k_{\mathrm{o}}^{\mathrm{l}}}\right\} - \{l \to r\} \end{aligned} \tag{6.34}$$

and the osmotic part is

$$\begin{aligned} \beta F_{\mathrm{o}}^{\mathrm{osm}} &= -N_{\mathrm{o}+}^{l}\left\{1+\ln\left[\frac{D_1}{N_{\mathrm{o}+}^{\mathrm{l}}}\left(\frac{m_+}{m_{\mathrm{p}}}\right)^{3/2}\lambda_{\mathrm{S}}^{-3}\right]\right\} - N_{\mathrm{o}-}^{\mathrm{l}}\left\{1+\ln\left[\frac{D_1}{N_{\mathrm{o}-}^{\mathrm{l}}}\left(\frac{m_-}{m_{\mathrm{p}}}\right)^{3/2}\lambda_{\mathrm{S}}^{-3}\right]\right\} \\ &\quad - N_{\mathrm{o}+}^{\mathrm{r}}\left\{1+\ln\left[\frac{D_{\mathrm{r}}}{N_{\mathrm{o}+}^{\mathrm{r}}}\left(\frac{m_+}{m_{\mathrm{p}}}\right)^{3/2}\lambda_{\mathrm{S}}^{-3}\right]\right\} - N_{\mathrm{o}-}^{\mathrm{r}}\left\{1+\ln\left[\frac{D_{\mathrm{r}}}{N_{\mathrm{o}-}^{\mathrm{r}}}\left(\frac{m_-}{m_{\mathrm{p}}}\right)^{3/2}\lambda_{\mathrm{S}}^{-3}\right]\right\} \end{aligned} \tag{6.35}$$

Here k_i, k_o^l and k_o^r are the moduli, and $N_{i\pm}$, $N_{o\pm}{}^l$ and $N_{o\pm}{}^r$ are the numbers of small ions in the regions R_i, R_o^l and R_o^r. Summing up these electric and nonelectric parts of the Helmholtz free energies of all regions, we obtain the adiabatic potential of the system in the form

$$V_N^F(d_1,\ldots,d_N;D_l,D_r)=\sum_{i=1}^{N}\left[F_i^{el}(d_i)+F_i^{osm}(d_i)\right]+F_o^{el}(D_l,D_r)+F_o^{osm}(D_l,D_r) \quad (6.36)$$

Equation 6.36 for the adiabatic potential is exact within the framework of the mean field description. However, the structure of the electric part F^{el} is too complex to disclose its analytic properties. Here we examine the adiabatic potential numerically following the Carlson theory of elliptic integrals [15–21]. To proceed with numerical computation, it is necessary to enter a set of parameters designed to describe an experimental situation. It will not surprise the reader who has made it this far that we use values of the chemically fixed parameters specified by the n-butylammonium vermiculite gels [22], namely $m_+ = 74\ m_p$ and $m_- = 36\ m_p$. The average density n_0 of the small ions is given by

$$n_0 = 6.02 \times 10^{-1} \times c\ [\text{nm}^{-3}] \quad (6.37)$$

with c (mol l^{-1}) being the salt concentration. We take $k_BT = \beta^{-1} = 0.025$ eV (1 eV $\approx 1.6 \times 10^{-12}$ erg), $\lambda_S^{-2} = 96$ nm^{-2} and $\lambda_B = e^2\beta/\varepsilon = 0.72$ nm. It is natural to assume that the physicochemical state is the same in all of the inner regions R_i ($i = 1, 2, \ldots, N$). Therefore, we examine mainly the two-plate system ($N = 1$) with a symmetric configuration ($D_l = D_r = (D - d_1)/2$, except for Figure 6.13 and Figure 6.14). To investigate the wall effect for such a system, the width D of the container is set to be 200 nm throughout (except for Figure 6.11 and Figure 6.12).

Let us first perform the analysis under the Dirichlet boundary condition, for which the value of the surface potential Φ_S is fixed. As we have seen in Chapter 3, analysis of the effect of uniaxial stress on the swelling of n-butylammonium vermiculite results in the average value of the surface potential $\varphi_S \approx -70$ mV being insensitive to electrolyte concentration [3], and the surface potential in the smectite minerals has also been found to be independent of electrolyte concentration and has the value $\varphi_S \approx -60$ mV [23]. Using these values and taking into account the fact that the surface potential Φ_S is as hard to measure as the charge densities Z_o^l and Z_o^r, we make numerical calculations of the thermodynamic quantities for the range of values $\Phi_S = \beta e\varphi_S = -2.0$ to -8.0, corresponding to $\varphi_S = -50$ to -200 mV. Numerical results are obtained for the plate system satisfying the conditions $Z_o \equiv Z_o^l = Z_o^r$.

Figure 6.2 and Figure 6.3 show the respective behaviors of the mean electric potential $\Phi(x)$ and the small ion distributions $n_\pm(x)$ for the distances $d_1 = 20$ nm and $D_l = D_r = 90$ nm, the salt concentration $c = 0.001$ mol l^{-1} and the surface potential $\Phi_S = -4.0$ (corresponding to the surface charge density $Z_o = -0.084$ nm^{-2} in the Neumann model). The origin of the coordinate x is taken at the midpoint between the container surfaces. Figure 6.2 shows that the potential satisfies approximately $\Phi(x) = 0$ on the surfaces ($x_c^l = -100$ nm and $x_c^r = 100$ nm) of the container. This indicates

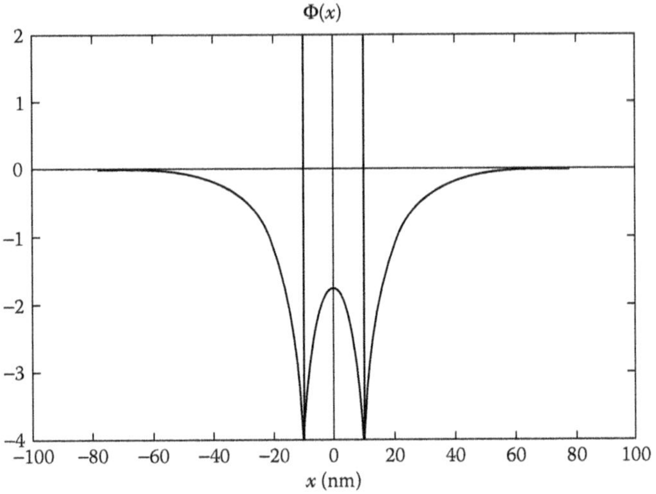

FIGURE 6.2 Variation of the dimensionless surface potential $\Phi(x)$ against the coordinate x whose origin is taken at the midpoint between plates with separation $d_1 = 20$ nm. The analysis is for the two-charged-plate system ($N = 1$). The concentration of the external soaking solution is $c = 0.001$ mol l^{-1}, and the surface potential is $\Phi_s = -4.0$ in the Dirichlet model, which corresponds to the surface charge density $Z_o = -0.084$ nm^{-2} in the Neumann model. The two plates are located at $x_1 = -10$ nm and $x_2 = 10$ nm.

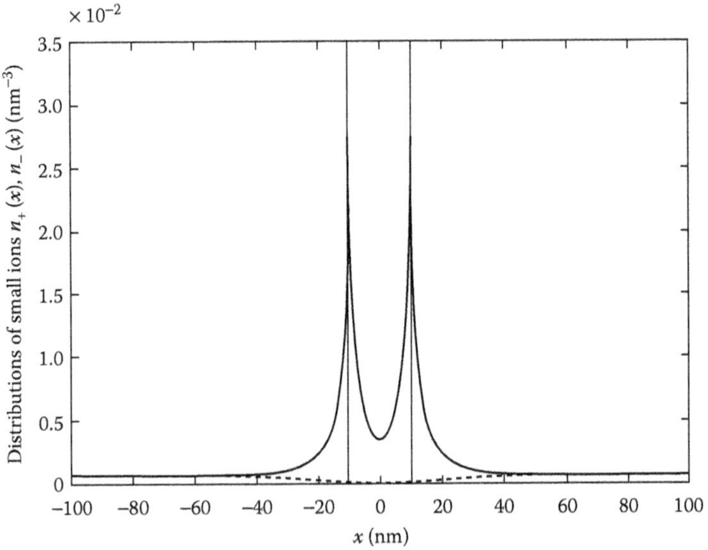

FIGURE 6.3 Distributions of small ions (n_+: solid line; n_-: dashed line) for the interplate distance $d_1 = 20$ nm, the electrolyte concentration $c = 0.001$ mol l^{-1} and the surface potential $\Phi_s = -4.0$ in the Dirichlet model (the surface charge density $Z_o = -0.084$ nm^{-2} in the Neumann model). The origin of the coordinate x is taken at the midpoint between the container surfaces.

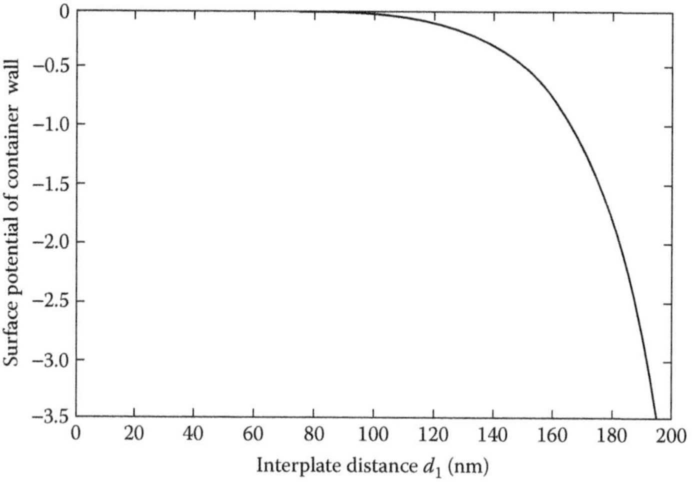

FIGURE 6.4 Variation of the surface potential of the container wall against the interplate distance d_1 for $c = 0.001$ mol l^{-1} and $\Phi_s = -4.0$.

the solution to be locally neutral at points far from the plates, and the condition of the infinite container (Equation 6 of reference [13])

$$\Phi(\pm\infty) = 0$$

$$\left.\frac{d\Phi(x)}{dx}\right|_{x=\pm\infty} = 0 \tag{6.38}$$

is also realized in the finite container. Figure 6.4 shows the variation of the surface potential of the container wall against the interplate distance d_1 for $c = 0.001$ mol l^{-1} and $\Phi_S = -4.0$ with the symmetric configuration $D_l = D_r$. If the plate is fully apart from the container wall, the wall potential becomes approximately zero. The potential of the container wall approaches Φ_S as $D_l, D_r \rightarrow 0$ ($d_1 \rightarrow 200$ nm in this case).

The behavior of the charge density Z_1 of the inner surface of the plate as a function of the interplate distance d_1 is drawn in Figure 6.5 for $\Phi_S = -4.0$ and -5.0 ($Z_o = -0.084$ and -0.140 nm^{-2} in the Neumann model) when $c = 0.001$ mol l^{-1}. The charge density Z_1 of the inner surface of the plate varies subject to the relations in Equation 6.19 and Equation 6.20: it rapidly approaches the charge density of the outer plate surface Z_o with increasing d_1, and it reduces rapidly to zero as $d_1 \rightarrow 0$. Such behavior occurs irrespective of the type of boundary condition employed; in the Neumann model for the two-plate system in the infinite container, the variation of Z_1 with d_1 (see Figure 4 of reference [13]) is essentially the same as that shown in Figure 6.5. Indeed, the disappearance of Z_1 as $d_1 \rightarrow 0$ is essential for the consistency of the theory. If Z_1 remained nonzero at $d_1 = 0$, the charge-balance condition (Equation 6.27) would break down because the numbers of small ions N_{1+} and N_{1-} in R_1 vanish in this limit. This reduction of the charge density Z_1 should be interpreted

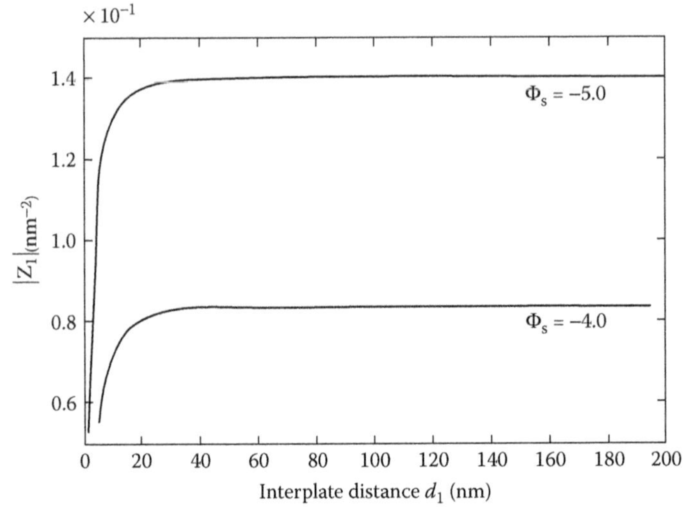

FIGURE 6.5 Dependence of the magnitude of the inner-surface charge density $|Z_i|$ on the interplate distance d_1 for $\Phi_S = -4.0$ and -5.0 in the Dirichlet model ($Z_o = -0.084$ and -0.140 nm^{-2} in the Neumann model) when $c = 0.001$ mol l^{-1}. $|Z_i|$ vanishes rapidly to zero as $d_1 \to 0$, and it saturates quickly to the outer-surface charge density $|Z_o|$.

as occurring through the recombination of surface charges with counterions. To describe such a phenomenon at the molecular level goes beyond the scope of mean field theory. Here, it is assumed that the plate changes its surface charge density to satisfy the condition of charge balance in each region.

Figure 6.6 shows the behavior of the electric parts F_o^{el} and F_i^{el} and osmotic parts F_o^{osm} and F_1^{osm} of the Helmholtz free energy as a function of the interplate distance d_1 for the salt concentration $c = 0.001$ mol l^{-1} and the surface potential $\Phi_S = -4.0$. While the osmotic pressure of the small ions in R_1 exerts a repulsion between the two plates, the electric contribution from R_1 and the osmotic contribution from $R_o^l \cup R_o^r$ give rise to an attraction. Figure 6.7 and Figure 6.8 show, respectively, the sum of the electric parts F_o^{el} and F_1^{el} and the sum of the osmotic parts F_o^{osm} and F_1^{osm} of the Helmholtz free energy for the salt concentration $c = 0.002$ mol l^{-1} and the surface potential $\Phi_S = -4.0$. Figure 6.7 shows the electric attraction between the two plates (and between a plate and the container surface). Figure 6.8 shows that the osmotic pressure is repulsive at medium range and attractive at long range between two plates (and between a plate and the container surface) and takes a minimum at three points. In the minimum point at the middle, the depth of the potential is shallow in comparison with the other two points and becomes shallower on taking the sum of the electric parts and the osmotic parts. As a result, the third minimum point cannot be found clearly in the total adiabatic potential (Equation 6.36). The behavior of the total adiabatic potential $V_I^F(d_1)$ is shown in Figure 6.9 for the concentrations $c = 0.001$ to 0.002 mol l^{-1}. It is repulsive at medium range and attractive at long range between two

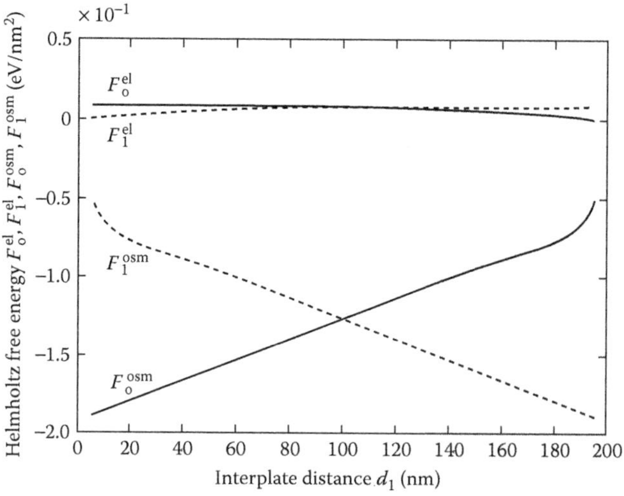

FIGURE 6.6 Electric parts F_o^{el} and F_1^{el} and osmotic parts F_o^{osm} and F_1^{osm} of the adiabatic pair potential vs. the interplate distance d_1 for $c = 0.001$ mol l^{-1} and $\Phi_S = -4.0$ in the Dirichlet model ($Z_o = -0.084$ nm^{-2} in the Neumann model). The solid lines and dashed lines represent the potentials for the region $R_o^l \cup R_o^r$ and the region R_1, respectively.

plates (and between a plate and the container surface). The potential V_1^F takes a minimum at $d_1 = 17.7$ and 182.3 nm for $c = 0.001$ mol l^{-1}, at $d_1 = 14.4$ and 185.6 nm for $c = 0.0015$ mol l^{-1}, and at $d_1 = 12.5$ and 187.5 nm for $c = 0.002$ mol l^{-1}. Figure 6.6 through Figure 6.9 are symmetric with respect to $d_1 = 100$ nm.

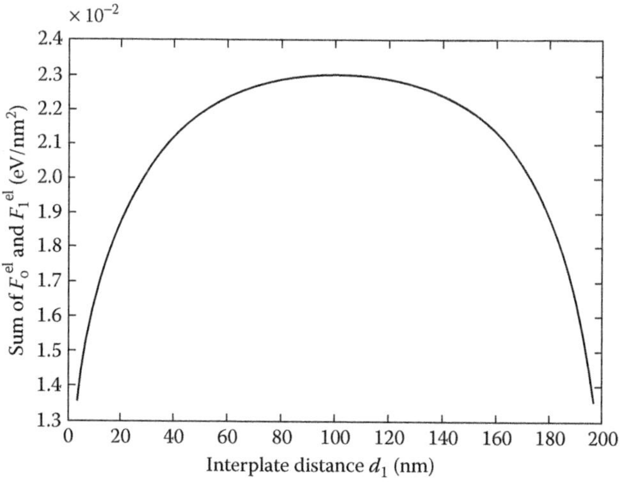

FIGURE 6.7 Sum of the electric parts F_o^{el} and F_1^{el} of the Helmholtz free energy vs. the interplate distance d_1 for $c = 0.002$ mol l^{-1} and $\Phi_S = -4.0$ in the Dirichlet model ($Z_o = -0.118$ nm^{-2} in the Neumann model).

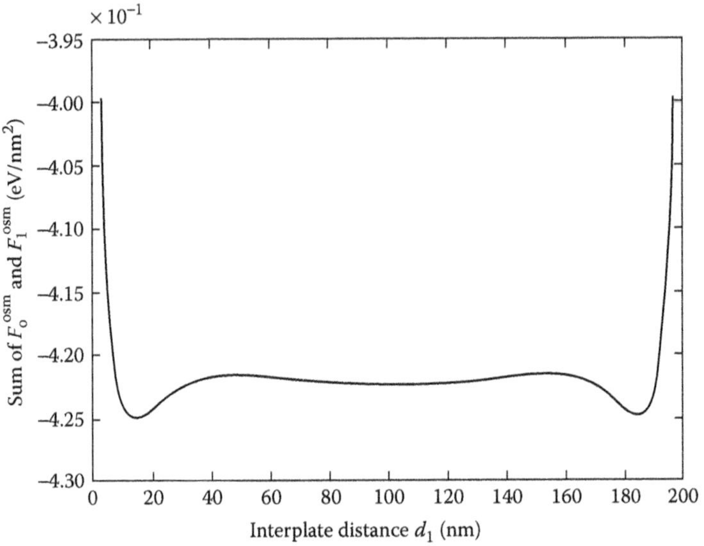

FIGURE 6.8 Sum of the osmotic parts F_0^{osm} and F_1^{osm} of the Helmholtz free energy vs. the interplate distance d_1 for $c = 0.002$ mol l^{-1} and $\Phi_S = -4.0$ in the Dirichlet model ($Z_0 = -0.118$ nm^{-2} in the Neumann model).

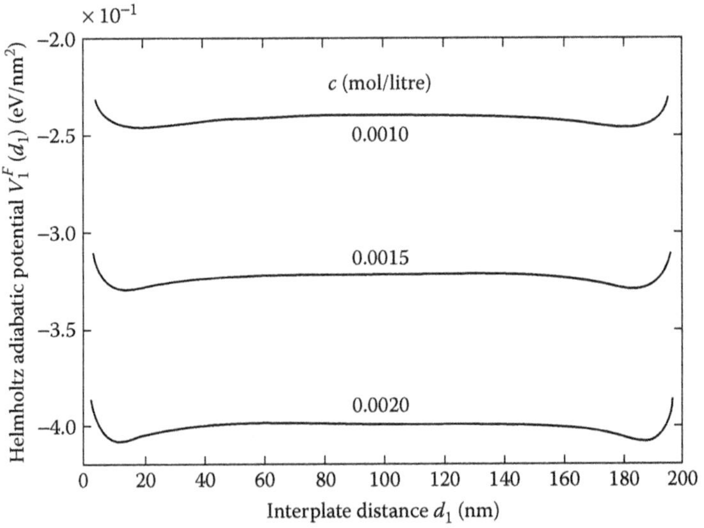

FIGURE 6.9 Adiabatic potentials $V_1^F(d_1)$ for three values of the electrolyte concentration c in the Dirichlet model. The surface potential of all plates is taken to be $\Phi_S = -4.0$ ($Z_0 = -0.084$, -0.102 and -0.118 nm^{-2} for $c = 0.0010$, 0.0015 and 0.0020 mol l^{-1} in the Neumann model). The potential V_1^F takes a minimum around $d_1 = 15$ nm and 185 nm.

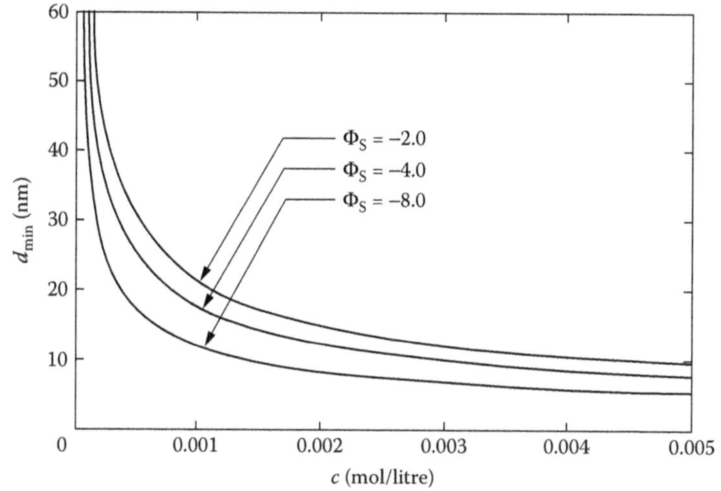

FIGURE 6.10 Trajectories of the stable interplate distance d_{min} plotted against the concentration c of the external soaking solution for three values of the surface potential Φ_S in the Dirichlet model. The width of the container is set to be 200 nm and $D_l = D_r$. The swelling is enhanced for small values of $|\Phi_S|$.

In Figure 6.10 the trajectories of the interplate distance d_{min} (the smaller one) for the potential minimum are drawn for different values of Φ_S, showing the effect of increasing the plate separation against the dilution of the salt concentration c. The interplate distance d_{min} approaches $D/2$ as $c \to 0$. We note that the swelling is enhanced for small values of $|\Phi_S|$. Figure 6.11 shows the trajectories of d_{min} for $c = 0.001$ mol l^{-1}, $\Phi_S = -4.0$ and $D_l = D_r$, changing the width D of the container. Until D reaches a specific value ($D \approx 58$ nm in Figure 6.11), the potential V_1^F takes a minimum at the one point $d_{min} = D/2$. After that, the potential V_1^F takes a minimum at two points, and the symmetry of the graph breaks "spontaneously" (see Figure 6.12, for example). Here, we choose to draw the graph for the smaller d_{min} of the two (the larger one is given by $D - d_{min}$). The potential minimum d_{min} becomes smaller as D increases, and d_{min} takes a constant (namely, asymptotic) value for sufficiently large D in comparison with d_{min}. This means that the contribution of the container wall disappears. The asymptotic value is in agreement with the value of d_{min} obtained in our previous article on the infinite container problem (see Figure 10 of reference [13]). Figure 6.12 shows the adiabatic potentials $V_1^F(d_1)$ for $D = 50$ nm (the left graph) and $D = 70$ nm (the right graph) in the Dirichlet model. The surface potential of all plates and the electrolyte concentration are taken to be $\Phi_S = -4.0$ and $c = 0.001$ mol l^{-1}. When $D = 50$ nm (<58 nm), the potential V_1^F takes a minimum at one point, $d_1 = 25$ nm. When $D = 70$ nm (>58 nm), the potential V_1^F takes a minimum at two points, $d_1 = 19.6$ and 50.4 nm.

Next, let us change D_l, in fixing d_i to d_{min}, the value at which the adiabatic potential takes its minimum in the case of $D_l = D_r$. Figure 6.13 shows the adiabatic potentials V_1^F for $c = 0.001$ mol l^{-1}, $D = 200$ nm and $d_1 = 17.7$ nm in the Dirichlet

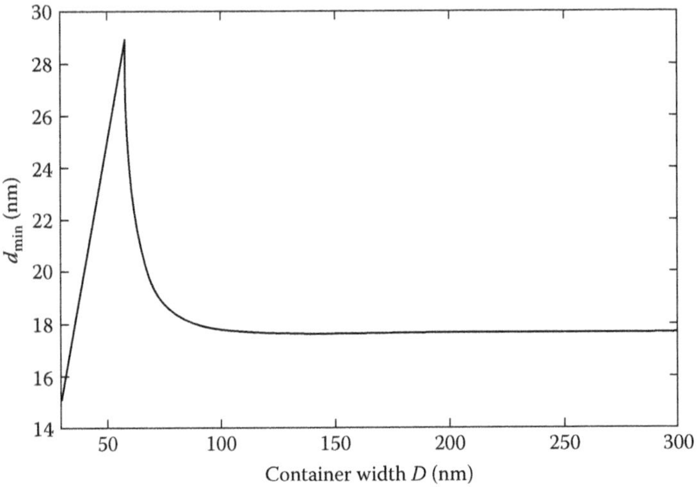

FIGURE 6.11 Trajectory of the stable interplate distance d_{min} plotted against the width D of the container for the surface potential $\Phi_S = -4.0$ of the plates in the Dirichlet model, when $c = 0.001$ mol l^{-1} and $D_l = D_r$. Until D reaches a specific value ($D \approx 58$ nm in this figure), the potential V_1^F takes a minimum at one point $d_{min} = D/2$. After that, the potential V_1^F takes a minimum at two points (see Figure 6.12, for example). Here, we choose to draw the graph for the smaller d_{min} of the two (the larger one is given by $D - d_{min}$).

model as a function of the central position between the two plates. The potential V_1^F takes a minimum at $d_1 = 17.7$ nm for $c = 0.001$ mol l^{-1} in Figure 6.9. The surface potential of all plates is taken to be $\Phi_S = -4.0$. The potential V_1^F takes a minimum at $D_l = d_{min}/2$ and $D_r = d_{min}/2$ in Figure 6.13.

In the case of the many-plate system ($N \geq 2$), we can obtain the adiabatic potential $V_N^F(d_1, \ldots, d_N; D_l, D_r)$ by summing up the contributions from all the regions. The potential takes a minimum at $d_1 = d_2 = \ldots = d_N$ if the plate system is symmetric with no defects. For example, Figure 6.14 shows the adiabatic potential V_6^F and the electrostatic potential $\Phi(x)$ for $N = 6$, $c = 0.001$ mol l^{-1} and $d_i = 17.7$ nm ($i = 1, 2, \ldots, 6$) in the Dirichlet model. The potential V_6^F is shown as a function of the central position between the first plate and the seventh plate in the upper graph in Figure 6.14. As seen from Figure 6.13 and the upper graph of Figure 6.14, the minimum point of the adiabatic pair potential exists in two symmetrical positions. Actually, left–right symmetry is "spontaneously" broken according to external initial conditions, and either one is realized. In the lower graph in Figure 6.14, we have chosen the potential minimum near the left-wall surface.

The lower graph of Figure 6.14 represents the behavior of the mean electric potential $\Phi(x)$ in the plate configuration at which the potential V_6^F takes its minimum ($2D_l = d_i = 17.7$ nm). The surface potential of all plates is taken to be $\Phi_S = -4.0$. The potential V_6^F takes a minimum at $d_i = 17.7$ nm, similar to the case of $D_l = D_r$. The value $d_i = 17.7$ nm is the same as d_{min} of Figure 6.9. In this sense, the solution of the many-plate problem does not really introduce any new physics beyond the two-plate problem, and we will not consider it further when we calculate the Gibbs free

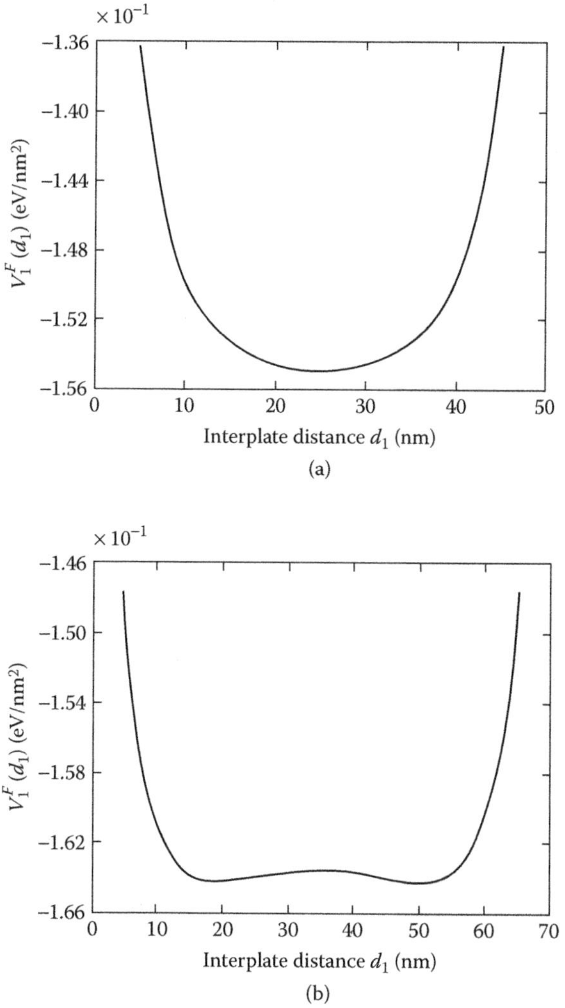

FIGURE 6.12 Adiabatic potentials $V_1^F(d_1)$ for $D = 50$ nm (a) and 70 nm (b) in the Dirichlet model. The surface potential of all plates and the electrolyte concentration are taken to be $\Phi_S = -4.0$ and $c = 0.001$ mol l^{-1}, respectively.

energy below. First, however, there are many noteworthy points about the preceding calculation of the Helmholtz free energy.

The most important general point is that there is a long-range attraction between the plates when we calculate the Helmholtz free energy of highly charged plates in a container with finite width and derive the exact adiabatic potential for the plates in mean field theory [22]. The same result was obtained previously for a container of infinite width [13]. In linearizing the problem, you clearly throw out the baby with the bathwater in the calculation of the Helmholtz free energy, and you need the solution of the nonlinear problem to demonstrate the long-range attraction.

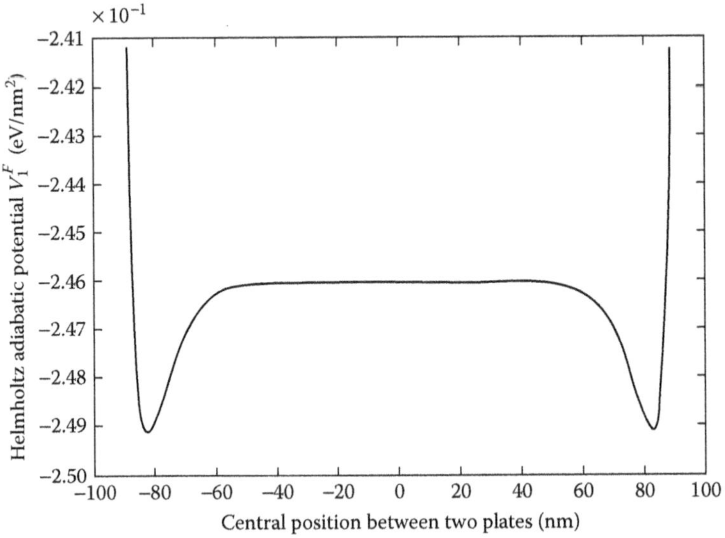

FIGURE 6.13 Adiabatic potentials V_1^F for $c = 0.001$ mol l^{-1} and $d_1 = 17.7$ nm in the Dirichlet model as a function of the central position between two plates. The origin of the coordinate is taken at the midpoint between the container surfaces. The surface potential of all plates is taken to be $\Phi_S = -4.0$. The potential V_1^F takes a minimum at $d_1 = 17.7$ nm for $c = 0.001$ mol l^{-1} in Figure 6.9.

The second most important point is that such exact calculations enable us to pinpoint the origin of the attraction. The adiabatic potential expressed in Equation 6.36 as the sum of the contributions from pairs of surfaces consists of an osmotic part and an electric part. The osmotic pressures arise from the random motions of the small ions trapped in the regions R_i. On the other hand, the electric part of the adiabatic potential represents the interaction mediated not by material entities, but by the electric field produced by the equilibrium distribution of the small ions and the plate charges. This is the "electric rope" that binds the plates together at long range. By attracting counterions and repelling co-ions, the fixed surface charges of the plates create asymmetric distributions of small ions both in number and in charge. From these asymmetries, the plates receive two kinds of reaction. First, the excess of small ions in number induces an osmotic repulsion between the plates. Second, the counterions that are shared by two plates give rise to an excess of charge opposite to that of the plates, and this induces an effective electric attraction between the plates. It is a delicate balance of these two diametrically opposed effects that leads to the osmotic and electric stability attained at the minimum of the adiabatic potential (Equation 6.36).

As is well known, the counterions distributed close to the plates work to shield their surface charges. Figure 6.3 shows, however, that there exist counterions at the middle of the inner region that do not belong to either of the plates. It is those counterions that result in an attraction, *a counterion-mediated attraction*, between

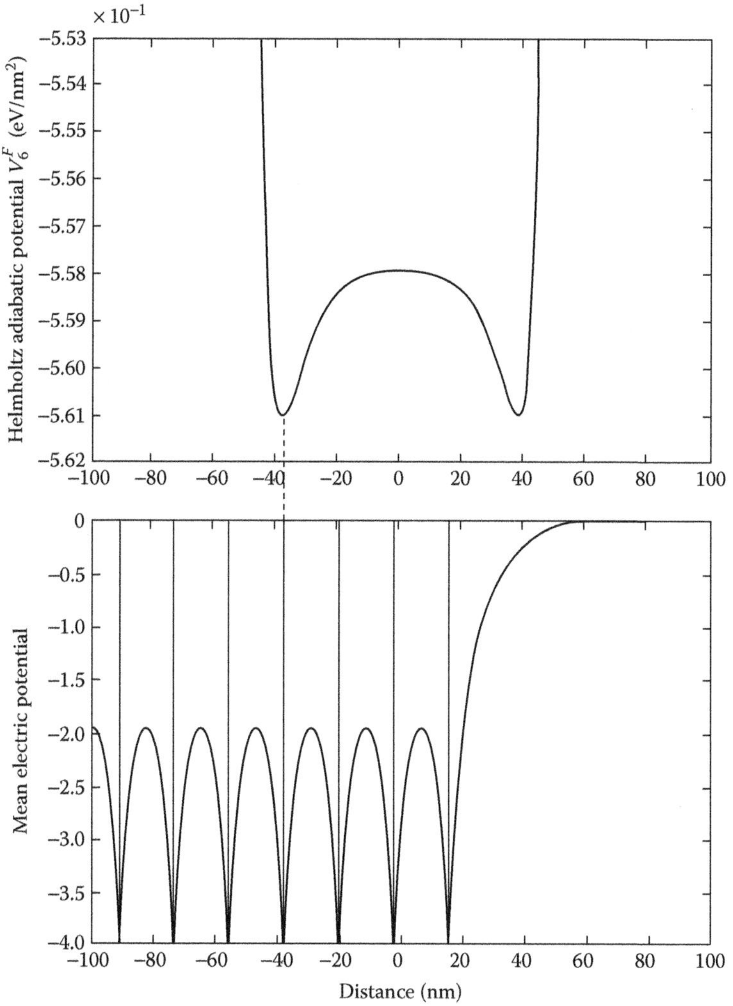

FIGURE 6.14 The case of the seven-plate system ($N = 6$). The adiabatic potential V_6^{F} and the electrostatic potential $\Phi(x)$ for $c = 0.001$ mol l^{-1} and $d_i = 17.7$ nm ($i = 1, 2, \ldots, 6$) in the Dirichlet model. The origin of the coordinate x is taken at the midpoint between the container surfaces. The potential V_6^{F} is shown as a function of the central position between the first plate and the seventh plate in the upper graph. The lower graph represents the behavior of $\Phi(x)$ at the minimum point of V_6^{F} ($2D_1 = d_i = 17.7$ nm). The surface potential of all plates is taken to be $\Phi_S = -4.0$. The potential V_6^{F} takes a minimum at $d_i = 17.7$ nm for $c = 0.001$ mol l^{-1}, $\Phi_S = -4.0$, and $D_1 = D_r$.

the plates of both sides. It is because both this attraction and the osmotic repulsion originate in the average distribution of ions that we are led to the well-known experimental fact stated by Hunter [24] that "the secondary minimum that occurs at larger distances ($\approx 7\kappa^{-1}$) is responsible for a number of important effects in colloidal suspensions." The approximation used in the DLVO theory [5, 6] failed to compute these competing effects correctly because it ignored the strong correlation between the plate charges and the counterions. Another radical departure of the new calculations from the DLVO theory is that, as shown in Figure 6.8, solely the osmotic part of the free energy can produce a medium-range repulsion and a long-range attraction. This is a result of the wall effect in a finite container. In the limit $D \to \infty$ (for the symmetric case $D_l = D_r$ and the two-plate system $N = 1$), the present model shifts to the model of the infinite container in reference [13]. In this limit, the osmotic part of the free energy turns out to lead to a repulsion, and the attraction is due to the electric rope alone.

Another point of interest is that we have fully shown that the models for $D \to \infty$ and $D = 200$ nm are in agreement. In the experimental system this corresponds to a sol concentration $r \approx 0.01$ for plates with crystalline spacing $2a \approx 2$ nm. From the point of view of the theory, the infinite-container limit is already effectively reached at $r = 0.01$. When we studied the sol-concentration effect in the n-butylammonium vermiculite system [25], we never noticed any difference between the data obtained at $r = 0.01$ and the data previously obtained at $r < 0.01$. On the theoretical side, the demonstration that our new solutions agree with the $D \to \infty$ solutions given previously will enable us to discuss objections to reference [13] in the light of the new results, since the physicochemical principles of references [13, 22] are the same. We reserve these discussions to the following chapter. Here, we note simply that if D becomes large enough, the analytical results will not depend on D. This means that the contribution of the container wall is lost. This can be understood also from Equation 6.38 and comparison of Figures 2, 3, 7 and 10 in reference [13] with Figure 6.2, Figure 6.3, Figure 6.15 and Figure 6.9 here, respectively.

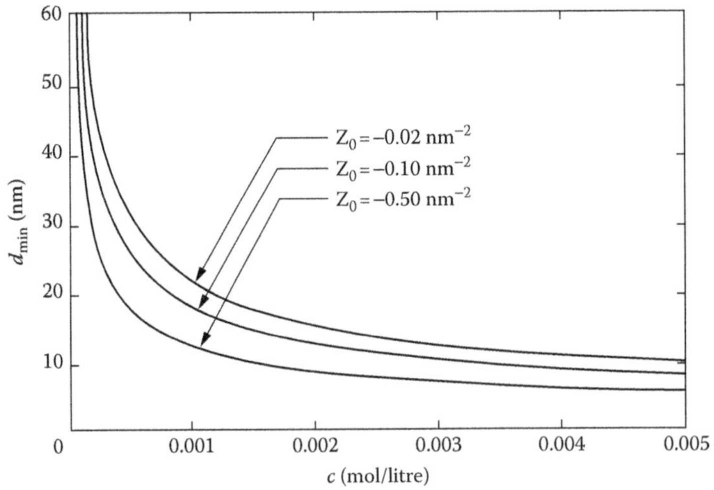

FIGURE 6.15 Trajectories of the stable interplate distance d_{min} plotted against the concentration c for three values of the surface charge density Z_o in the Neumann model. The width of the container is set to be 200 nm and $D_l = D_r$. The swelling is enhanced for small values of $|Z_o|$.

Figure 6.15 shows the trajectories of the stable interplate spacing $d_{\min}$ plotted against the concentration c for three values of the surface charge density Z_o in the Neumann model. It has been included for the sake of completeness. Because all the experimental evidence in Chapters 1 to 5 points us in the direction of trying to understand swollen clays as a constant surface potential system, it is natural to concentrate on solutions of the Dirichlet model. The details of the calculation used to obtain Figure 6.15 are given in the appendix of reference [22]. It can be said that the attraction between the plates works strongly, in either model, insofar as the surface charge is large (see Figure 6.10 and Figure 6.15).

We now wish to investigate the Gibbs free energy of the system based on the formalism of references [10, 26] and compare the results with those from the Helmholtz free energy. In reference [10], the Helmholtz and Gibbs free energies of macroionic suspensions were proved to have the following integral representation as

$$\beta G = \beta F - \beta F_0 + \frac{1}{8\pi\lambda_{\mathrm{B}}}\int_V (\nabla\Phi)^2 dV + \beta G_0 \tag{6.39}$$

where Φ is an average electric potential satisfying the PB equation, and F_0 and G_0 are, respectively, the nonelectric parts of the Helmholtz and Gibbs free energies. This integral representation is exact in the sense of the mean field description. In the nonelectric parts F_0 and G_0 of the free energy, both the effect of the solvent and the contribution of the nonelectric charge limit of the charged plate are also included in addition to the osmotic effect of the small ions. Here, these effects are regarded as a common background contribution working uniformly for all regions, and we treat only the osmotic parts of the free energies. Accordingly, we use the expressions F^{osm} and G^{osm} instead of F_0 and G_0. Substitution of the osmotic part (Equation 6.30) of the Helmholtz free energy into Equation 6.39 yields

$$\begin{aligned}\beta G &= \beta F - \beta F^{\mathrm{osm}} + \beta E^{\mathrm{el}} + \sum_j N_j \left(\frac{\partial \beta F^{\mathrm{osm}}}{\partial N_j}\right)_{d,T,Z_n} + \sum_n Z_n \left(\frac{\partial \beta F^{\mathrm{osm}}}{\partial Z_n}\right)_{d,T,N_j} \\ &= \beta F + \beta E^{\mathrm{el}} + N_+ + N_-\end{aligned} \tag{6.40}$$

where

$$\begin{aligned}\beta E^{\mathrm{el}} &= \frac{1}{8\pi\lambda_{\mathrm{B}}}\int_{-l_{\mathrm{L}}}^{l_{\mathrm{R}}} \left(\frac{d\Phi(x)}{dx}\right)^2 dx = \frac{2\kappa^2}{8\pi\lambda_{\mathrm{B}}}\int_{-l_{\mathrm{L}}}^{l_{\mathrm{R}}} (\cosh\Phi(x) - \cosh(0))\, dx \\ &= N_+ + N_- - n_0 d\frac{1+k^2}{k}\end{aligned} \tag{6.41}$$

We can divide this Gibbs free energy G into an electric part G^{el} and an osmotic part G^{osm}, namely

$$G = G^{\mathrm{el}} + G^{\mathrm{osm}} \tag{6.42}$$

Here, so as to satisfy the relation

$$\beta G - \beta G^{\text{osm}} = \beta F - \beta F^{\text{osm}} + \beta E^{\text{el}} \tag{6.43}$$

that was obtained in references [7, 8], we define G^{osm} as

$$\begin{aligned}\beta G^{\text{osm}} &= \sum_j N_j \left(\frac{\partial \beta F^{\text{osm}}}{\partial N_j}\right)_{d,T,Z_n} + \sum_n Z_n \left(\frac{\partial \beta F^{\text{osm}}}{\partial Z_n}\right)_{d,T,N_j} \\ &= \beta F^{\text{osm}} + N_+ + N_-\end{aligned} \tag{6.44}$$

corresponding to Equation 6.30. As a result, we get the electric part

$$\begin{aligned}\beta G^{\text{el}} = &-\left\{2\,|Z_{\text{L}}|\ln\left[\sqrt{k}\,\text{sn}\left(-\frac{\kappa}{2\sqrt{k}} l_{\text{L}} + K(k), k\right)\right]\right\} \\ &-\{L \to R\} - N_+ \ln\left(\frac{N_+}{n_0 d}\right) - N_- \ln\left(\frac{N_-}{n_0 d}\right)\end{aligned} \tag{6.45}$$

and the osmotic part

$$\beta G^{\text{osm}} = -N_+ \ln\left[\frac{d}{N_+ \lambda_{\text{S}}^3}\left(\frac{m_+}{m_{\text{p}}}\right)^{\frac{3}{2}}\right] - N_- \ln\left[\frac{d}{N_- \lambda_{\text{S}}^3}\left(\frac{m_-}{m_{\text{p}}}\right)^{\frac{3}{2}}\right] \tag{6.46}$$

For the sake of simplicity, let us examine the two-plate system ($N = 1$) with a symmetric configuration ($D_{\text{l}} = D_{\text{r}} = (D - d_1)/2$). We first apply the result obtained above to the inner region R_1. The electric part of the Helmholtz free energy in the region R_1 sandwiched between the two plates is given by making the substitution $i = 1$ in Equation 6.32. The osmotic part is represented likewise in Equation 6.33. Similarly, from Equation 6.45 and Equation 6.46, the electric part and the osmotic part of the Gibbs free energy in the inner region R_1 are determined as

$$\begin{aligned}\beta G_1^{\text{el}}(d_1) = &-4\,|Z_1|\ln\left[\sqrt{k_1}\,\text{sn}\left(-\frac{\kappa}{4\sqrt{k_1}} d_1 + K(k_1), k_1\right)\right] \\ &-N_{1+}\ln\left(\frac{N_{1+}}{n_0 d_1}\right) - N_{1-}\ln\left(\frac{N_{1-}}{n_0 d_1}\right)\end{aligned} \tag{6.47}$$

and

$$\beta G_1^{\mathrm{osm}}(d_1) = -N_{1+} \ln\left[\frac{d_1}{N_{1+}\lambda_{\mathrm{S}}^3}\left(\frac{m_+}{m_{\mathrm{p}}}\right)^{\frac{3}{2}}\right] - N_{1-} \ln\left[\frac{d_1}{N_{1-}\lambda_{\mathrm{S}}^3}\left(\frac{m_-}{m_{\mathrm{p}}}\right)^{\frac{3}{2}}\right] \tag{6.48}$$

respectively.

For the outer region $\mathrm{R}_{\mathrm{o}}^{\mathrm{l}} \cup \mathrm{R}_{\mathrm{o}}^{\mathrm{r}}$ outside the plates, the electric part and the osmotic part of the Helmholtz free energy are given by Equation 6.34 and Equation 6.35, respectively. Similarly, from Equation 6.45 and Equation 6.46, the electric part and the osmotic part of the Gibbs free energy in the outer region $\mathrm{R}_{\mathrm{o}}^{\mathrm{l}} \cup \mathrm{R}_{\mathrm{o}}^{\mathrm{r}}$ are determined as

$$\beta G_{\mathrm{o}}^{\mathrm{el}}(D_1, D_{\mathrm{r}}) = -\left\{2\left|Z_{\mathrm{o}}^{\mathrm{l}}\right| \ln\left[\sqrt{k_{\mathrm{o}}^{\mathrm{l}}}\,\mathrm{sn}\left(-\frac{\kappa}{2\sqrt{k_{\mathrm{o}}^{\mathrm{l}}}} D_1 + K\left(k_{\mathrm{o}}^{\mathrm{l}}\right), k_{\mathrm{o}}^{\mathrm{l}}\right)\right]\right.$$

$$\left. + N_{\mathrm{o}-}^{\mathrm{l}} \ln\left(\frac{N_{\mathrm{o}-}^{\mathrm{l}}}{n_0 D_1}\right) + N_{\mathrm{o}+}^{\mathrm{l}} \ln\left(\frac{N_{\mathrm{o}+}^{\mathrm{l}}}{n_0 D_1}\right)\right\} - \{l \to r\} \tag{6.49}$$

and

$$\beta G_{\mathrm{o}}^{\mathrm{osm}}(D_1, D_{\mathrm{r}}) = -N_{\mathrm{o}+}^{\mathrm{l}} \ln\left[\frac{D_l}{N_{\mathrm{o}+}^{\mathrm{l}}\lambda_{\mathrm{S}}^3}\left(\frac{m_+}{m_{\mathrm{p}}}\right)^{\frac{3}{2}}\right] - N_{\mathrm{o}-}^{\mathrm{l}} \ln\left[\frac{D_1}{N_{\mathrm{o}-}^{\mathrm{l}}\lambda_{\mathrm{S}}^3}\left(\frac{m_-}{m_{\mathrm{p}}}\right)^{\frac{3}{2}}\right] - \{l \to r\} \tag{6.50}$$

respectively.

We now have the free energy of each region expressed as a function of the width of that region. Summing up the electric and osmotic parts of the free energies of all regions, we get the total free energy of the system. This total free energy is the arrangement function for two plates laid in the container, and it is interpreted as the adiabatic potential with regard to the interplate distances. The Helmholtz adiabatic pair potential is given by making the substitution $i = 1$ in Equation 6.36. The Gibbs adiabatic potential is obtained as

$$V_1^{\mathrm{G}}(d_1; D_1, D_{\mathrm{r}}) = G_1^{\mathrm{el}}(d_1) + G_1^{\mathrm{osm}}(d_1) + G_{\mathrm{o}}^{\mathrm{el}}(D_1, D_{\mathrm{r}}) + G_{\mathrm{o}}^{\mathrm{osm}}(D_1, D_{\mathrm{r}}) \tag{6.51}$$

The formula for the adiabatic Gibbs potential (Equation 6.51) is exact within the framework of the mean field description. As with the adiabatic Helmholtz potential, we examine it numerically following the Carlson theory of elliptic integrals [15–21]. In order to make clear comparisons between the two types of free energy, the width D of the container is set to be 200 nm throughout (except for Figure 6.21).

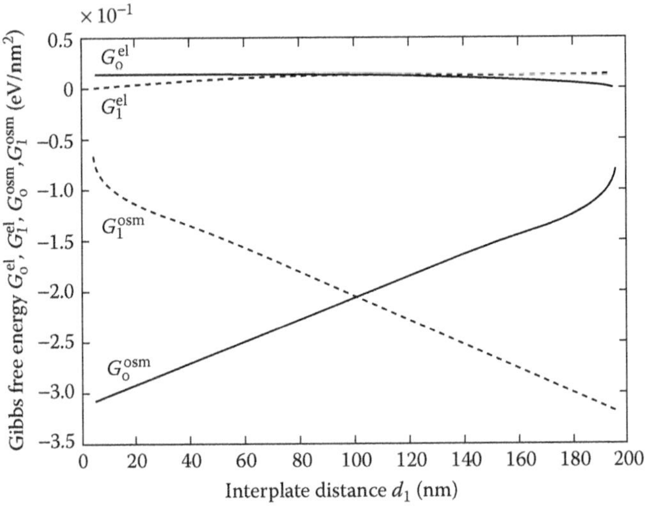

FIGURE 6.16 Electric parts G_o^{el} and G_1^{el} and osmotic parts G_o^{osm} and G_1^{osm} of the Gibbs free energy vs. the interplate distance d_1 for $c = 0.002$ mol l^{-1} and $\Phi_S = -4.0$ in the Dirichlet model ($Z_0 = -0.118$ nm^{-2} in the Neumann model). The solid lines and dashed lines represent the free energies for the region $R_o^l \cup R_o^r$ and the region R_1, respectively.

Figure 6.16 shows the behavior of the electric parts G_o^{el} and G_1^{el} and osmotic parts G_o^{osm} and G_1^{osm} of the adiabatic potential as a function of the interplate distance d_1 for the salt concentration $c = 0.002$ mol l^{-1} and the surface potential $\Phi_S = -4.0$ in the Dirichlet model. While the osmotic pressure of the small ions in R_1 exerts a repulsion between the two plates, the electric contribution from R_1 and the osmotic contribution from $R_o^l \cup R_o^r$ give rise to an attraction. Figure 6.17 and Figure 6.18 show respectively the sum of the electric parts G_o^{el} and G_1^{el} and the sum of the osmotic parts G_o^{osm} and G_1^{osm} of the adiabatic potential for the salt concentration $c = 0.002$ mol l^{-1} and the surface potential $\Phi_S = -4.0$. Figure 6.17 shows the electric attraction between the two plates (and between a plate and the container surface). Figure 6.18 shows that the osmotic pressure is repulsive at medium range and attractive at long range between two plates (and between a plate and the container surface) and takes a minimum at three points. In the minimum point at the middle, the depth of the potential is shallow in comparison with the other two points and becomes shallower on taking the sum of the electric parts and the osmotic parts. As a result, the third minimum point cannot be found clearly in the total adiabatic potential (Equation 6.51). The behavior of the total adiabatic potential $V_1^G(d_1)$ is shown in Figure 6.19 for the concentrations $c = 0.001$ to 0.002 mol l^{-1}. It is repulsive at medium range and attractive at long range between two plates (and between a plate and the container surface). This potential characteristic does not depend on change of salt concentration ($c = 0.001$, 0.0015, 0.002 mol l^{-1}). The potential V_1^G takes a minimum at $d_1 = 16.2$ and 183.8 nm for $c = 0.001$ mol l^{-1}, at $d_1 = 13.2$ and 186.8 nm for $c = 0.0015$ mol l^{-1}, and at $d_1 = 11.4$ and 188.6 nm for $c = 0.002$ mol l^{-1}.

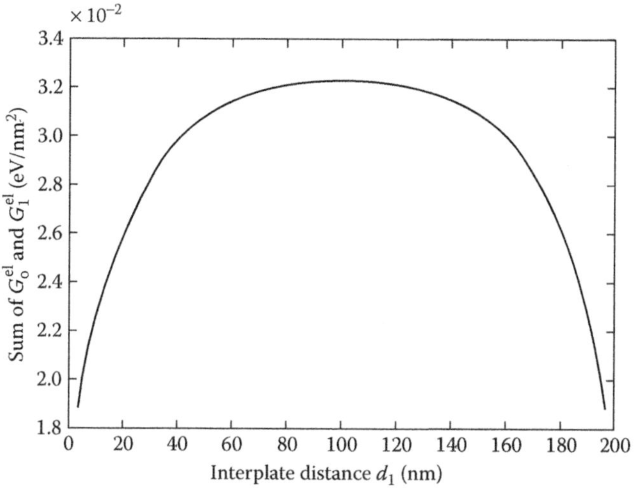

FIGURE 6.17 Sum of the electric parts G_o^{el} and G_1^{el} of the Gibbs free energy vs. the interplate distance d_1 for $c = 0.002$ mol l^{-1} and $\Phi_S = -4.0$ in the Dirichlet model ($Z_0 = -0.118$ nm^{-2} in the Neumann model).

In Figure 6.20, the trajectories of the interplate distance d_{min} (the smaller one) for the potential minimum are drawn for different values of Φ_S, showing the increase of the plate separation against the dilution of the salt concentration c. The interplate distance d_{min} approaches $D/2$ as $c \to 0$. We note that the swelling is enhanced for

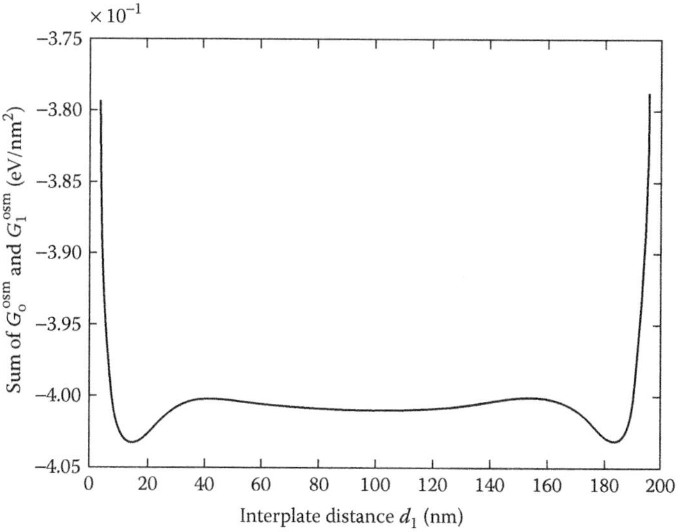

FIGURE 6.18 Sum of the osmotic parts G_o^{osm} and G_1^{osm} of the Gibbs free energy vs. the interplate distance d_1 for $c = 0.002$ mol l^{-1} and $\Phi_S = -4.0$ in the Dirichlet model ($Z_0 = -0.118$ nm^{-2} in the Neumann model).

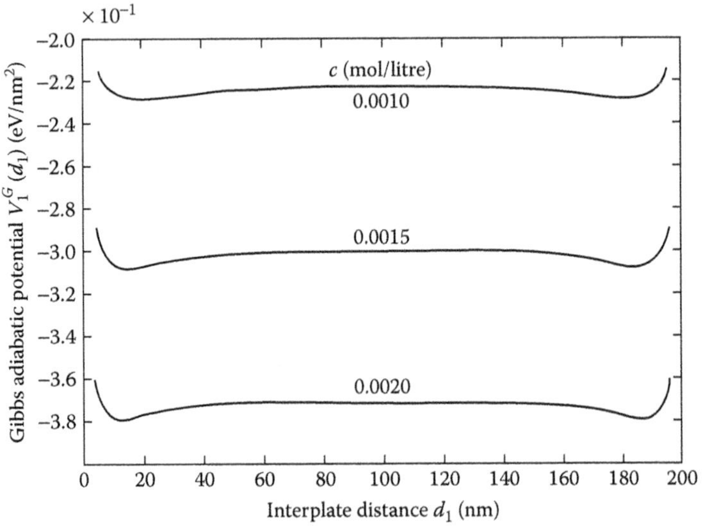

FIGURE 6.19 Gibbs adiabatic potentials $V_1^G(d_1)$ for three values of the electrolyte concentration c in the Dirichlet model. The surface potential of all plates is taken to be $\Phi_S = -4.0$ ($Z_0 = -0.084$, -0.102 and -0.118 nm^{-2} for $c = 0.0010$, 0.0015 and 0.0020 mol l^{-1} in the Neumann model). The potential V_1^G takes a minimum around $d_1 = 15$ nm and 185 nm.

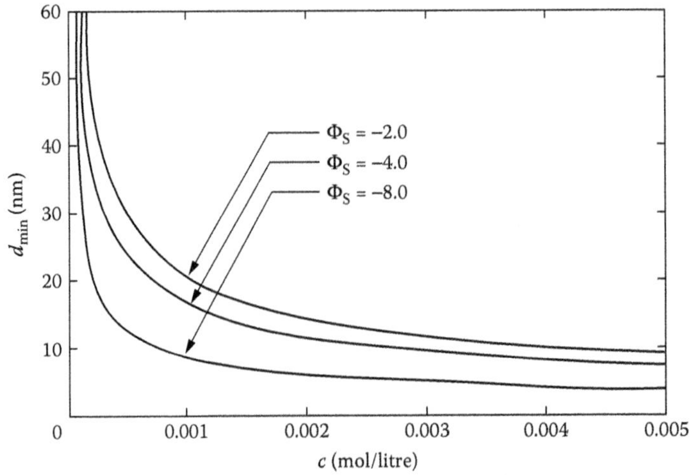

FIGURE 6.20 Trajectories of the stable interplate distance d_{min} of the Gibbs adiabatic potential V_1^G plotted against the concentration c of the external soaking solution for three values of the surface potential Φ_S in the Dirichlet model. The width of the container is set to be 200 nm and $D_l = D_r$. The swelling is enhanced for small values of $|\Phi_S|$.

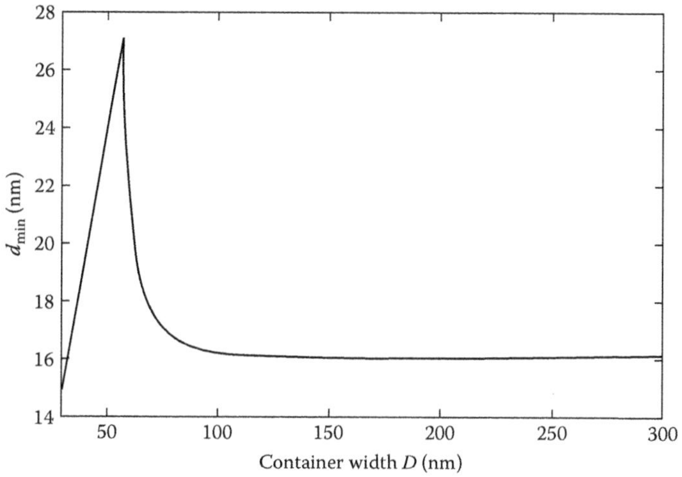

FIGURE 6.21 Trajectory of the stable interplate distance d_{min} of the potential $V_1{}^G$ plotted against the width D of the container for the surface potential $\Phi_S = -4.0$ of the plates in the Dirichlet model, when $c = 0.001$ mol l^{-1} and $D_l = D_r$.

small values of $|\Phi_S|$. Figure 6.21 shows the trajectory of d_{min} for $c = 0.001$ mol l^{-1}, $\Phi_S = -4.0$ and $D_l = D_r$, changing the width D of the container. Until D reaches a specific value ($D \approx 54.9$ nm in Figure 6.21), the potential $V_1{}^G$ takes a minimum at the one point $d_{min} = D/2$. After that, the potential $V_1{}^G$ takes a minimum at two points, and the symmetry of the graph breaks "spontaneously" (see Figure 6.19, for example). Here, we choose to draw the graph for the smaller d_{min} of the two (the larger one is given by $D - d_{min}$). The potential minimum d_{min} becomes smaller as D increases, and d_{min} takes a constant (namely, asymptotic) value for sufficiently large D in comparison with d_{min}. This means that the contribution of the container wall disappears.

In Figure 6.22, the two kinds of adiabatic potential $V_1{}^F$ and $V_1{}^G$ are drawn on the same energy scale for $c = 0.003$ mol l^{-1} and $\Phi_S = -4.0$. As shown by this figure, both adiabatic potentials take a maximum at $d_1 = 100$ nm. The difference $V_1(100\text{ nm}) - V_1(d_{min})$ therefore measures the strength of the plate binding. Figure 6.23 shows the potential difference $V_1(100\text{ nm}) - V_1(d_{min})$ plotted against the salt concentration c for $\Phi_S = -4.0$. In Figure 6.24, the values of the interplate distance d_{min} for the two adiabatic potentials are compared changing the salt concentration c for $\Phi_S = -4.0$. Figure 6.22 shows that $V_1{}^G(d_1)$ is larger than $V_1{}^F(d_1)$ for all d_1, owing to the average energy $E^{el} = G^{el} - F^{el}$ of Equation 6.43 and the van't Hoff osmotic pressure, which is derived from the relation $G^{osm} = F^{osm} + (N_+ + N_-)k_BT$ of Equation 6.44. However, by comparison between the potentials with these contributions deducted, it turns out that the potential minimum of $V_1{}^G(d_1)$ is deeper than that of $V_1{}^F(d_1)$ (see Figure 6.23) and that the interplate distance d_{min} of $V_1^G(d_1)$ is smaller than that of $V_1^F(d_1)$ (see Figure 6.24), though their differences are small. Consequently, the two adiabatic potentials have qualitatively the same structure of a repulsive component and an attractive component, but the Gibbs adiabatic potential shows the stronger attractive

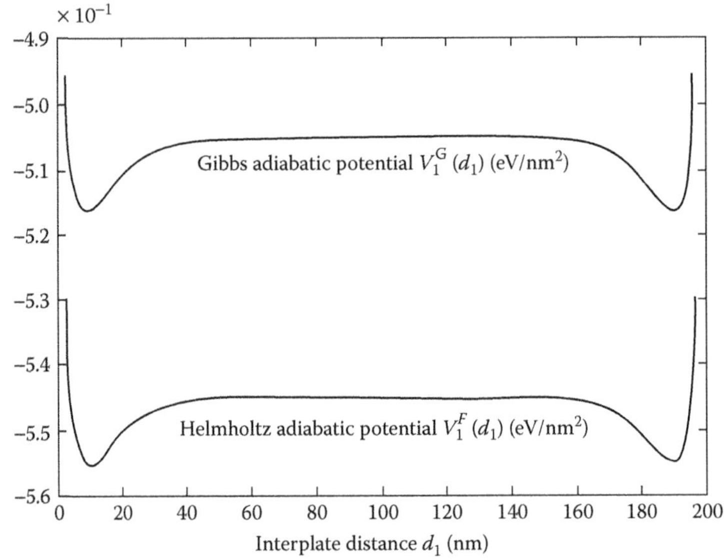

FIGURE 6.22 Comparison between the Helmholtz adiabatic potential $V_1^F(d_1)$ and Gibbs adiabatic potential $V_1^G(d_1)$ for the salt concentration $c = 0.003$ mol l^{-1}, the container width $D = 200$ nm ($D_l = D_r$) and the surface potential $\Phi_S = -4.0$ in the Dirichlet model.

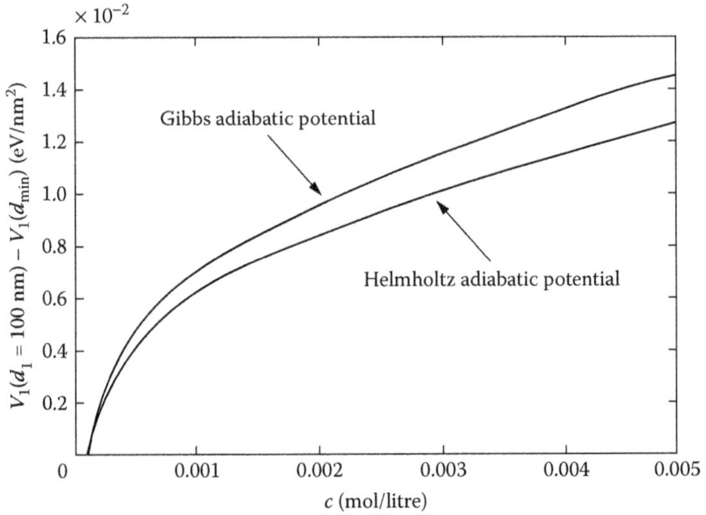

FIGURE 6.23 Potential differences V_1 (100 nm) − V_1 (d_{min}) for the Helmholtz and Gibbs adiabatic potentials plotted against the concentration c of the external soaking solution for the surface potential $\Phi_S = -4.0$ in the Dirichlet model. The width of the container is set to be 200 nm and $D_l = D_r$. The potential difference measures the strength of the binding between the plates.

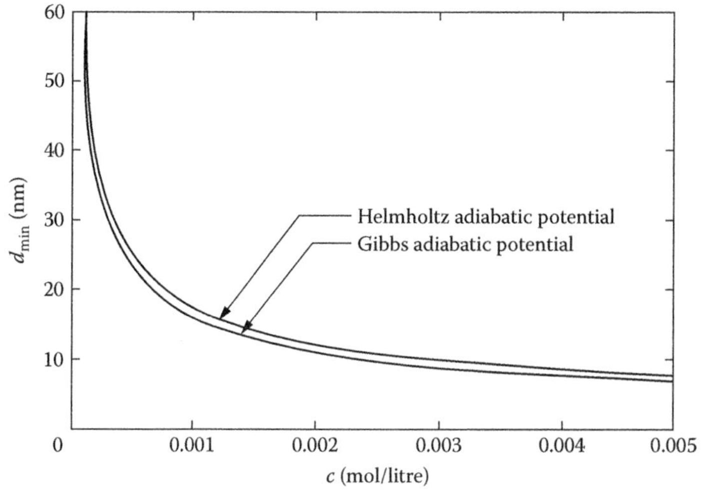

FIGURE 6.24 Trajectories of the stable interplate distance d_{min} of the Helmholtz and Gibbs adiabatic potentials plotted against the concentration c of the external soaking solution for the surface potential $\Phi_S = -4.0$ in the Dirichlet model. The width of the container is set to be 200 nm and $D_l = D_r$.

effect. Both adiabatic potentials have repulsive components for small d_l and weak attractive components for large d_l (< $D/2$).

Let us consider why $V_l^G(d_l)$ has the stronger attractive effect than $V_l^F(d_l)$. In order to examine it, the average energy E^{el} of the solution is shown in Figure 6.25

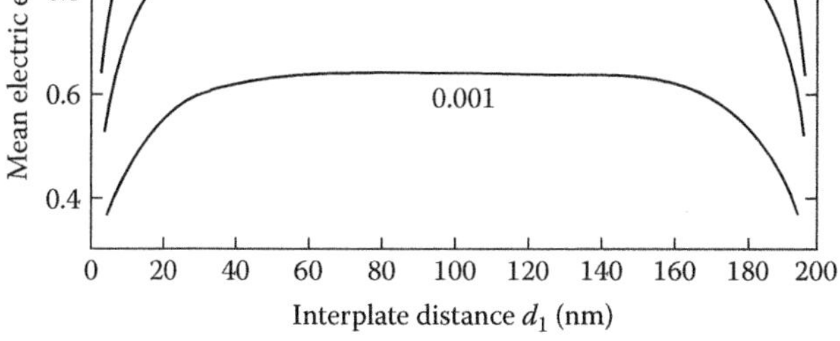

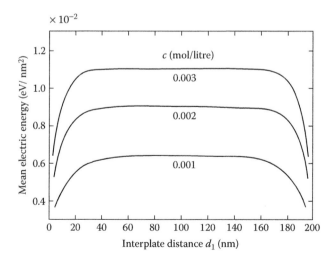

FIGURE 6.25 Average electric energy E^{el} for three values of the electrolyte concentration $c = 0.001$, 0.002 and 0.003 mol l^{-1} in the Dirichlet model. The surface potential of the plates is taken to be $\Phi_S = -4.0$.

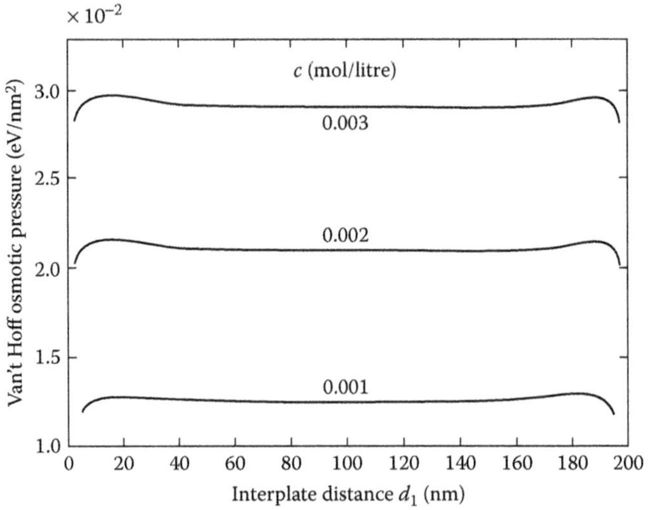

FIGURE 6.26 Van't Hoff osmotic pressure $(N_+ + N_-)k_BT$ for three values of the electrolyte concentration $c = 0.001$, 0.002 and 0.003 mol l^{-1} in the Dirichlet model. The surface potential of the plates is taken to be $\Phi_S = -4.0$.

for the concentrations $c = 0.001$, 0.002 and 0.003 mol l^{-1} and $\Phi_S = -4.0$, and the van't Hoff osmotic pressure $(N_+ + N_-)k_BT$ is shown in Figure 6.26 under the same conditions. According to these figures, E^{el} and $(N_+ + N_-)k_BT$ have a short-range attractive effect. On the other hand, from the general Equation 6.40, the Gibbs free energy adds the average electric energy and van't Hoff osmotic pressure term to the Helmholtz free energy and, as a result, the attractive effect of $V_1^G(d_1)$ becomes stronger.

REFERENCES

1. Krishnan, C.V. and Friedman, H.L., *J. Phys. Chem.,* 74, 3900, 1979.
2. Desnoyers, J.E. and Arel, M., *Can. J. Chem.,* 45, 359, 1967.
3. Crawford, R.J., Smalley, M.V., and Thomas, R.K., *Adv. Colloid Interface Sci.,* 34, 537, 1991.
4. Debye, P.J.W. and Hückel, E., *Physik. Zeits.,* 24, 185, 1923.
5. Derjaguin, B.V. and Landau, L., *Acta Physicochimica,* 14, 633, 1941.
6. Verwey, E.J.W. and Overbeek, J.Th.G., *Theory of the Stability of Lyophobic Colloids,* Elsevier, Amsterdam, 1948.
7. Sogami, I., *Phys. Lett. A,* 96, 199, 1983.
8. Sogami, I. and Ise, N., *J. Chem. Phys.,* 81, 6320, 1984.
9. Smalley, M.V., *Mol. Phys.,* 71, 1251, 1990.
10. Sogami, I.S., Smalley, M.V., and Shinohara, T., *Prog. Theor. Phys.,* 113, 235, 2005.
11. Erdélyi, A., Magnus, W., Oberhettinger, F., and Tricomi, F.G., *Higher Transcendental Functions,* Vol. 2, McGraw-Hill, New York, 1953.

12. Byrd, P.E. and Friedman, M.D., *Handbook of Elliptic Integrals for Engineers and Scientists,* 2nd ed., Springer, Berlin, 1971.
13. Sogami, I., Shinohara, T., and Smalley, M.V., *Mol. Phys.,* 76, 1, 1992.
14. Landau, L.D. and Lifshitz, E.M., *Statistical Physics,* Pergamon, New York, 1959.
15. Carlson, B.C., *Special Functions of Applied Mathematics,* Chap. 9, Academic Press, New York, 1977.
16. Carlson, B.C., SIAM *J. Math. Anal.,* 8, 231, 1977.
17. Carlson, B.C., SIAM *J. Math. Anal.,* 9, 524, 1978.
18. Zill, D.G. and Carlson, B.C., *Math. Comput.,* 24, 199, 1970.
19. Carlson, B.C., *Numer. Math.,* 33, 1, 1979.
20. Press, W.H. and Teukolsky, S.A., *Comput. Phys.,* 4, 92, 1990.
21. Press, W.H., Teukolsky, S.A., Vetterling, W.T., and Flannery, B.P., *Numerical Recipes in Fortran 77, The Art of Scientific Computing,* Vol. 1, Fortran Numerical Recipes, 2nd ed., Cambridge University Press, Cambridge, 1996.
22. Shinohara, T., Smalley, M.V., and Sogami, I.S., *Mol. Phys.,* 101, 1883, 2003.
23. Low, P.F., *Langmuir,* 3, 18, 1987.
24. Hunter, R.J., *Foundations of Colloid Science,* Clarendon Press, Oxford, 1987.
25. Williams, G.D., Moody, K.R., Smalley, M.V., and King, S.M., *Clays Clay Min.,* 42, 614, 1994.
26. Shinohara, T., Smalley, M.V., and Sogami, I.S., *Int. J. Mod. Phys.* B, 19, 3217, 2005.

7 Criticisms and Refutations

To clarify the different roles played by the Helmholtz and Gibbs free energies of ionic solutions, it is relevant to reconsider the derivation of these thermodynamic quantities in the original Debye-Hückel theory [1–4].

In their investigation of dilute solutions of simple ions, Debye and Hückel regarded the jth small ion as a spherical particle with radius a such that the small ion has an electric charge $z_j e$ at its spherical center and the dielectric constant ε inside the sphere. In this case, the total electrostatic energy of the system is obtained as

$$E^{\mathrm{el}} = -\frac{1}{2}\frac{e^2\kappa}{\varepsilon}\frac{1}{1+\kappa a}\sum_j z_j^2 N_j = -\frac{1}{2}k_{\mathrm{B}}TV\frac{1}{4\pi}\frac{\kappa^3}{1+\kappa a} \tag{7.1}$$

Here the Debye screening parameter κ is defined by

$$\kappa^2 = \frac{4\pi e^2}{\varepsilon k_{\mathrm{B}}TV}\sum_j z_j^2 N_j \tag{7.2}$$

in terms of the particle number of the jth small ion N_j and the total volume of the system V. Debye's charging-up method results readily in the Helmholtz free energy as

$$F = F_0 + F_{\mathrm{el}} = F_0 - \frac{1}{12\pi}k_{\mathrm{B}}TV\kappa^3\tau(\kappa a) \tag{7.3}$$

Where the size effect of the ion is represented by the function

$$\tau(\kappa a) = \frac{3}{(\kappa a)^3}\int_0^{\kappa a}\frac{t^2}{1+t}dt = \frac{3}{(\kappa a)^3}\left\{\ln(1+\kappa a) - \kappa a + \frac{1}{2}(\kappa a)^2\right\} \tag{7.4}$$

When κa is sufficiently small, this function is written in series expansion form as

$$\tau(\kappa a) \cong 1 - \frac{3}{4}\kappa a + \frac{3}{5}(\kappa a)^2 + \cdots \tag{7.5}$$

By summing up the chemical potentials

$$\mu_j^{\mathrm{el}} = \left(\frac{\partial F_{\mathrm{el}}}{\partial N_j}\right)_{T,V} = -\frac{\kappa e^2}{2\varepsilon(1+\kappa a)}z_j^2 \tag{7.6}$$

calculated from the Helmholtz free energy (Equation 7.3), the Gibbs free energy is determined to be

$$G = G_0 - \frac{1}{8\pi} k_B T V \frac{\kappa^3}{1+\kappa a} \tag{7.7}$$

Taking the difference between the Gibbs and Helmholtz free energies, we obtain the equation of state for the osmotic pressure of the small-ion gas as follows:

$$P = k_B T \left\{ \frac{1}{V} \sum_j N_j - \frac{\kappa^3}{24\pi} \left[\frac{3}{1+\kappa a} - 2\tau(\kappa a) \right] \right\} \tag{7.8}$$

In this way, the net electric effects of the ions reduce all of the thermodynamic energies E^{el}, F and G of the system. It is essential to note that the reduction appears more strongly in G than in F and, as a result, the osmotic pressure decreases. It is readily proved that the function inside the square brackets in Equation 7.8 is positive definite for arbitrary κa j the electric interaction acts to decrease the osmotic pressure in the solution, regardless of the size of the ions. It was Langmuir who first noticed the importance of the decrease of the osmotic pressure and interpreted this effect as a manifestation of the attractive nature of the net electric interactions among all ionic species in the solution. Applying this result directly to macroionic systems, Langmuir [5] suggested that similar effects also appear in solutions including macroions.

In the original theory of Langmuir [5], all the ionic species are treated symmetrically so that small ions must have surrounding clouds of macroions as well as macroions having clouds of small ions and small ions themselves having mutual atmospheres of small ions with opposite charges. To remedy this apparent drawback of Langmuir's treatment, Sogami and Ise [6, 7] adopted a viewpoint where the size and the timescale of motion of the macroions are considered to be asymmetrically larger than those of the small ions.

In a macroionic suspension where the number and configuration of the particles are fixed, this condition requires that the effective valency of the particles Z_n must change against a variation of the number of small ions N_i. Therefore the chemical potential of the small ions that is obtained by $\partial F/\partial N_j$ must necessarily be correlated with the derivative of F with respect to Z_n. It is reasonable to interpret the derivative of the Helmholtz free energy $\partial F/\partial Z_n$ as the *chemical potential of effective valency of the* n*th particle* [6, 7]. The Gibbs free energy of the system G is obtained by summing up the chemical potentials of the small ions of all kinds and the chemical potentials of the effective valencies of all particles.

To take derivatives of F with respect to N_j and Z_n and to sum up the component chemical potentials without contradiction to the constraint condition [8, 9], we introduce the following differential operator D_{NZ} as

$$D_{NZ} = \sum_j N_j \frac{\partial}{\partial N_j} + \sum_n \frac{\partial}{\partial Z_n} \tag{7.9}$$

It is readily confirmed that, when applied to the neutrality condition (Equation 6.27), this operator reproduces the condition again without resulting in any additional constraints [8, 9]. Therefore, this operator D_{NZ} can be interpreted as a kind of covariant derivative that preserves invariance of a surface spanned by the constraint in a fictitious space $(N_1, \ldots, N_s, Z_1, \ldots, Z_M)$.

For application of the covariant derivative D_{NZ}, it is convenient to use the Helmholtz free energy of the region R in the form [10]

$$\beta F = -\frac{1}{8\pi\lambda_{\mathrm{B}}}\int_{-l_{\mathrm{L}}}^{l_{\mathrm{R}}}\left[\frac{d\Phi(x)}{dx}\right]^2 dx + \sum_n Z_n\Phi(x_n) - \sum_j N_j \ln\left[\frac{1}{d}\int_{-l_{\mathrm{L}}}^{l_{\mathrm{R}}} e^{-z_j\Phi(x)}dx\right] + \beta F_0 \tag{7.10}$$

according to the notation of Chapter 6. Here, Z_n is the relevant surface charge density on the plate, $\Phi(x_n)$ is the surface potential at one of the boundaries $x_n = -l_{\mathrm{L}} + 0, l_{\mathrm{R}} - 0$ of the region R and d is the height of the region R. Noting that the average potential Φ as the solution of the Poisson-Boltzmann (PB) equation depends on N_j and Z_n, we find

$$D_{NZ}(\beta F) = \sum_j N_j\left(\frac{\partial\beta F}{\partial N_j}\right)_{Z_n,\Phi} + \sum_n Z_n\left(\frac{\partial\beta F}{\partial Z_n}\right)_{N_j,\Phi} + D_{NZ}(\Phi)\left(\frac{\delta\beta F}{\delta\Phi}\right)_{N_j,Z_n} \tag{7.11}$$

Then, using the extremal condition as

$$\delta\beta F(\Phi)/\delta\Phi = 0 \tag{7.12}$$

and the definition $G = D_{NZ}(F)$, the Gibbs free energy of the macroionic suspension has the integral representation

$$\beta G = \sum_n Z_n\Phi(x_n) - \sum_j N_j \ln\left[\frac{1}{d}\int_{-l_L}^{l_R} e^{-z_j\Phi(x)}dx\right] + \beta G_0 \tag{7.13}$$

where $G_0 = D_{NZ}(F_0)$ is the nonelectric part of the Gibbs free energy. Comparison of Equation 7.13 and Equation 7.10 reveals that we have in this way proved the relation in Equation 6.39, derived originally in reference [11] and used to obtain the Gibbs free energy from the Helmholtz free energy in Chapter 6.

When I first adapted the SI theory to clays, as described in Chapter 2, the first criticism of the SI theory, by Overbeek [12], was in the pipeline. As we have seen, in the linearized coulombic attraction theory, the Helmholtz free energy F is purely repulsive, but the Gibbs free energy G calculated from F shows a minimum in the pair interaction for large distances. Overbeek found that "it is quite unexpected that such a large, qualitative difference is found in a condensed virtually incompressible system." I answered this point at the time by pointing out that in the n-butylammonium vermiculite gels and other swollen clays, the phase bounded by the macroions is highly compressible [13], as illustrated in Figure 2.1. Note again the crucial difference between V and V^* in this argument. Overbeek went on to conjure up a solvent term that exactly cancelled the attractive branch in the Gibbs pair potential and claimed (I am sorry to use the polemical word "claimed," but I cannot bring myself to write "proved" here) that $F = G$. If I had known then about the simple proof given above (see Equation 7.3 and Equation 7.7), I could have demonstrated that this "$F = G$" claim is clearly contradicted by the Debye-Hückel theory [1–4]. At the time, I used the principle of virtual work to show that if $F - G$ were equal to zero, as claimed by Overbeek [12], then any mechanical work done on the system would simply disappear.

Overbeek's criticism is based on a misunderstanding of the operator D_{NZ} (see Equation 7.9) in Sogami-Ise theory. Overbeek criticized the variation of the macroionic charge number Z_n "since in the original equations for F, the particles are assumed to have a constant charge and no mechanism is included for charge variation by adsorption or dissociation with the concomitant changes in non-electrical free energy." This type of idea will resurface in criticisms of the SSS (Sogami, Shinohara, Smalley) theory [10, 14] described below. We can now point out how any other choice violates the neutrality condition [8, 9]: Overbeek's criticism is refuted by the law of conservation of charge. We reiterate that the effective valency of the particles Z_n must change against a variation in the number of small ions N_i. Changing the number of the small ions independently is not a meaningful process. There must be a concomitant change in the valency of the macroions to preserve charge neutrality. It is a crucial error to omit the *chemical potential of effective valency of the particles* [6, 7, 13]. We emphasize that, while the total number of counterions in the macroionic phase is fixed, these ions can be either adsorbed onto the surface of the particles or can be in the diffuse double layers around the particles. There is no magical energy of adsorption associated with this process, as discussed in detail below when we consider criticisms of SSS theory.

The calculations described in the preceding Chapter 6 were by Shinohara, Sogami and myself [15, 16]. For those readers who skipped the slog through Chapter 6, I recall that we gave the exact mean field theory solution to the one-dimensional colloid problem in terms of both the adiabatic Gibbs pair potential and the adiabatic Helmholtz pair potential for plates confined in a *finite container*. Obviously, we need these results to make a sensible discussion of clay swelling under conditions when the sol concentration is significantly greater than 0.01 ($r > 0.01$), but, needless to say, the calculations are more difficult than those when the container walls go to infinity ($D \to \infty$). In fact, we gave the exact adiabatic Helmholtz pair potential for plates in an *infinite container* over a decade earlier [10]. One of the features of the ion distributions (see Figure 3 of reference [10] or Figure 2 of reference [15]) in the exact solution is that in the region sandwiched by the plates, the distribution for counterions becomes dominant and co-ions are almost completely excluded. If we omit the contribution of the co-ions altogether (i.e., if we make the approximation of counterion dominance), we can solve the problem analytically in terms of elementary functions rather than elliptic integrals. We solved this easier problem earlier [14], and references [14] and [10] became known as the SSS theory (at least, to people who criticized it). Although reference [14], which provided the solution under the approximation of counterion dominance, has been surpassed by the exact solutions, it has the merit of being much easier to understand mathematically while retaining the essential physics of the problem, in particular the "electric rope" (see Chapter 6).

Overbeek [17] published a criticism of the SSS theory in 1993. His arguments were largely based on how we took the $D \rightarrow \infty$ limit in reference [10]. Sogami and I [18] refuted Overbeek's criticism in 1995. Here I note that by having dealt with the Helmholtz free energy of the finite system [15], as described in Chapter 6, we have offered a different and yet more compelling rebuttal of Overbeek's objections. I also wish to note here that I have a high regard for Overbeek's early work and consider the *Theory of the Stability of Lyophobic Colloids* [19] to be a great book. You cannot have a paradigm shift without first having a paradigm.

One of the great achievements of DLVO (Derjaguin-Landau-Verwey-Overbeek) theory [19,20] was in recognizing that the attraction of plates with like charges that results from the Levine and Dube treatment [21] is spurious. In the 1930s, Levine supported a long-range attraction between charged colloidal particles upon the false basis of the minimum in U^{E}_{mn} (see Equation 2.1). He failed to realize that the entropy associated with the thermal motion of the simple ions works to cancel out completely the effective attraction in the electric pair potential U^{E}_{mn}; as we have seen in Chapter 2, in the linearized theory, the Helmholtz pair potential U^{F}_{mn} (see Equation 2.2) turns out to be purely repulsive for all plate distances. This had already been pointed out by Frumkin [22] in 1938, the same year as Langmuir's seminal paper [5], and by Derjaguin the following year [23]. Derjaguin and Landau direct some violent polemic at Levine in their classic 1941 paper [20]. Referring also to Derjaguin's work the previous year [24], they state clearly that their disagreement with Levine arises from the latter's erroneous expression for the free energy: "... in particular it has been shown that the expression for the free energy found by S. Levine by a statistical method incorrectly applied contradicts thermodynamics. Expressions equivalent to that employed by S. Levine and, consequently, also erroneous..."

An amusing historical irony arose when Levine criticized the SSS theory in the 1990s [25]. In the 1930s, he had supported a long-range attraction on a false basis, and in the 1990s he opposed the long-range attraction in the SSS theory on the false basis of a conceptual error compounded by a mathematical blunder [26]. Levine's ability to contradict thermodynamics on the basis of an incorrectly applied method had not diminished over the intervening half century.

The most interesting criticism of the SSS theory was made by Ettelaie [27]. It is based on the idea that, although our individual expressions for the electric and osmotic free energy terms in the inner and outer regions are themselves correct, we "do not correctly deal with a simple but important term in the Helmholtz free energy." Ettelaie works under the assumption that the adsorption of an ion from the solution onto the solid surface involves a change in the environment of that ion. He refers to μ_g as "the chemical potential of an ion in the solution" (the subscript "g" indicates a gaslike phase), to μ_s as "the chemical potential of the same ion on the surface" (the subscript "s" indicates a solidlike phase), and to ϕ_s as the "electric potential difference between the surface and the solution." He writes $\Delta\mu_c = \mu_s - \mu_g$ and states that "when equilibrium between the plate and solution is finally established, the chemical potential difference between the two phases becomes zero, leading to the important relation"

$$\phi_s = \Delta\mu_c/e \tag{7.14}$$

He refers to Equation 7.14 (Equation 2 in his paper) as the Nernst equation and states his approach explicitly as follows: "The only important point is that the Nernst equation should correctly be obeyed, if one is considering a constant surface potential problem." His criticism is clearly based on the idea that "there exists a chemical potential difference $\Delta\mu_c$ between the two phases." Many colloid scientists think in this way, but few express it so elegantly. The idea resurfaces in Overbeek's criticism of SSS theory [17, 18].

It is a misinterpretation of the concept of a phase to consider these two possible sites for a counterion as separate phases. This confuses a macroscopic property of a well-defined phase (a solid) in equilibrium with another well-defined phase (a homogeneous electrolyte solution) with an internal property of an inhomogeneous single phase, the macroionic (or colloidal or gel) phase. It divides the macroionic phase into two regions that have no physical counterpart.

The Nernst equation applies (if we neglect the activity coefficients of the ions, in keeping with PB theory) to the emf (electromotive force) of an electrochemical cell. The emf of such a cell and the surface potential of a colloidal particle are quantities of quite different kinds. It is not possible to measure ϕ_s for a colloidal particle with a potentiometer (where would we place the electrodes?), and even if we could, we have no reason to expect that it would obey the Nernst equation. We have been at pains to point out that all the experimental evidence on the n-butylammonium vermiculite system is consistent with the surface potential being roughly constant over two decades of salt concentration. This is clearly incompatible with the Nernst equation, and so are results on the smectite clays [28]. Furthermore, if the zeta potential can be related to the electrical potential difference ϕ_s, there are many known cases showing deviations from Nernst behavior, as discussed by Hunter [29]. This is why Ettelaie was wrong to use the Nernst equation as a first principle in colloid science.

The application of the Nernst equation has no basis. However, because the underlying idea of a discontinuous change in chemical potential near the macroion surface is so prevalent, I wish to prove that it leads to the system not being at thermal equilibrium.

Firstly, the two "phases" considered by Ettelaie do not have the same chemical composition. One phase (the surface-adsorbed phase) contains the macroion, and the other phase (the solution) does not contain the macroion. The condition for thermal equilibrium in these circumstances has been clearly stated by Guggenheim [30] as follows: "For the distribution of the ionic species i between two phases α, β of different chemical composition, the equilibrium condition is equality of the electrochemical potential μ_i, that is to say

$$\mu_i^\alpha = \mu_i^\beta \tag{7.15}$$

Any splitting of $\mu_i^\beta - \mu_i^\alpha$ into a chemical part and an electrical part is in general arbitrary and without physical significance."

There is therefore no basis for Ettelaie's version of this equation, expressing the electrochemical potential as $\mu_s = \mu_g + e\phi_s$. The mathematical equation corresponding to the statement that "the chemical potential difference between the phases becomes zero" is in fact

$$\Delta\mu_c = 0 \tag{7.16}$$

and not Equation 7.14. The statement corresponding to Equation 7.14 is that the chemical potential difference between the two phases becomes equal to a finite potential energy term. In Ettelaie's two-state model for the simple ion system, the boundary between the "solidlike" and "gaslike" regions is a plane where the chemical potential gradient is infinite due to a discontinuous change in μ_i.

The SSS theory is based on the PB equation. An advantage of considering the $D \to \infty$ limit is that we can immediately identify the normalization constant n_0 in the Boltzmann distribution with n_∞, the number density far from the plates. Pursuing our usual notation where z_i is the valency, we can then calculate the simple ion distributions $n_+(x)$ and $n_-(x)$ as

$$n_i(x) = n_\infty \exp(-z_i e\phi(x)/k_B T) \tag{7.17}$$

In Equation 7.17, it must be clearly recognized that n_∞, which is common to R_o and R_i, is not a function of d. The quantity n_∞ (the salt concentration in the infinite reservoir of soaking solution) is experimentally defined precisely because it is independent of d. Fixing n_∞ in accordance with the choice $\phi(x \to \infty) = 0$ at the outset corresponds to the limit of the infinite container, i.e., it corresponds to the limit of infinite dilution of the plates in the electrolyte solution ($r < 0.01$). *Equation 7.17 expresses the condition of thermal equilibrium over all regions.* It does not matter whether a counterion is adsorbed onto the surface of the plates, whether it is in the double layers trapped between the plates, whether it is close to the external surfaces of the plates, or whether it is in the infinite reservoir of electrolyte solution far from the plates: if the system is at thermal equilibrium, the electrochemical potential of the simple ions is constant everywhere.

It is easy to prove that the SSS system is at thermal equilibrium by taking logarithms of the Boltzmann equation (Equation 7.17) and rearranging as follows

$$k_B T \ln n_i(x) + e\phi(x) = k_B T \ln n_\infty \tag{7.18}$$

The left-hand side of Equation 7.18 is the electrochemical potential of the ion i and can be written as $\mu_i(x)$. In this form

$$\mu_i(x) = k_B T \ln n_\infty \tag{7.19}$$

Equation 7.19 shows that the electrochemical potential of the ion is constant and equal to its value at an infinite distance from the plates, as it must be at thermal equilibrium in the infinite dilution limit.

To prove that the system treated by Ettelaie is not at thermal equilibrium, consider the consequences of adding on the $e\phi_s$ term to Equation 7.19. Because the "gas" of small ions has been treated by the PB equation, Equation 7.19 can be written as

$$\mu_g = k_B T \ln n_\infty \tag{7.20}$$

There is nothing exceptional in this equation: because the electrochemical potential is everywhere constant within the "gaslike phase," $\mu_i(x)$ can be replaced by the constant μ_g. Inserting the Nernst equation (Equation 7.14) to obtain the electrochemical potential in the "solidlike phase"

$$\mu_s = k_B T \ln n_\infty + e\phi_s = k_B T \ln\left[n_\infty \exp\left(\frac{e\phi_s}{k_B T}\right)\right] \tag{7.21}$$

now begs the question, "What is the surface-adsorbed phase in equilibrium with?" The immediately apparent answer is that the surface-adsorbed phase is in equilibrium with a simple ion gas of number density $n_\infty \exp(e\phi_s/k_B T)$. In other words, it is not in equilibrium with the experimentally measurable and controllable number density of ions n_∞. The application of the Nernst equation (Equation 7.14) leads to the system not being at thermal equilibrium.

It was perceptive of Ettelaie to recognize that SSS theory is incompatible with the Nernst equation, but he failed to realize that it is the SSS theory that is correct, not the Nernst equation. The Ettelaie [27], Overbeek [17] and Levine and Hall [25] criticisms were "head on" attacks on the SSS theory, all based on the idea that we missed nonelectrical terms in our calculations. As Sogami and I pointed out in 1995, the reason we did not include these terms is that they do not exist [18]. The previous year, Ise and I had pointed out a more subtle kind of logical or dielectical error among workers still using the DLVO theory [31]. When it is shown that a purely repulsive potential can account for one particular phenomenon, and when the counterinterpretation in terms of a potential with both attractive and repulsive components is ignored, the reader is left with the impression that the repulsive DLVO (or, rather, Yukawa) potential is the only correct one. We chose the example of the thermal compression of colloidal crystals to illustrate the point that, in some circumstances, similar results can be obtained using either the repulsive DLVO potential or the Sogami potential [31]. We now make our second diversion into the world of colloidal spheres to demonstrate the point that it is only when we take a global view of the properties of a system that the advantage of the Sogami potential becomes apparent.

Rundquist et al. [32] studied the thermal compression of regular crystalline arrays of dyed sulfonated polystyrene spheres. They used absorption of high-intensity radiation by the dye to induce local heating of the crystalline array and probed the variation in lattice parameter by Kossel line analysis [33, 34]. Following the traditional method of analysis, they interpreted the observed compression in terms of the repulsive DLVO potential. The essential experimental facts were as follows: (a) low ionic strengths ($<10^{-5}$ M) were used; (b) the sphere diameter was 830 Å; (c) the (bare) surface charge of the particles (Z) was 2370e, where e is the electronic charge; (d) the structure was body-centered cubic (bcc) (e) the sphere volume fraction was 2% (it was noted that higher volume fractions led to the face-centered cubic (fcc) structure); and (f) the interparticle separation was 2450 Å.

We first consider the geometric facts (b) and (d–f) and ask the question, "Is the structure space filling?" We denote the nearest-neighbor separation by b and the length of the usual cubic (nonprimitive) unit cell by c. By Pythagoras's theorem, $c = 2b/\sqrt{3}$. Assuming the Kossel line analysis to be accurate to three significant figures [35], b = 2450 Å implies c = 2830 Å. Let V_{cell} be the volume of the unit cell, given by $V_{cell} = c^3 = 2.27 \times 10^{10}$ Å^3. The volume occupied by one sphere, V_p, the volume of the primitive unit cell, is given by $V_p = 1.13 \times 10^{10}$ Å^3. Let V_s be the volume of a sphere (radius a). Because a = 415 Å, $V_s = 2.99 \times 10^8$ Å^3. The volume fraction in the bcc structure (r_c) is therefore given by $r_c = V_s/V_p = 0.0264$. The spheres occupy 2.64% of the dispersion volume within the bcc structure. Comparing this with the global volume fraction (r_b) of 2%, the answer to our question is no. The bcc structure must coexist with a less dense phase, that is, it is part of a two-state structure. Unless we are to accept the mysterious many-body interaction (MMBI) described in Chapter 3 [36–38], this proves the existence of a net attractive force at fixed interparticle distance [39] and undermines the use of a purely repulsive potential.

Let us now apply the Sogami potential to the problem. As with the one-dimensional problem, we get a good qualitative insight into the behavior of the system by simply equating the observed interparticle separation (b) with the position (R_m) of the minimum in $U^G(R)$, the Gibbs pair potential for spheres with separation R, namely [7]

$$R_m = \{\kappa a \coth \kappa a + 1 + [(\kappa a \coth \kappa a + 1)(\kappa a \coth \kappa a + 3)]^{1/2}\}/\kappa \qquad (7.22)$$

With a = 415 Å, we set $R_m = b$ = 2450 Å. Then we find κ from Equation 7.22 to be 2.19×10^{-3} Å^{-1}. This solves the geometric problem easily: the spheres are constrained to sit at their observed interparticle separation by the long-range attractive tail in the Sogami potential and so, quite naturally, do not adopt a space-filling structure.

As this is the only occasion in this book that I am going to make any detailed use of the Sogami potential for spheres, I wish to show how easy it is to solve Equation 7.22 over the experimentally accessible range. Noting that the numerator in Equation 7.22 is a function of the product of a with κ only, we write

TABLE 7.1
Sogami Minimum (R_m) in Terms of the Number of Debye Screening Lengths (κR_m) and the Number of Particle Radii (R_m/a)

$a\kappa$	$f(a\kappa) = \kappa R_m$	R_m/a
0	4.828	•
0.1	4.833	48.3
0.5	4.997	9.99
1.0	5.471	5.47
1.5	6.175	4.12
2.0	7.025	3.51
2.5	7.956	3.18
3.0	8.929	2.98
3.5	9.920	2.83

$R_m = f(a\kappa)/\kappa$ and construct Table 7.1. The second column of the table expresses R_m as the number of Debye screening lengths at the potential minimum, which lies between 4 and 10. Note that experimentally, the secondary minimum between colloidal particles always lies between 4 and 10 Debye lengths, with Hunter quoting $7/\kappa$ as the typical separation [29]. We recall that neither DLVO theory nor the volume-term theories have any explanation to offer for this important global fact. By contrast, the Sogami theory, in which both the repulsion and attraction are electrical in origin and expressed as a pair potential, naturally predicts that colloidal particles must sit at a roughly constant number of Debye lengths, whether for plates or for spheres. However, since κ is *a priori* unknown and because the radius of the particles is much easier to measure, it is better to convert R_m into a number of particle radii by dividing the second column of Table 7.1 by the first. The results are given in the third column of Table 7.1. Returning to the example case, $R_m/a = 5.9$, the calculation gives $a\kappa = 0.91$, from which $\kappa = 2.19 \times 10^{-3}$ Å^{-1}, as stated.

We now ask the question, "Is this a reasonable value for κ in view of the previously stated essential experimental facts (a) and (c)?" The relationship between κ and the electrolyte concentration c for a univalent electrolyte in water at 25°C is given by Equation 7.23,

$$\kappa^2 = 0.107c \tag{7.23}$$

where κ is expressed in Å^{-1}, and c is in moles per liter. In the example case, $\kappa = 2.19 \times 10^{-3}$ Å^{-1} corresponds to $c = 4.49 \times 10^{-5}$ M. This only fits with fact (a), which states that the background ionic impurity concentration is less than 10^{-5} M if the counterions make the dominant contribution to κ. Fact (c) is easiest to address if we convert c into a number density n (Å^{-3}) using

$$n = 6.02 \times 10^{-4} c \tag{7.24}$$

giving $n = 2.70 \times 10^{-8}$ Å^{-3}. If we use the analytic surface charge to calculate n, which we represent by n_a, then fact (c) tells us that there are 2370 charges in 1.13×10^{10} Å^3, that is, $n_a = Z/V_p = 2370/1.13 \times 10^{10} = 2.09 \times 10^{-7}$ Å^{-3}. If the dispersion is perfectly deionized, then $n/n_a = 0.13$ to two significant figures, the limit of experimental accuracy in such determinations.

This value can be compared with the number of effective charges on latex spheres with SO_3H groups, determined by transference measurements [40]. The ratio n/n_a was 0.10 for $Z = 17{,}000$ for particles similar to those studied by Rundquist et al. [32]. The value $n/n_a = 0.13$ for $Z = 2370$ calculated above fits in perfectly sensibly with this result. The renormalized charge used by Rundquist et al. [32], $Z_{eff} = 1150$, gives $n/n_a = 0.49$. Such a value is appropriate for soluble polyelectrolytes, but it is not applicable for polystyrene sulfonate latex spheres [46]. We note that the interpretation in terms of the repulsive DLVO potential cannot explain the observed interparticle separation, and it does not give a reasonable value for the effective surface charge on the spheres. In contrast, the Sogami potential has now given a consistent explanation of all the previously identified essential experimental facts (a) to (f).

Now let us consider the photothermal compression data. Rundquist et al. [32] do not state the temperature at which $b = 2450$ Å was observed. To proceed, we choose $T = 25°\text{C}$ without affecting our qualitative conclusions. The essential equation is

$$\kappa^2 \propto \frac{1}{\varepsilon T} \tag{7.25}$$

and the basic facts that we need are in Table 7.2 [41], which suffices to calculate the T dependence of the Sogami minimum over the experimentally accessible range. Note that it is only the product εT that enters into the calculation, given in the third column of the table. Application of Equation 7.25 enables us to construct the fourth column of the table, but we should note that κ^2 is also proportional to c (or n), so we had to choose a temperature dependence for c, which in turn depends on the effective surface charge. We made the simplest possible choice, namely $c(T)$ = constant. This is a weak point, but it makes the T dependence of b directly comparable with that of Rundquist et al. [32], who made the same approximation. Fixing $\kappa = 2.19 \times 10^{-3}$ Å^{-1} at 25°C then gives us the T dependence of κ and hence the T dependence of R_m, shown in Figure 7.1. To compare with the calculation of Rundquist et al. [32], we have assumed that their ΔT (Kelvin) is equal to the temperature in degrees centigrade.

TABLE 7.2
Inverse Debye Screening Length κ between 0 and 50°C

T (°C)	ε	εT (K)	κ/κ_{298}
0	87.90	24,000	0.988
10	83.96	23,760	0.993
20	80.20	23,500	0.998
25	78.55	23,410	1.000
30	76.60	23,210	1.004
40	73.17	22,900	1.011
50	69.88	22,570	1.018

With respect to the Sogami calculation, we first note that the total thermal compression between 0 and 50°C is 59 Å, which is a 2.4% compression, easy to measure by Kossel line analysis. Secondly we note that two competing factors are involved. The Debye screening length contracts by approximately 3.0% across the range — the major factor leading to the thermal compression — but this is partially offset by the fact that the number of Debye screening lengths

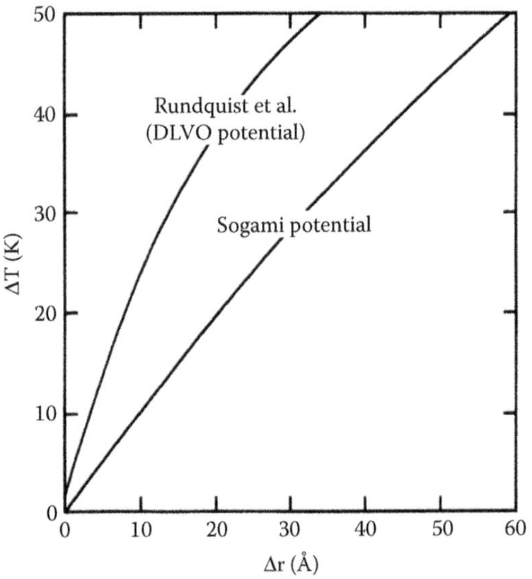

FIGURE 7.1 Comparison of the Sogami prediction for the thermal compression of colloidal crystals with that of Rundquist et al. [32] (DLVO prediction).

at the Sogami minimum increases by approximately 0.6%, leading to an overall contraction of 2.4%. The second factor is clear from the second column of Table 7.1.

Although Figure 7.1 makes the deviation between the two calculations seem quite pronounced, it is worth noting that the plots are qualitatively similar (both nearly linear with a slight concavity with respect to the temperature axis). From 0 to 50°C, the contraction is 1.3% according to the repulsive DLVO potential and 2.4% according to the Sogami potential. In the former case, the "repulsion" between the particles becomes weaker; in the latter case, the attraction becomes stronger. If we take one single fact (the photothermal compression of colloidal crystals) in isolation, we cannot distinguish between the two potentials.

In this case, the nonspace-filling nature of the ordered structure is the first fact that must be taken into account. The second important fact is that Ise et al. [42] reported a qualitatively similar temperature dependence of the interparticle separation in a latex dispersion for the case when the whole dispersion was heated. The repulsive DLVO potential offers only a superficially plausible explanation for photothermal compression when there are external "cold" regions to compress the laser-heated region (a nonequilibrium situation). In the case of thermal compression (an equilibrium situation), the DLVO theory would predict no contraction at all. This is because the particles are always repelling each other (the van der Waals forces are negligible at separations on the order of 0.1 to 1 μm), so they must fill the dispersion homogeneously irrespective of the temperature and so should show no variation of lattice parameter in the case where the whole dispersion is heated. We emphasize that the claim in the paper of Rundquist et al. [32] — that the lattice is compressed by repulsions arising from the surrounding unheated parts — is unwarranted, despite the qualitatively correct result of their calculation, namely, that the lattice is compressed on heating.

The Rundquist et al. paper [32] is typical of many papers in colloid science that find an individual phenomenon that can be interpreted in terms of a purely repulsive potential (the Yukawa potential, invariably named the DLVO potential in such papers); the possibility that the individual phenomenon has an alternative interpretation in terms of the Sogami potential is ignored. I have shown above how both potentials can be used to give a satisfactory explanation of some photothermal compression data on colloidal spheres, but the advantage of the Sogami potential becomes apparent when we consider the *global* properties of the system. Similarly, we saw in Chapter 3 that the force–distance curves obtained by Crocker and Grier [43, 44] can be interpreted in terms of either the Yukawa potential or the Sogami potential. These papers created the misleading impression that the coulombic attraction theory had been experimentally disproved, so let us finish by comparing the context of their experimental results on spheres with our results on clays, as shown in Table 7.3. We reemphasize that it is only when we take a global view of the properties of a system that the advantage of the Sogami potential becomes apparent.

TABLE 7.3
Comparison of Two Investigations in Colloid Science

Data on Vermiculite System in this Book	Crocker and Grier Data on Colloidal Spheres [43, 44]
1. Clay system studied rigorously as a function of a controlled salt concentration c over a range of two orders of magnitude between 0.001 and 0.1 M	Spheres studied at one unknown salt concentration (an attempt was made to estimate c from the κ values of the plots themselves)
2. Clay system studied rigorously as a function of the volume fraction r over a range spanning an order of magnitude in r, between 0.01 and 0.1	Spheres studied at one unknown (low) volume fraction
3. Clay system studied rigorously over a wide range of temperature T, between 0 and 50°C, including an investigation of the temperature-induced reversible phase transition between the gel and crystalline states	Spheres studied at room temperature, $T = 25°C$
4. Clay system studied rigorously as a function of hydrostatic pressure P over a wide range of pressures between 1 and 2000 atm, including an investigation of the pressure-induced reversible phase transition between the gel and crystalline phases	Spheres studied at atmospheric pressure, $P = 1$ atm
5. Clay system studied rigorously as a function of uniaxial stress p over a wide range of well-defined salt concentrations between 0.001 and 0.1 M	Spheres studied as a function of an effective pressure at one unknown salt concentration
6. Analytic surface charge of clay plates known accurately from the chemical formula of the clay; surface charge *in situ* measured qualitatively by neutron diffuse scattering (see Chapter 8)	Analytic surface charge unknown; no attempt made to measure *in situ* surface charge

REFERENCES

1. Debye, P.J.W. and Hückel, E., *Physik. Zeits.*, 24, 183, 1923.
2. Debye, P.J.W. and Hückel, E., *Physik. Zeits.*, 25, 97, 1924.
3. Debye, P.J.W., *The Collected Papers of Peter J. W. Debye,* Interscience Publishers, New York, 1954.
4. McQuarrie, D.A., *Statistical Mechanics,* Harper Collins Publishers, New York 1973.
5. Langmuir, I., *J. Chem. Phys.*, 6, 873, 1938.
6. Sogami, I., *Phys. Lett. A*, 96, 199, 1983.
7. Sogami, I. and Ise, N., *J. Chem. Phys.*, 81, 6320, 1984.
8. Dirac, P.A.M., *Can. J. Math.*, 2, 129, 1950.
9. Dirac, P.A.M., *Lectures on Quantum Mechanics,* Academic Press, New York, 1964.
10. Sogami, I., Shinohara, T., and Smalley, M.V., *Mol. Phys.*, 76, 1, 1992.

11. Sogami, I.S., Smalley, M.V., and Shinohara, T., *Prog. Theor. Phys.,* 113, 235, 2005.
12. Overbeek, J.T.G., *J. Chem. Phys.,* 87, 4406, 1987.
13. Smalley, M.V., *Mol. Phys.,* 71, 1251, 1990.
14. Sogami, I., Shinohara, T., and Smalley, M.V., *Mol. Phys.,* 74, 599, 1991.
15. Shinohara, T., Smalley, M.V., and Sogami, I.S., *Mol. Phys.,* 101, 1883, 2003.
16. Shinohara, T., Smalley, M.V., and Sogami, I.S., *Int. J. Mod. Phys. B,* 19, 3217, 2005.
17. Overbeek, J.T.G., *Mol. Phys.,* 80, 685, 1993.
18. Smalley, M.V. and Sogami, I.S., *Mol. Phys.,* 85, 869, 1995.
19. Verwey, E.J.W. and Overbeek, J.Th.G., *Theory of the Stability of Lyophobic Colloids,* Elsevier, Amsterdam, 1948.
20. Derjaguin, B.V. and Landau, L., *Acta Physicochimica URSS,* 14, 633, 1941.
21. Levine, S. and Dube, G., *Trans. Faraday Soc.,* 35, 1125, 1939.
22. Frumkin, A. and Gorodetzkaja, A., *Acta Physicochimica URSS,* 9, 327, 1938.
23. Derjaguin, B., *Acta Physicochimica URSS,* 10, 333, 1939.
24. Derjaguin, B., *Trans. Faraday Soc.,* 36, 203, 1940.
25. Levine, S. and Hall, D.G., *Langmuir,* 8, 1090, 1992.
26. Smalley, M.V., *Langmuir,* 11, 1813, 1995.
27. Ettelaie, R., *Langmuir,* 9, 1888, 1993.
28. Low, P.F., *Langmuir,* 3, 18, 1987.
29. Hunter, R.J., *Foundations of Colloid Science,* Clarendon Press, Oxford, 1987.
30. Guggenheim, E.A., *Thermodynamics,* North Holland, Amsterdam, 1957.
31. Ise, N. and Smalley, M.V., *Phys. Rev. B,* 50, 16722, 1994.
32. Rundquist, P.A., Jagannathan, S., Kesavamoorthy, R., Brnadic, C., Xu, S., and Asher, S.A., *J. Chem. Phys.,* 94, 711, 1991.
33. Kossel, W., Loeck, V. and Voges, H., *Zeits. Physik,* 94, 139, 1935.
34. Kossel, W. and Voges, H., *Ann. Phys.,* 23, 677, 1937.
35. Sogami, I.S. and Yoshiyama, T., *Phase Transitions,* 21, 171, 1990.
36. van Roij, R. and Hansen, J-P., *Phys Rev. Lett.,* 79, 3082, 1997.
37. Warren, P.B., *J. Chem. Phys.,* 112, 4683, 2000.
38. Chan, D.Y.C., Linse, P., and Petris, S.N., *Langmuir,* 17, 4202, 2001.
39. Kamenetzsky, E.A., Magliocco, L.G., and Panzer, H.P., *Science,* 263, 207, 1994.
40. Ito, K., Ise, N., and Okubo, T., *J. Chem. Phys.,* 82, 5732, 1985.
41. Archer, D.G. and Wang, P., *J. Phys. Chem. Ref. Data,* 19, 371, 1990.
42. Ise, N., Ito, K., Okubo, T., Dosho, S., and Sogami, I., *J. Am. Chem. Soc.,* 107, 8074, 1985.
43. Crocker, J.C. and Grier, D.G., *Phys. Rev. Lett.,* 73, 352, 1994.
44. Crocker, J.C. and Grier, D.G., *Phys. Rev. Lett.,* 77, 1897, 1996.

8 The Structure of a Dressed Macroion in Solution

In Chapter 3, we made a passing reference to the diffuse neutron scattering from the n-butylammonium vermiculite gels, just evident on the scale of Figure 3.3b. Although we first saw these beautiful ripples in June 1987, I can remember the morning as if it were yesterday. In one of those marvelous moments of serendipity, I accidentally blew up the ordinate scale when I was looking at the data from the overnight scan by an order of magnitude. Instead of seeing what I had expected — some well-defined low-angle peaks (the sample was our old favorite, a 0.1 M gel at 10°C, with a *d*-value of 122 Å) and a flattish background at higher angles — I was greeted by the sight shown in Figure 8.1a. We realized that the wide-angle scattering had something to do with the details of the scattering-length density in and around the clay plates. We therefore swelled up a crystal that we had prepared in deuterated n-butylammonium chloride, $C_4D_9ND_3Cl$, in a deuterated 0.1 M solution. The wide-angle scattering that we saw 24 hours later from the swollen D-salt sample is shown in Figure 8.1b. I need hardly tell you that Figure 8.1b, in which the first broad maximum around $Q = 0.4$ Å^{-1} observed with the H-salt is entirely absent, is different from Figure 8.1a.

We reckoned that above $Q = 0.2$ Å^{-1} the interparticle scattering should be completely damped out because the first-order peak from the sample was at $Q = 0.05$ Å^{-1}, and we had never seen any unstressed gel sample give a fourth-order peak, with even third-order peaks quite rare. (This figures with the Sogami minimum being quite shallow, such that the gels are one-dimensional paracrystals with a range of spacings around the minimum, rather than perfect crystals with a single *d*-value.) The remaining pattern, the ripples of diffuse scattering, must therefore depend only on the Fourier transform of the unit cell contents. We knew that the clay plates must have an average positive scattering-length density and that each plate has a thickness of approximately 10 Å. Likewise, the D_2O should also be fairly uniform for most of the distance between the clay plates, giving a block of positive scattering-length density for distances greater than about 20 Å from the clay layers. We also knew that the nine protons in the protonated n-butylammonium ion would give a concentration of negative scattering-length density at the counterion positions. Our first guess was that the first broad maximum around $Q = 0.4$ Å^{-1} was some kind of Young's slits effect with neutrons rather than light and that we were observing the so-called Stern layer of adsorbed counterions [1]. Thus far, as the following detailed analysis will prove, we had the right idea [2], but it was a full decade after the first observation before Jan Swenson [3, 4] was able to make a correct interpretation of the data, and his careful

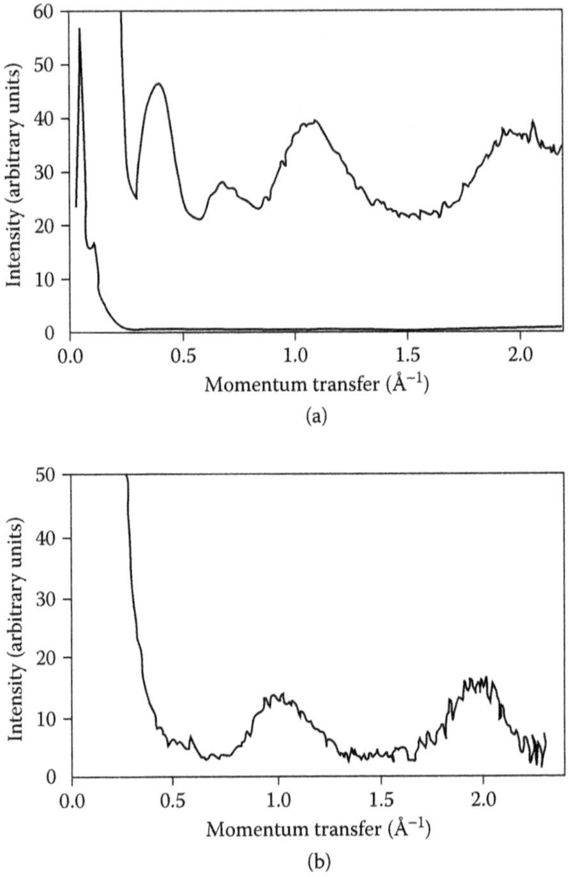

FIGURE 8.1 The ripples of diffuse scattering observed in the first D16 experiment. The gels had been soaked in 0.1 M n-butylammonium chloride solutions in D_2O: (a) with the usual protonated n-butylammonium ions; (b) with deuterated n-butylammonium ions. In (a) the sharp peak at Q = 0.05 Å^{-1} is the first-order diffraction peak; the low angle scattering has been scaled down by a factor of 15 with respect to the ripples.

and accurate analysis, using the pair correlation function method, is described in this chapter.

The samples were oriented such that the clay plates were perpendicular to the momentum transfer Q, described as $Q = Q_z$. Spectra were recorded separately for each scattering angle, which was increased from 6 to 100° in steps of 2° (the detector covered about 9°), and also for monitors in the incident and transmitted beams. The incident wavelength was 4.53 Å. The data from different scattering angles were then combined and corrected for background and container scattering and absorption. They were then normalized to a proper structure factor for the z-direction (perpendicular to the clay plates), $S(Q_z)$. For each Q-value, we used only the detector cells that gave mutually consistent results from the overlapping frames. The structure factor obtained from each sample was Fourier transformed to obtain the total neutron-weighted pair correlation function along the z-axis as

$$G(r_z) = \frac{2\sum_{i=1}^{n} c_i b_i^2}{\pi\rho^0\left(\sum_{i=1}^{N} c_i b_i\right)^2} \int_0^\infty (S(Q_z - 1)\sin(Q_z r_z) dQ_z + 1 \tag{8.1}$$

where ρ° is the average number density and c_i and b_i are the concentration and neutron-scattering length of atom i, respectively.

The c_i and b_i values are known from the structural formula of the clay and the d-values. As we saw in Chapter 1, the former is given by $Si_{6.13}\,Mg_{5.44}\,Al_{1.65}\,Fe_{0.50}\,Ti_{0.13}\,Ca_{0.13}\,Cr_{0.01}\,K_{0.01}\,O_{20}\,(OH)_4\,(C_4H_9ND_3)_{1.29}\cdot nD_2O$ for the samples prepared with H-salt, where n is the number of water molecules per structural unit of the vermiculite plate. As we emphasized at the end of the preceding chapter (see Table 7.3), the extent of swelling and whether or not it occurs depends on many thermodynamic variables and has been characterized as a function of hydrostatic pressure P [5], temperature T [6], uniaxial stress along the swelling axis p [7], the volume fraction of the clay in the condensed matter system r [8] and the electrolyte concentration c in the aqueous solution [5–8]. The ripples experiments took several hours each, compared with a few minutes for the measurement of the d-values from the more intense low-angle peaks, so it was not possible to cover such a wide range of conditions in these experiments. We begin with two experiments in which T was held constant at 10°C, P was held constant at 1 atm, $p = 0$ and $r < 0.01$, which describes the situation where, even after swelling, the clay occupies only a small volume of the condensed-matter system. The electrolyte concentrations studied were $c = 0.1$ M and $c = 0.01$ M. For the 0.01 M sample, the equilibrium d-value was about 300 Å, and for the 0.1 M samples, it was about 120 Å, as described previously [5, 6]. These values correspond to $n \cong 400$ and $n \cong 150$ in the structural formula given above. It is worth remembering that the majority of the scattering observed is from the water, which was fully deuterated in all cases.

The total $G(r_z)$ can be expressed as a neutron-weighted sum (dependent on the scattering lengths of the constituent atoms) of the partial pair correlation functions $g_{ij}(r_z)$ according to

$$G(r_z) = \frac{\sum_{i,j=1}^{n} c_i c_j \langle b_i \rangle \langle b_j \rangle g_{ij}(r_z)}{\left(\sum_{i=1}^{n} c_i b_i\right)^2} \tag{8.2}$$

Using the method of isotope substitution, it is possible to obtain partial pair correlation functions experimentally. Two isotopically different samples were prepared at $c = 0.1$ M, one where all the hydrogen in the butylammonium chains was ordinary ^1H and one where they were entirely deuterated, to make use of the large difference in scattering length between H and D (b_H = −3.74 fm and b_D = 6.67 fm). In this case, all correlations containing the butyl chains, including the ion–ion correlation, are changed by the H/D substitution of the butyl chains, providing us with the sum of the $g_{ij}(r_z)$ involving those atoms.

The structure factors of the 0.1 M H-butylammonium chloride solution without clay and the vermiculites in 0.01 and 0.1 M H-butylammonium chloride solutions and 0.1 M D-butylammonium chloride solution are shown in Figure 8.2 in the Q-range 0.2 to 2.1 Å^{-1}. The Bragg peaks due to the interplate correlations, which are located at lower Q-values (see Figure 8.1a), are not shown in this figure. It should be noted that the total scattering is strongly dominated by correlations within the deuterated water, since even in the more concentrated salt solution of 0.1 M, the distance between the 10 Å thick clay plates is about 110 Å, which means that about 90% of the total volume is occupied by water molecules. Considering this, the overall differences between the structure factors are interestingly large.

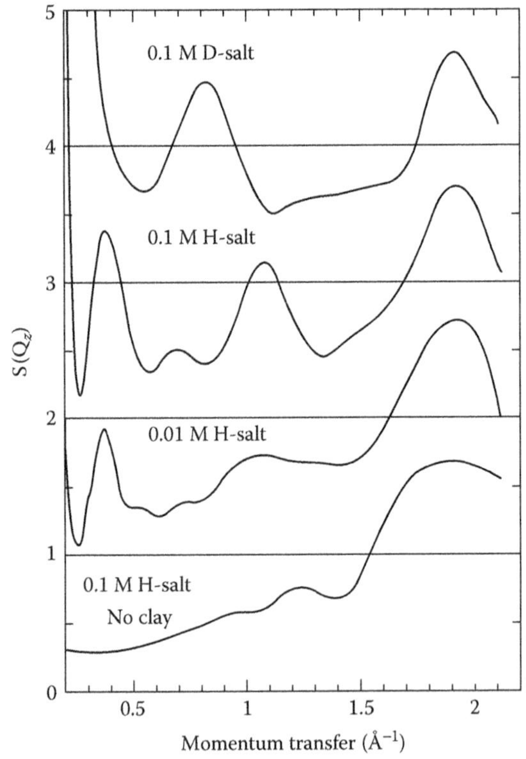

FIGURE 8.2 Structure factors $S(Q_z)$ obtained from neutron-scattering patterns of butylammonium vermiculite gels (upper panels) and from a 0.1 M protonated butylammonium salt solution with no clay (lowest panel). The upper panels show $S(Q_z)$ for gels prepared in a 0.1 M deuterated salt solution and in 0.1 and 0.01 M protonated salt solutions. The momentum transfer Q was perpendicular to the clay plates, and the structure factor $S(Q_z)$ has been normalized after correction for background scattering and absorption.

The pure 0.1 M salt solution (with no clay) shows only one pronounced structural feature in the Q-range studied: a broad peak at about 1.9 Å^{-1}. However, even the vermiculite gel in the lower concentration H-salt solution shows a new, clearly pronounced peak at about 0.4 Å^{-1} and a weak broad peak around 1.05 Å^{-1}, which both grow in intensity with increasing salt concentration. One can also observe a weak peak or shoulder at about 0.7 Å^{-1} for both H-salt concentrations. The position of the peak Q_1 at 0.4 Å^{-1} corresponds to a real-space characteristic length of about $2\pi/Q_1 = 16$ Å. The structure factor of the vermiculite in the 0.1 M D-salt solution is completely different, except for the peak at about 1.9 Å^{-1}, which is due to correlations within the deuterated water. The peak at about 0.4 Å^{-1} is totally absent for the vermiculite in the D-salt solution. Instead, a relatively strong peak is observed at about 0.8 Å^{-1}. It is also interesting to note that the vermiculite in the D-salt solution shows a much higher intensity in the region around 0.3 Å^{-1} than the corresponding vermiculite in the H-salt solution.

To simplify the interpretation of the structural features observed in reciprocal space, the structure factors were Fourier transformed to atomic pair correlation functions, $G(r_z)$, measured in the direction perpendicular to the clay plates (in the z-direction). The structural functions $r(G(r_z) - 1)$ derived from $G(r_z)$ are shown in Figure 8.3 for the same vermiculites as shown in Figure 8.2. It should be noted that

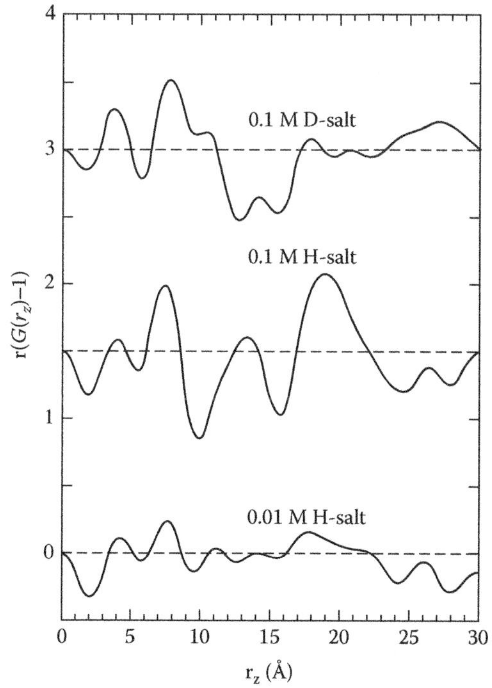

FIGURE 8.3 Atomic pair correlation functions $G(r_z)$ obtained from the structure factors shown in Figure 8.2, presented as $r(G(r_z) - 1)$, where r_z is the distance along the swelling axis of the clay. Each function oscillates around zero and has been displaced in the upper panels to show the functions obtained from gels prepared in a 0.1 M deuterated salt solution and in 0.1 and 0.01 M protonated salt solutions.

the structural features in the region $0 < r_z < 5$ Å are very approximate due to the low maximum Q-value of 2.13 $Å^{-1}$ in the D16 experiments and will therefore not be discussed in this chapter. All the vermiculites show a strong peak at 7 to 7.5 Å, which due to its generality and r-value we interpret as mainly arising from the correlation between the two surface oxygen layers within the clay plates (see Figure 1.1). At $c = 0.01$ M the interlayer distance is two to three times greater than that at $c = 0.1$ M, so the reduced intensity of this feature in the 0.01 M solution is due to a lower clay concentration in the path of the neutron beam. A common feature for the vermiculites in H-salt solutions is that they show a relatively strong peak at about 18 Å and show dips at about 9.5 and 16 Å, which increase in magnitude with increasing salt concentration. The vermiculite in the D-salt solution shows a shoulder at 10 to 11 Å, a broad dip in the region $11 < r_z < 17$ Å, and a broad peak in the region $23 < r_z < 30$ Å. Before we interpret these features, it is instructive to examine two difference functions in real space.

First, the function $G(r_z)$ for the 0.1 M H-butylammonium chloride solution (with no clay) has been subtracted from the function $G(r_z)$ for the vermiculite in the same solution. Let us call this function $\Delta G(r_z)(H)$, since it should give us the correlations around a clay plate relative to a uniform 0.1 M H-salt solution. This function will show peaks for clay–clay and clay–water correlations and dips for clay–ion correlations. We have chosen to take the difference between the normalized $G(r_z)$ values, since that will give positive and negative values in $\Delta G(r_z)$ where we have "excess" water and butyl chain correlations, respectively. Second, the function $G(r_z)$ for the vermiculite in the 0.1 M H-butylammonium chloride solution has been subtracted from the function $G(r_z)$ for the vermiculite in the corresponding D-salt solution. Let us call this function $\Delta G(r_z)(D)$, since it should give us correlations due to the replacement of H-ions by D-ions. In $\Delta G(r_z)(D)$, the sign of all the correlations should be the reverse of those observed in $\Delta G(r_z)(H)$; the correlations involving butyl chains give positive contributions to $\Delta G(r_z)(D)$, while the correlations involving D_2O give negative contributions.

Figure 8.4 shows the difference functions $\Delta G(r_z)$, with the (H) and (D) functions represented by the dotted and full lines, respectively. The function $\Delta G(r_z)(H)$ (between the vermiculite and its H-salt solution) shows peaks at about 7 (the intraclay plate correlation), 13 and 18 Å and dips at about 9.5, 16 and 22 to 29 Å. The function $\Delta G(r_z)(D)$ (between the vermiculites in D- and H-salt solutions) shows peaks at about 9.5, 16 and 22 to 29 Å and dips at about 13 and 19 Å. The coincidence of the dips in one function with the peaks in the other is very noticeable and remarkable in view of the fact that the original scattering patterns giving rise to the function $\Delta G(r_z)(D)$ were obtained from different pieces of the naturally occurring Eucatex mineral. The observation of this antiphase correlation between the two difference functions gives us great confidence that all the features observed in the raw Fourier transforms of the data are structurally meaningful.

By doing a simultaneous analysis of the $\Delta G(r_z)$ presented in Figure 8.4 and the $G(r_z)$ used in presenting Figure 8.3, it is evident that we see butyl chain–clay, butyl chain–butyl chain and water–water correlations. The peaks or dips at about 13 and 18 Å in $\Delta G(r_z)$ or $r(G(r_z) - 1)$ must be due to water–water correlations. It is likely

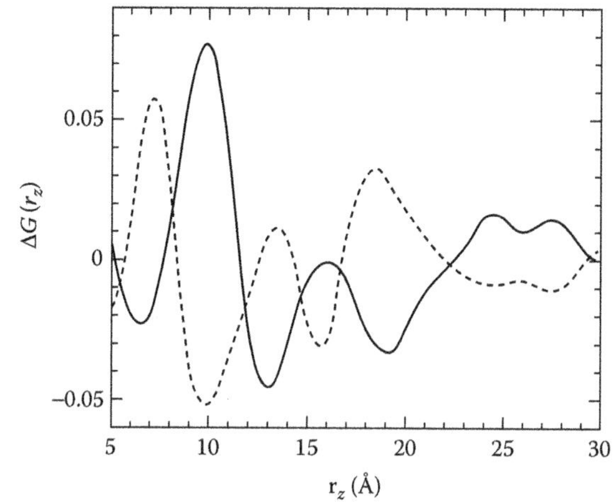

FIGURE 8.4 Two difference pair-correlation functions $\Delta G(r_z)$. The dotted line shows $\Delta G(r_z)$(H), obtained from the difference between the vermiculite and its H-salt solution. The full line shows $\Delta G(r_z)$(D), obtained from the difference between the vermiculites in D- and H-salt solutions.

that we have a broad water–water correlation in the range 11 to 21 Å. It therefore follows that this broad peak in $\Delta G(r_z)$(H) (between the vermiculite in 0.1 M H-salt solution and its pure solution) is "interrupted" in the middle by a butyl chain–clay correlation, which produces a "negative peak" or dip at about 16 Å. Butyl chain–butyl chain correlations can be seen in both $\Delta G(r_z)$(H) and $\Delta G(r_z)$(D) in the region $22 < r < 29$ Å, and relatively strong correlations between the surface oxygen layers within the clay plates and butyl chains are evident at about 9.5 Å (the distance to the nearest oxygen layer) and 16 Å (the distance to the oxygen layer on the other side of the clay plate). Thus, there is a high concentration of butyl chains about 11 to 15 Å from the center of the clay plates.

The results mean that we obtain a structural picture where every clay plate is coated with about a 6 Å thick water layer and where the highest concentration of butyl chains is located next to the water layer in a layer about 4 Å thick. The situation is illustrated in Figure 8.5, which gives us a crude picture of a "dressed macroion" in solution. The positively charged ammonium ions may be directed toward the clay plate and stick into the neighboring water layer. However, the ions cannot be directly bound to the clay plate, since the thickness of the water layer means that they must be located at least 4 Å from the surface of the clay plate. Furthermore, it is interesting to note that the density of the water layer closest to the clay plate seems to be slightly less than the density of bulk water. This can mainly be seen from the function $r(G(r_z) - 1)$ for the vermiculite in the 0.1 M D-salt solution (see the upper panel of Figure 8.3).

Before moving on to how the structure of the dressed macroion (see Figure 8.5) depends on uniaxial stress and sol concentration as well as salt concentration, and

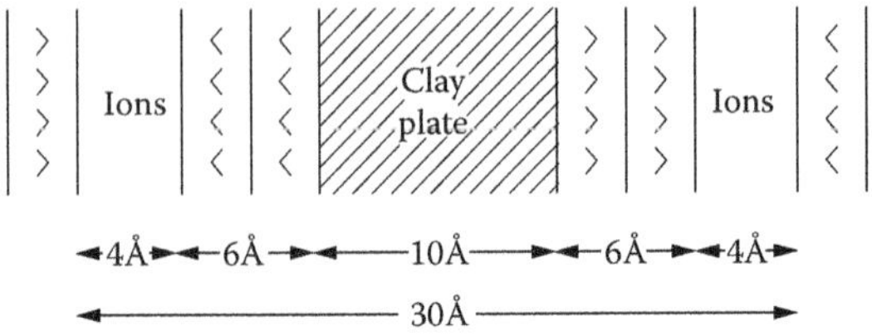

FIGURE 8.5 Schematic structure of a dressed macroion in solution. The arrowheads represent partially ordered water molecules.

how our picture came to be extended to include a further layer of water molecules on the outer side of the layers of n-butylammonium ions, I wish to pay tribute to the intuition of some of the earlier workers in the field. First, our discovery that the charged clay plates are covered with two layers of water molecules is reminiscent of Garrett and Walker's statement [9] that "the initiation of swelling is thought to involve the development of 'icebergs.'" The image is also an appealing one in the sense that the water layer closest to the clay surface seems to have a lower density than that of bulk liquid water. What is clear is that the ions are not directly bound to the surface. This constitutes a counterexample to the Stern layer picture [1], in which the majority of the counterions are supposed to be bound directly to the surface. If we wish to retain the concept of an "outer Helmholtz plane" at which we can define an effective surface electric potential [10], we must look farther out, beyond the first adsorbed layer of counterions. These two counterion layers lie between 22 and 30 Å apart, much thicker than a clay plate, which is usually taken to have a thickness of 10 Å [11]. Any continuum electrical theory, such as the coulombic attraction theory or the DLVO (Derjaguin-Landau-Verwey-Overbeek) theory, can only apply outside this 30 Å thick region. It is remarkable that a similar thickness of 28 Å for a system of a clay plate with adsorbed butylammonium ions was obtained by an independent method by Rausell-Colom [12]. In reference [12], the d-values in butylammonium vermiculite gels were measured as a function of applied uniaxial stress along the swelling axis, and linear plots of $\ln p$ vs. d were obtained. The intercepts of these plots extrapolated to $d = 28$ Å, which was interpreted as the incompressible plate thickness.

We now wish to examine how the structure of the dressed macroion, as determined by our diffuse neutron scattering experiments, varies with applied uniaxial stress. The best data was obtained at $c = 0.03$ M, with $r < 0.01$, when the uniaxial pressure p was varied between 0 and 0.2 atm; a single experiment conducted over two days on D16 gave clear ripples at different applied stresses, as described below. As in the experiments at $c = 0.1$ and 0.01 M described above, T and P were held constant at 10°C and 1 atm, respectively. The details of the experiment and the small-angle patterns observed are given in reference [4]. Here we concentrate on how applied uniaxial pressure affects the higher Q-range of the structure factors.

In Figure 8.6 we show the structure factors in the Q-range 0.2 to 2.1 Å^{-1} for a vermiculite in 0.03 M D-salt solution at zero pressure and for a sample in 0.03 M

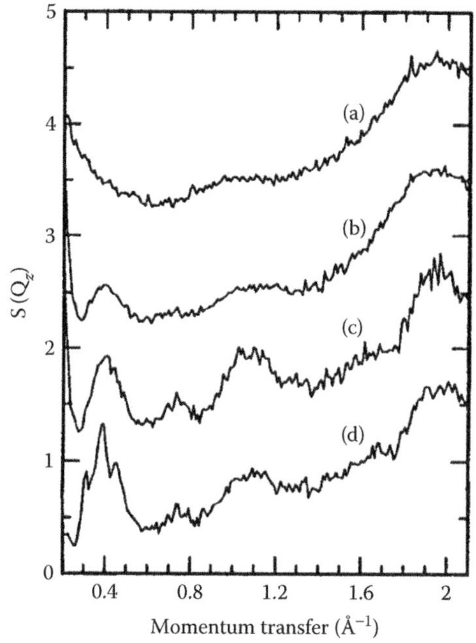

FIGURE 8.6 Structure factors $S(Q_z)$ in the range $0.2 < Q < 2.1$ Å^{-1} obtained from neutron scattering patterns of 0.03 M butylammonium vermiculite gels at $r = 0.0065$ and $T = 10$°C. (a) D-salt at $p = 0$; (b), (c) and (d) H-salt at p = 0, 0.04 and 0.07 atm, respectively. Consecutive curves are shifted vertically for clarity.

H-salt solution at the applied pressures 0, 0.04 and 0.07 atm. Of course, the Bragg peaks due to the interplate correlations are located at lower Q-values than are shown in Figure 8.6; the d-values obtained from these were 169, 120 and 104 Å for curves (b), (c) and (d), respectively [4]. As expected, the qualitative features of both the D-salt and H-salt butylammonium vermiculites are as described above for the $c = 0.1$ M case. The applied uniaxial stress has almost no effect on the positions of the peaks for the sample in the H-salt solution. The intensity of the strong peak at about 2.0 Å^{-1} shows only a minor stress dependence (it is mainly the width of the peak that seems to change), while the intensities of the other peaks change significantly with the applied stress. The intensity of the peak at about 0.4 Å^{-1} increases with increasing stress, and at a pressure of 0.07 atm, the top of the peak shows a remarkable sharpness. For the peaks located at about 0.7 and 1.1 Å^{-1} the intensity is largest at the intermediate pressure of 0.04 atm. The structure factor of the vermiculite in the 0.03 M D-salt solution seems rather featureless, but one should note its much higher intensity compared with the vermiculite in the 0.03 M H-salt solution in the Q-range around 0.2 to 0.3 Å^{-1}. This is evident in Figure 8.7, which shows the difference curve $\Delta S(Q_z)$ between the structure factors of the vermiculites in the 0.03 M D- and H-salts. The corresponding difference curve for the vermiculites in the 0.1 M salt solutions is

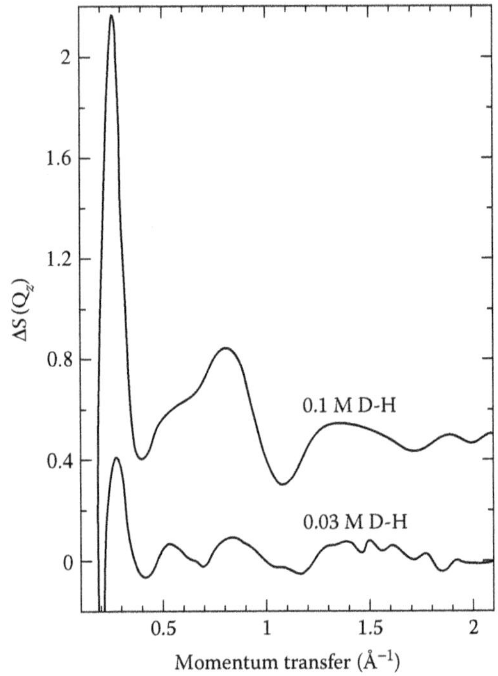

FIGURE 8.7 Difference curve $\Delta S(Q_z)$ between the vermiculites in D- and H-salt solutions at $c = 0.03$ M (lower curve) and at $c = 0.1$ M (upper curve).

also shown for comparison. Both difference curves show a relatively strong peak at about 0.25 Å^{-1} and smaller peaks (or shoulders) at about 0.55 and 0.8 Å^{-1}. The overall shapes of the two curves are similar. For $Q < 0.2$ Å^{-1}, the curves are strongly negative due to the fact that the Bragg peaks arising from the interplate correlations are much stronger for the vermiculites in H-salt, where the scattering contrast between the clay layers and the interlayer solution is greater.

As with the data at $c = 0.1$ M, the structure factors were Fourier transformed to atomic pair correlation functions, $G(r_z)$, using Equation 8.1. Difference pair correlation functions $\Delta G(r_z)$ were obtained by Fourier transformation of the difference structure factors $\Delta S(Q_z)$ shown in Figure 8.7. These agreed well with the same functions calculated by subtracting the $G(r_z)$ obtained by Fourier transformation of the raw structure factors, again confirming that the $G(r_z)$ are structurally meaningful. The structural correlation functions $r(G(r_z) - 1)$ derived from $G(r_z)$ are shown in Figure 8.8 for the same vermiculites and the same uniaxial stresses as shown in Figure 8.6.

The main feature of Figure 8.8 is that the curves (b), (c) and (d), measured at $p = 0$, 0.04 and 0.07 atm, respectively, are very similar to each other. Comparison with Figure 8.3 also reveals that they are very similar to the corresponding curve obtained with $c = 0.1$-M H-salt, and the same structural analysis applies in each case. Before considering the details, it is clear that the dressed macroion structure is basically unaffected by uniaxial stress over the pressure regime studied.

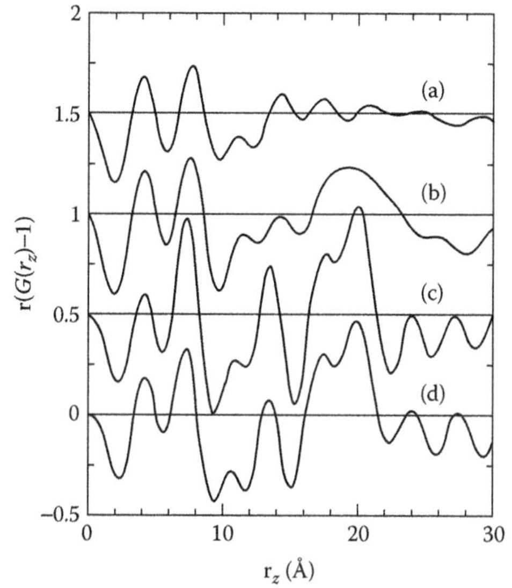

FIGURE 8.8 Atomic one-dimensional pair-correlation functions $G(r_z)$ derived from the structure factors shown in Figure 8.6, presented as $r(G(r_z) - 1)$. (a) to (d) are as described in the caption to Figure 8.6. Consecutive curves are shifted by 0.5 for clarity.

The structure factors shown in Figure 8.6 and the real-space correlation functions $r(G(r_z) - 1)$ shown in Figure 8.8 do indicate an increased ordering with increased pressure (at least for pressures up to about 0.04 atm) in the solution closest to the clay plates. The correlation between the water molecules bound to the butylammonium ions (on the "inner" side closest to a clay plate) on each side of a clay plate gives rise to the peak at about 0.4 Å^{-1} in the structure factors of the vermiculite in H-salt solution and the broad peak in the range $17 < r_z < 22$ Å of the corresponding $r(G(r_z) - 1)$ functions. These peaks get narrower in real space and their intensities increase with increasing pressure, which indicates that the distance from the center of the clay plate to the water layer on each side gets slightly more well-defined. The average distance is, however, maintained. It also seems that the low-density water layer closest to the clay surface retains its low density and, therefore, probably also its structure, even at the highest pressure. Because the water–water correlation gets slightly more well-defined, it is also likely that the layer of butylammonium ions gets slightly narrower. The main conclusion to be drawn is that the dressed macroion structure of the clay plate, plus its two adsorbed layers of water molecules and one adsorbed layer of counterions, is constant with respect to uniaxial stress.

From the correlation functions $r(G(r_z) - 1)$ shown in Figure 8.8, it seems likely that while the density of the water molecules closest to the clay plates is less than the density of bulk water, the water molecules closest to the butylammonium ions seem to have a slightly higher density and to be more orientationally ordered than bulk water. A natural interpretation of this analysis is that one layer of water molecules is adsorbed on the clay plate and that the second layer is strongly bound to the butylammonium ions. Thus, the structural ordering and density of these two water layers may differ significantly. Within this schema, it would also be expected that there would be a strongly bound water layer on the solution side of the counterion layer, and there were indications in the data that this is indeed the case [4]. This gives us a structural picture of the dressed macroion in which the bare clay platelet thickness is extended to approximately 35 Å by successive adsorbed layers of water molecules, counterions and further partially ordered water molecules. It therefore seems that primitive theories, treating the solvent as a dielectric continuum, can only be valid outside a block of a thickness of at least 35 Å, farther out than the 30 Å suggested above. It is noteworthy that the discovery of an effective clay layer thickness of about 35 Å improves the agreement between the coulombic attraction theory and the observed *d*-values in the butylammonium vermiculite system. It has been shown previously that when the salt fractionation effect is taken into account, the theory predicts spacings of seven Debye lengths under low-*r*, low-*c* conditions, giving results systematically lower than the observed values. At $c = 0.03$ M, for example, the Debye length is 18 Å and the prediction is $d = 126$ Å. If we add to this the invariant effective plate thickness, we obtain $d \cong 160$ Å, in good agreement with the data. We therefore believe that continuum electrical theory should be retained as the central theory of clay swelling, but with the proviso that one must use an effective clay layer thickness of about 35 Å rather than a "naked" clay layer thickness of 10 Å. It is at the surface of the dressed macroion rather than the naked macroion that we can retain the concept of an effective surface potential.

The fraction of the total number of butylammonium ions located within the layers about 11 to 15 Å from the center of the clay plates is difficult to estimate accurately, mainly due to the limited *Q*-range and the partly unknown fraction of the gel in the neutron beam. However, with the assumption that 100% of the sample scattering (the container scattering is subtracted) is from the gel, it is possible to make a rough estimation of the number of butylammonium ions within the layer; the fractions of butylammonium ions have been estimated to be 46% and 39% for the gels in 0.03 and 0.1 M salt solutions, respectively [4]. If some of the sample scattering was from the external solution, these values would have been larger. Thus, the calculated values are very approximate, and it is likely that the real fraction of butylammonium ions located within these layers is larger than the given values. In this context, it is worth noting that the intrasolution and solution–clay correlations giving rise to the observable peaks in $G(r_z)$ are very weak compared with the total scattering, reflecting the disordered nature of the samples. The absolute values are a factor of two lower than those predicted from the Yukawa fits to the uniaxial stress results, which give fractions of 90% and 80% for the gels in the 0.03 and 0.1 M solutions, respectively (see Table 3.4). However, it is important to note that the adsorption does proceed in the direction predicted by electrical theory, with fewer ions adsorbed at the higher salt concentration.

We also investigated how the structure factor changes with the sol concentration *r* of the clay in the condensed matter system at $p = 0$. Figure 8.9 shows the low-*Q*-range ($0.02 < Q_z < 0.15$ Å^{-1}) of the structure factor of vermiculites in a 0.03 M H-butylammonium chloride solution for $r = 0.0065$, $r = 0.062$ and $r = 0.39$. In Figure 8.10, the higher *Q*-range of the same structure factors is shown. Our prediction for the phase boundary at $c = 0.03$ M is $r^* = 0.24$ (see Table 4.4), so the point $r = 0.39$ should

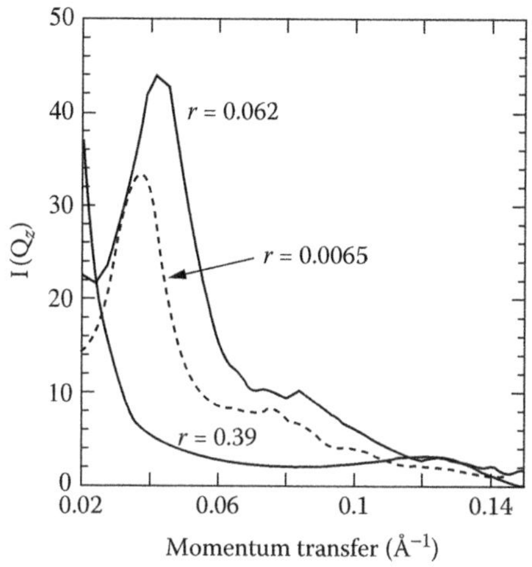

FIGURE 8.9 Low-Q neutron data $I(Q_z)$ obtained at $c = 0.03$ M and $p = 0$ as a function of the sol concentration r.

lie within the one-phase region of colloid stability. In fact, there is a weak first-order gel peak between 0.12 $Å^{-1}$ and 0.13 $Å^{-1}$ in Figure 8.9, corresponding to a d-value of 50 Å at $r = 0.39$, which is what we expect if the clay has soaked up all of the available solution homogeneously. However, the wide-angle diffuse scattering was contaminated by peaks from the 19.4 Å crystalline phase, as shown in Figure 8.10b. Although this made the subtraction necessary for calculating the dressed macroion structure difficult in the case of the highest sol concentration, the diffuse scattering in Figure 8.10b shows no evidence for a change in the structure pictured in Figure 8.5. For the points within the two-phase region of colloid stability, the ripples were clearly of the same qualitative form, as shown in Figure 8.10a; most of the differences in the data for $r = 0.0065$ and $r = 0.062$ can be explained by the larger volume of vermiculite gel in the neutron beam in the latter case. It can be concluded that the dressed macroion structure is unaffected by the sol concentration, as with the salt concentration and uniaxial pressure. It seems to be a universal structure in the vermiculite gels.

The main result to emerge from the study of the ripples of diffuse scattering is the structure of the dressed macroion and its invariance with respect to changes in the variables r, c and p. In every case studied, the clay surface is covered by two layers of water molecules, then a layer of counterions, giving a block of 30 Å thickness. If we interpret the "outer" water layer as being associated with the counterions, it is logical to expect a further partially ordered water layer on the solution side of the counterion layer, and although the evidence is inconclusive on this point, there are indications in the data that this is indeed the case. This gives rise to a structural block 35 Å thick, which probably represents the smallest d-value possible for the gel phase. At higher r, c and p values, we might expect it to be possible to produce such a spacing, but then to produce collapse into the $d = 19.4$ Å crystalline phase.

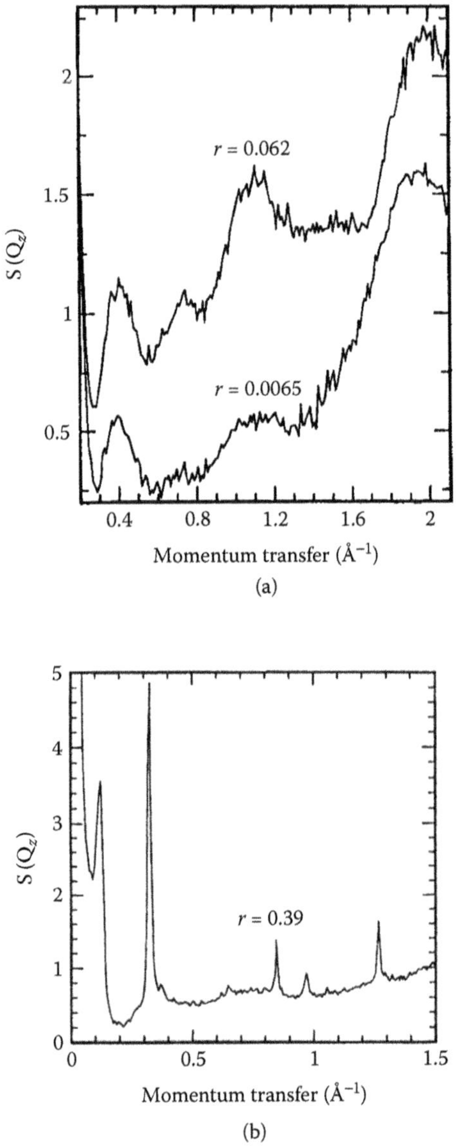

FIGURE 8.10 Structure factors $S(Q_z)$ obtained at $c = 0.03$ M and $p = 0$ as a function of the sol concentration r in the higher Q-range. The upper figure (a) shows the data for $r = 0.062$ and $r = 0.0065$, and the lower figure (b) shows the data for $r = 0.39$. In (b), the scale has been expanded to show the more intense peaks due to the $d = 19.4$ Å crystalline phase of the vermiculite.

REFERENCES

1. Stern, O., *Z. Elektrochem.,* 30, 508, 1924.
2. Braganza, L.F., Crawford, R.J., Smalley, M.V., and Thomas, R.K., *Prog. Colloid Polymer Sci.,* 81, 232, 1990.
3. Swenson, J., Smalley, M.V., Thomas, R.K., Crawford, R.J., and Braganza, L.F., *Langmuir,* 13, 6654, 1997.
4. Swenson, J. Smalley, M.V., Thomas, R.K., and Crawford, R.J., *J. Phys. Chem. B,* 102, 5823, 1998.
5. Smalley, M.V., Thomas, R.K., Braganza, L.F., and Matsuo, T., *Clays Clay Min.,* 37, 474, 1989.
6. Braganza, L.F., Crawford, R.J., Smalley, M.V., and Thomas, R.K., *Clays Clay Min.,* 38, 90, 1990.
7. Crawford, R.J., Smalley, M.V., and Thomas, R.K., *Adv. Colloid Interface Sci.,* 34, 537, 1991.
8. Williams, G.D., Moody, K.R., Smalley, M.V., and King, S.M., *Clays Clay Min.,* 42, 614, 1994.
9. Garrett, W.G. and Walker, G.F., *Clays Clay Min.,* 9, 557, 1962.
10. Low, P.F., *Langmuir,* 3, 18, 1987.
11. van Olphen, H., *An Introduction to Clay Colloid Chemistry,* 2nd ed., Wiley, New York, 1977.
12. Rausell-Colom, J.A., *Trans. Faraday Soc.,* 60, 190, 1964.

9 The Counterion Distribution between Charged Plates in Solution

So, what are we to make of the story that emerges from Chapters 1 to 8? The central result that emerged from Chapter 8 is that in a model clay system, the "naked" clay particle (of a thickness of about 10 Å) is covered by two ordered layers of water molecules on each side, followed by a layer of counterions and another layer of partially ordered water molecules, to produce a "dressed" clay particle of a thickness of about 35 Å. Within this "dressed macroion," short-range molecular forces are dominant. We can interpret these as giving rise to an effective clay plate thickness of about 35 Å in a swollen clay.

The theories we discussed prior to the discovery of its structure apply outside of the dressed macroion for separations of the interplate distance (center-to-center distance of the clay plates) greater than about 35 Å. When in Chapter 3 we used the uniaxial stress results on the model clay system to calculate $\psi_0 \cong 70$ mV for the surface potential of the particles, we should assume that this is the value of an effective surface potential at about 15 to 20 Å from the center of the macroion. If you like, this is a natural way of extending the Stern layer picture of colloidal particles in solution. To round numbers, under typical conditions in clay science, with interplate separations in the range from 100 Å ($c = 0.1$ M) to 200 Å ($c = 0.03$ M), roughly half the counterions are "adsorbed" within the dressed macroion and roughly half are in the diffuse layers around the effective particles. We should definitely envisage these two regions as being in dynamic equilibrium. I also did neutron scattering experiments, not described so far, in which the D-salt was substituted into H-salt butylammonium vermiculites *in situ*. The exchange of the patterns was always rapid, showing that the ions adsorbed to the plates were rapidly replaced by ions from a fresh solution. Complete deliberate substitution to 50% H-salt/50% D-salt always produced a stable pattern that we can understand from a random distribution of the H-ions and D-ions between the surface sites and the double layers. There is no magical "energy of adsorption" involved in this process. The two types of site are in thermodynamic equilibrium, and the system adjusts to maintain a constant surface potential as a function of electrolyte concentration.

The main fact of the book so far is that the clay particles sit at about 7 Debye lengths from each other when the clay is in its swollen gel state. The interplate separation in the gel state is therefore inversely proportional to the square root of the electrolyte concentration, a fact noted by Walker in 1960 [1]. Such a separation of 7 Debye lengths is no special feature of clay science associated with the large aspect ratio of the clay particles: it is a typical separation in colloid science, a fact noted by Hunter in his *Foundations of Colloid Science* [2]. The fact that the interplate spacing is controlled in this way by the electrolyte concentration points strongly toward an electrostatic origin for the interaction.

If you have come this far with an open mind, you will realize that all the basically correct ideas for handling this interaction were described in a seminal paper by Sogami and Ise in 1984 [3]. I hope you will agree from the results in Chapters 1 to 7 that I have adapted this theory faithfully to plate macroion interactions [4] and have used it to develop a useful theory of clay swelling [5]. I also tried to look at the implications of applying the coulombic attraction theory to the case of ψ_0 approximately constant with respect to the salt concentration c, and especially to the example case $\psi_0 \cong 70$ mV observed in the model system. I came up with a constant salt fractionation factor s that was equal to 2.8 for $\psi_0 \cong 70$ mV. This is also approximately the case for the n-butylammonium vermiculites, as reported by Williams et al. [6], where the interplate spacings measured as a function of a wide range of sol concentrations r ($0.01 \leq r \leq 0.1$) and salt concentrations c (0.001 M $\leq c \leq$ 0.1 M) are in broad agreement with the predictions of the coulombic attraction theory (CAT). If we now envisage CAT as applying outside a 35 Å thick dressed particle, I think we have a broadly accurate picture of a swollen clay. Although I have perhaps banged on about it too much already, I still think its bears reemphasizing that the reversible phase transition with respect to both temperature [7] and hydrostatic pressure [8] in the model system are natural consequences of CAT, and are inexplicable in terms of the previous DLVO (Derjaguin-Landau-Verwey-Overbeek) theory of colloid stability [9, 10]. A more advanced formalism of the coulombic attraction theory has been forwarded by Sogami, Shinohara and myself [11–13], and this has placed CAT on a firmer theoretical basis. From now on, I regard the matter of the electrostatic interactions in clay swelling and colloid stability as closed. This does not mean that we have solved all the interesting problems in clay and soil science. Far from it.

It became fashionable among theoreticians to refer to the basic model in any field as "vanilla flavored," the version with no fancy flavorings. Looked at in this light, we might be tempted to label the diffuse scattering in Chapter 8 as "vanilla ripples," as they gave us a first basic picture of the ion and solvent distribution around a colloidal particle in solution. In that case, the ripples we are now about to discuss are definitely of the raspberry variety.

We saw in Chapter 3 that we could measure interparticle spacings as low as 45 Å; not so much greater than the 20 Å spacing in the crystalline phase. For such a system, surely it would be possible to make a complete structural determination along the c-axis? In such a case, the 35 Å thick macroions are not so far off touching, as envisaged in Figure 9.1, which is for illustrative purposes only. In the intervening 10 Å, we could easily envisage the remaining (approximately half) of the counterions jostling with three or four layers of water molecules in a liquidlike arrangement. The vanilla ripples of Chapter 8 were from spacings in the range $d \approx 120$ Å ($c = 0.1$ M) to $d \approx 180$ Å ($c = 0.03$ M). In such cases, the system was 90% or so aqueous, and we analyzed the scattering patterns in a way that would be familiar to people interested in liquids and amorphous materials, but perhaps not to those interested in

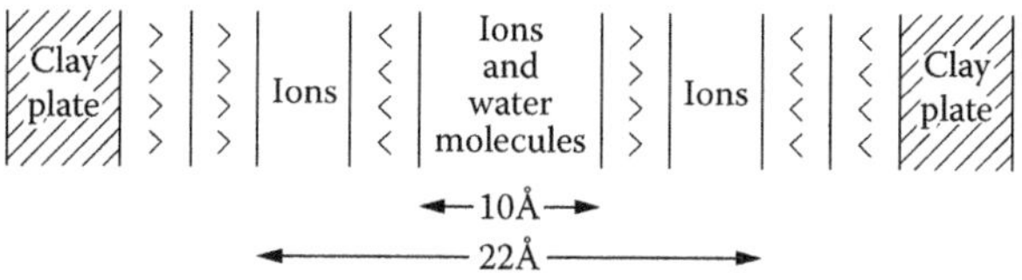

FIGURE 9.1 Schematic illustration of a possible structure of a clay with d = 45 Å. If the dressed macroion structure remains constant, there is only a 10 Å intermacroionic gap for the remaining counterions and water molecules.

crystallography. We are now going to adopt the crystallographic approach and look at the raspberry ripples that were mainly the work of Neal Skipper, Alan Soper and Graham Williams [14–18].

Even at d = 45 Å, water forms more than half the volume occupied by the clay, so the majority of the scattering is still from water molecules, and we should clearly expect a liquidlike component in the scattering. Our aim was to determine the complete scattering length density profile along the c-axis for $d \approx$ 50 Å, and so determine the complete ion distribution in a swollen clay. In the vanilla ripples on D16, the resolution was limited by the low maximum Q, and we were only able to make very rough estimates of the numbers of ions adsorbed to the plates. To determine the complete structure of the double layer throughout the region between two particles in a colloidal (gel) state, we will try to understand the structure from a crystallographic point of view; this will be on the understanding that we will be averaging over a lot of liquidlike motion in the interlayer region.

First of all, if you are going to understand the crystallography of the gels, you have to understand the crystallography of the crystalline phase. We started out along this path with a study of calcium vermiculite crystals in 1994 [14]. The choice of calcium ions as the interlayer cations in a crystalline vermiculite was not arbitrary. Calcium ions play a central role in many important biochemical processes. For example, they act as a cofactor for a variety of extracellular enzymes, and they are used to maintain the transmembrane potentials responsible for the contraction of muscle fibers and the transmission of nerve impulses [19–21]. To understand these biochemical reactions, which take place in an aqueous environment, it is clearly necessary to have a detailed knowledge of the hydration properties of calcium ions. The hydration of a calcium ion in simple ionic solutions of $CaCl_2$ had already been established by previous neutron diffraction experiments [22]. These showed that in very concentrated (4.5 M) solutions the calcium ions were surrounded by about six water molecules, whereas the "hydration shell" around calcium ions in less concentrated (1.0 M) solutions contained about ten water molecules. We may expect to see the latter type of coordination in dilute aqueous solutions such as cell fluids [19], but this leaves open the question of how calcium ions will hydrate in more confined geometries, such as those encountered at the surface of a protein or enzyme molecule, or within the pore of a cell membrane. Due to the complex nature of polypeptides and biomembranes, it would be extremely difficult to measure the hydration of calcium ions *in situ* in these systems. However, we may get some insight into their behavior by studying the hydration of calcium ions intercalated into a calcium-substituted vermiculite clay.

As we wanted to make a complete structural determination of the mineral, we used a more highly oriented vermiculite from Llano, Texas. This was a special clay (VTx-1) from the Clay Minerals Society's Source Clays Repository; it had the structural formula [23]

$$(Mg_{2.81}Al_{0.08})(Si_{2.89}Al_{1.11})O_{10}(OH)_2Ca_{0.465}\cdot nH_2O$$

where we have used the half-cell labeling common in clay crystallography. The crystal used in our experiments had approximate dimensions 15 × 10 × 2 mm and was prepared from its natural form (which had mostly Mg^{2+} as the interlayer cation) by repeated soaking in 1 M $Ca(NO_3)_2$, at about 50°C, over two years. The sample was very similar to the calcium-substituted smectite clays that make up a large portion of many soils and sedimentary rocks [23–25], but in the form of a large, highly oriented crystal. Our experiments were conducted using the time-of-flight liquids and amorphous materials diffractometer (LAD) on the ISIS source [26].

A big advantage of LAD is that you can measure down to a sufficiently low Q to accurately measure interlayer spacings up to about 50 to 60 Å (this would be needed for the later gel experiments), but up to a sufficiently high Q to reveal structural details at the molecular level. Using our usual method of orienting the sample such that the c-axis (actually the c*-axis, if you are a crystallographer) was parallel to the scattering vector Q, we measured the first 27 (00*l*) Bragg reflections of the calcium vermiculite, which had a c-axis *d*-value of 15.05 Å. We used isotope substitution of D_2O for H_2O among the interlayer water molecules, in conjunction with Monte Carlo difference analysis, to determine the number and position of water molecules in the interlayer region, and the results are described in detail in reference [14]. It turned out that there were six water molecules in each calcium-water complex, more like the case in a very concentrated (4.5 M) salt solution than in a molar solution. If we generalize this conclusion to other confined environments, we would expect hydrated cadmium and mercury ions to compete with hydrated calcium ions at surfaces of proteins or within biomembrane pores, which may go some way to explaining why these heavy-metal ions can act as poisons in physiological processes [19].

The LAD data on the calcium vermiculite were sufficiently accurate to show the orientations of individual water molecules, with four water molecules hydrogen bonded to a clay layer [14]. We wanted to study a swelling vermiculite with similar precision, and originally chose a lithium vermiculite in order to take advantage of the spherical nature of the cation in this case (in contrast to the other swelling cations, the alkylammoniums and the amino acids). The experiments were a success in the sense that the neutron diffraction experiments on the crystalline phase went well [15]. We again measured the first 27 (00*l*) Bragg reflections, again using the Llano mineral, this time for samples in which the *d*-value was 14.67 Å. A series of experiments were performed using isotopic substitution of ^6Li for ^7Li and D for H, and difference analysis then allowed us to establish the interlayer counterion and water distributions. We found that the lithium ions are located midway between the

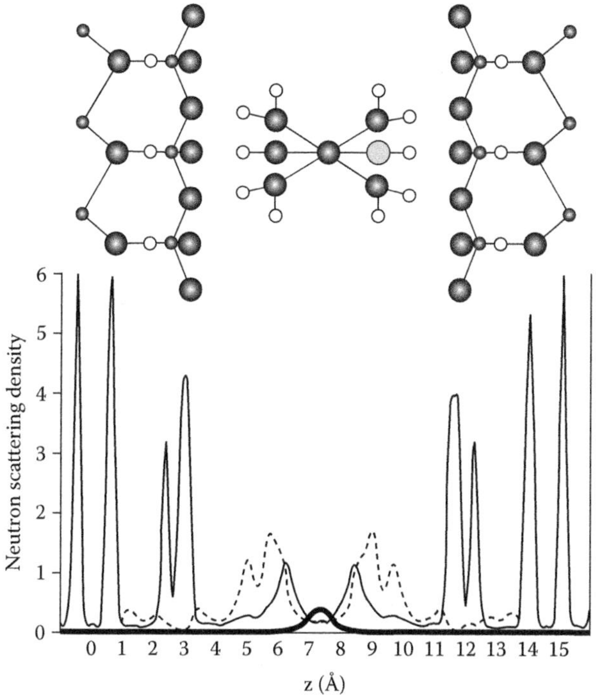

FIGURE 9.2 Neutron-scattering-density profiles $\rho(z)$ for a lithium-substituted vermiculite. Oxygen plus clay layer is the solid line, hydrogen is the dashed line, and lithium is the stars. The molecular model above shows two sections of clay surface and an undistorted octahedral $Li^{+}(H_2O)_6$ complex. In this model all six water molecules are hydrogen bonded directly to the clay plate; in practice we find that, on average, two of the six water molecules are less strongly oriented toward the plate.

clay plates and form octahedral hydration complexes with six water molecules. The structure is shown schematically in Figure 9.2, together with the neutron scattering density profiles used to fit the data. The analysis of the peaks in the simulated density profiles is given in Table 9.1. Our data show clearly that the ions are located midway between the clay sheets, giving rise to a peak at 7.335 Å. If only we could study the double layer in a swollen lithium vermiculite gel with such precision.

But it was not to be. Try as we might, the difference in scattering lengths between the ^{6}Li and ^{7}Li isotopes was too small to permit us to measure the lithium ion distribution in the swollen state. We had to content ourselves with the results for the crystalline phase, where the behavior of the lithium ions is different from that of the larger alkali metal cations [27]. Potassium and cesium ions bind directly to vermiculite clay surfaces rather than hydrating fully. Because only lithium-substituted vermiculites of the alkali metal series will swell macroscopically when soaked in water, it seems that interlayer cations must form fully hydrated ion–water complexes if the particles are to expand colloidally. This conclusion has since been supported

TABLE 9.1
Analysis of Peaks in the Simulated Density Profiles of 14.67 Å Lithium Vermiculite

Peak Position (Å)	Assignment of Chemical Species to Each Peak	Unnormalized Area	Normalized Area (barns str^{-1} $nucleus^{-1}$)	Chemical Equivalent
0.0	Octahedral cations	60.50	1.50	$0.95(Mg_{2.81}Al_{0.08}Fe_{0.07})$
1.04	Apical oxygen	70.20	1.74	3.00 O
2.75	Tetrahedral cations	31.87	0.79	$Si_{2.89}Al_{1.11}$
3.28	Basal plane oxygen	73.21	1.81	3.12 O
1.0–4.4	Hydrogen	27.75	0.65	1.19 H
7.34	Lithium	14.96	0.37	0.64 Li
6.13	Oxygen	87.62	2.17	3.74 O
5.16–5.88	Hydrogen	174.10	4.32	7.43 H

by the results in Chapter 8 and also holds true for the results on the gel phase of n-propylammonium vermiculite, to be discussed below.

In an ideal world, we might have studied the crystalline phases of all the n-alkylammonium vermiculites up to, say, six carbon atoms in the chain. It still remains a challenge to theory to understand why only the $n = 3$ and $n = 4$ chains lead to macroscopic swelling in this series. We started out with the methylammonium vermiculite [16, 17], reverting to the Eucatex system (with about 1.3 cations per structural unit), which swells more readily than the Llano system (with about 1.5 cations per structural unit). Understanding the effect of the naked surface charge on the swelling also remains a challenge to theory. Garrett and Walker [28] had previously noted that the swelling seems to be most pronounced for vermiculites with about 1.3 cations per cell (in their case, a Kenya vermiculite). These authors also gave the interlayer spacings for all the alkylammonium vermiculites between $n = 1$ (methylammonium) and $n = 8$ (octylammonium). These values included 12.3 Å for methylammonium and 19.2 Å for butylammonium, in good agreement with our results [16, 17]. The fitted neutron scattering length density profiles to our LAD data on methylammonium vermiculite are given in Figure 9.3. We used H/D substitution on both the water molecules and counterions, and $^{15}N/^{nat}N$ substitution on the counterions. By simulating two data sets simultaneously, we fitted common and difference components of the density profile for each substitution, and so found the position and orientation of both the counterions and water molecules in the system. It is worth noting that the isotopic substitution technique, which enables us to locate the hydrogen atoms unambiguously, leads to excellent fits with the data. For example, we obtained an R-factor of 3.7% in our calcium vermiculite experiments [14], compared with an R-factor in excess of 15% for a model proposed to fit X-ray scattering data from a calcium vermiculite [29].

Our data for a methylammonium vermiculite with $d = 12.3$ Å (Figure 9.3) show a maximum in the unsubstituted scattering density at 6.15 Å, which we attribute to

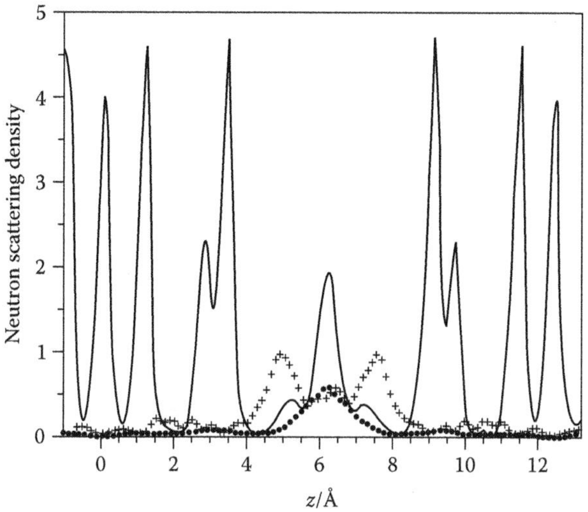

FIGURE 9.3 Neutron scattering density profiles $\rho(z)$ along the c* clay stacking axis for a hydrated methylammonium vermiculite crystal with a clay-layer spacing of 12.3 Å. Three profiles are shown, representing the scattering density due to the methyl group protons (solid dots), the water and ammonium group protons (crosses), and all remaining atoms (continuous line). A molecular graphics snapshot representation of this sample is shown in Figure 9.4.

the oxygen atoms of interlayer water molecules. The small shoulder at 5.1 Å is probably due to nitrogen atoms of ammonium groups, though we had no significant neutron difference data to confirm this. Peaks in the substitutable hydrogen density are found at 4.9 and 7.4 Å. These positions are consistent with each water molecule forming hydrogen bonds to only one clay surface, as is the case in the single-layer hydrate of sodium vermiculite [27]. These proton peaks also contain the hydrogen atoms of ammonium groups. Alkyl group protons are located in a broad peak at 6.15 Å, completing a picture of a rather weakly structured interlayer region. Figure 9.4 shows a molecular graphics snapshot obtained from Monte Carlo simulations of methylammonium vermiculite [30].

Our data for crystalline n-butylammonium Eucatex vermiculite, the swelling system studied in the earlier chapters, are shown in Figure 9.5 [16]. They point to an interlayer region that is, if anything, even less structured than in the methylammonium samples. From Figure 9.5 we see unsubstituted density peaks at 5.0, 6.12, 7.14 and 9.44 Å, substitutable interlayer proton peaks at 4.94 and 8.86 Å, and an alkyl proton peak at 8.21 Å. Again, nitrogen substitution was unsuccessful, resulting in a uniform distribution throughout the interlayer region. You cannot win them all. The observed peaks in real space are consistent with rather weak hydrogen bonding of water to the clay plates and with a tendency for the alkyl groups to occupy the center of the interlayer region. We saw no evidence for the formation of any micelle-like structure or of strong bonding of the ammonium groups to the negatively charged clay plates. It is interesting to compare this liquidlike structure for the interlayer

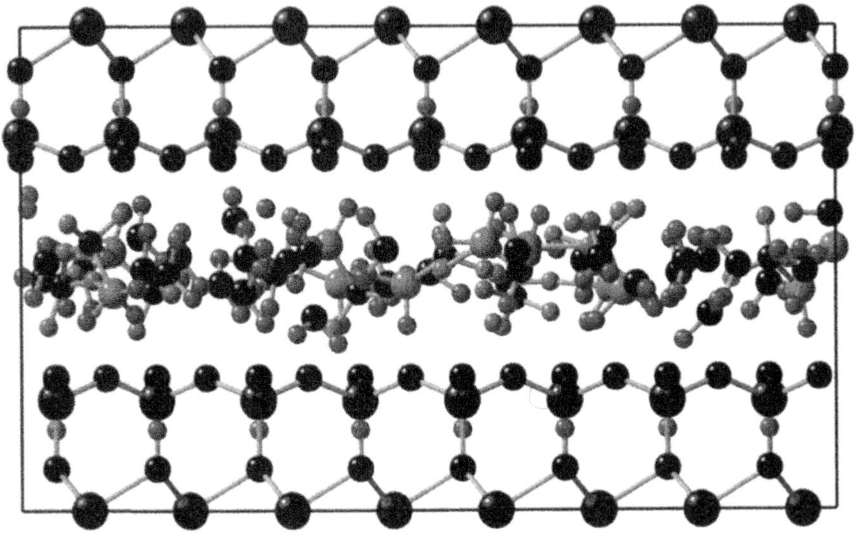

FIGURE 9.4 Molecular graphics snapshot of hydrated methylammonium vermiculite, with color coding to show the species that we target by using isotopic substitution in conjunction with neutron diffraction. Red = hydrogen; blue = nitrogen; green = methyl; black = unsubstituted species (oxygen, magnesium, silicon and carbon). In the system illustrated, the water content is 2.5 molecules per methylammonium counterion, and the clay layer spacing along the c*-axis is 12.3 Å. During the neutron diffraction experiments, the samples were aligned so that the clay layers were perpendicular to the scattering vector Q. See color insert following page 76.

region inside the crystals with the dressed macroion structure obtained in the gel phase [31, 32]. It is obvious that large structural changes take place adjacent to the naked clay plate when the swelling transition occurs.

The problem with measuring the double layer (by this, I now mean determining the counterion distribution as a function of distance between the plates) on a butylammonium vermiculite is that, if the sol concentration is kept low and uniaxial stress is not applied, it is only possible to produce gels with spacings of greater than 80 Å in the gel phase [32]. Such a spacing would be too large for us to see the first-order Bragg peak of the gel phase, a crucial component of the scattering, on LAD. We needed to go down to 50 Å or less. If we increased the sol concentration to the point necessary to decrease the spacing in this way in the n-butylammonium system, we would obtain mixtures of crystalline and gel phase (see Figure 8.10b) and be unable to perform accurate subtractions. Likewise, for the delicate raspberry ripples probed here, having a movable object in the cell, with the possibility of squeezing gel out of the beam, seemed to rule out uniaxial stress as a way of producing the required spacing. Luckily, we had also been working on the propylammonium vermiculites and found them to have a qualitatively similar but quantitatively very different phase diagram than their butylammonium counterparts.

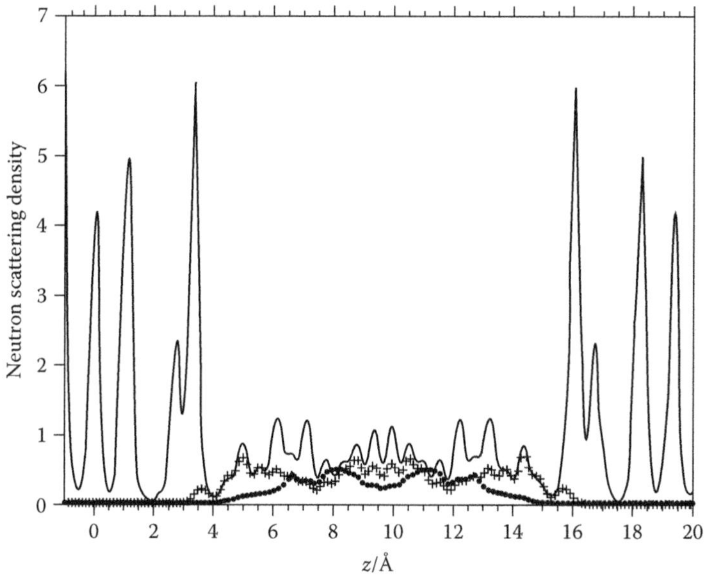

FIGURE 9.5 Neutron scattering density profiles $\rho(z)$ along the c* clay-stacking axis for a hydrated butylammonium vermiculite crystal with a clay layer spacing of 19.8 Å. Three profiles are shown, representing the scattering density due to the butyl group protons (solid dots), the water and ammonium group protons (crosses) and all remaining atoms (continuous line).

The phase diagram was mapped out on the LOQ instrument at ISIS, Didcot, U.K. [16], just as for the butylammonium vermiculite system [6]. Typical scattering patterns from an $r = 0.01$, $c = 0.25$ M propylammonium vermiculite gel are shown at $T = 36$, 38, 40 and 42°C in Figure 9.6. The story is by now a familiar one: a gel with a well-defined d-value at low temperatures (in this case $d \cong 60$ Å) collapses as the temperature increases at a well-defined phase transition temperature (in this case $T_c = 39$°C). Note that the butylammonium system will not swell at $c > 0.2$ M, whereas here we have colloidal swelling at $c = 0.25$ M. The complete c, T phase diagrams at low r for both the propylammonium and butylammonium systems, taken in contiguous experiments, are shown in Figure 9.7. It is clear that the propylammonium vermiculites will swell in salt concentrations up to about 0.5 M in cold water. In these circumstances, their d-values decrease below 50 Å, and so could be measured on LAD.

Before moving on to our attempt to measure the complete double layer in a swollen propylammonium vermiculite with $d = 43.6$ Å [18], we pause to note that (a) at ionic strengths relevant to cell fluids, namely $c \approx 0.12$ M [19], the phase-transition temperature in the propylammonium vermiculite system is not so far away from our body temperature and (b) similar temperature-induced gel-crystal transitions are observed in many biochemical systems. An example is the deoxyhemoglobin molecule that causes sickle cell anemia [33]. We also note that with both counterions, T_c decreases linearly with the logarithm of the salt concentration.

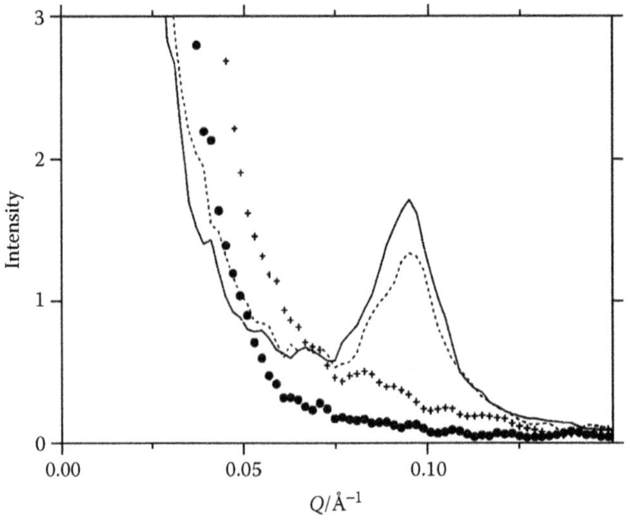

FIGURE 9.6 Low-angle neutron scattering intensity as a function of scattering vector Q (Å^{-1}) for propylammonium vermiculite immersed in a 0.25 M propylammonium chloride solution at four different temperatures, T = 309 K (continuous line), 311 K (dashed line), 313 K (crosses) and 315 K (solid circles). The low-angle peak at 0.1 Å^{-1} is due to a gel phase with a clay layer spacing of 60 Å. The data therefore show the temperature-induced gel-crystal phase transition occurring between 311 and 313 K. Note that body temperature is 311 K and that cell fluids are typically 0.2 M.

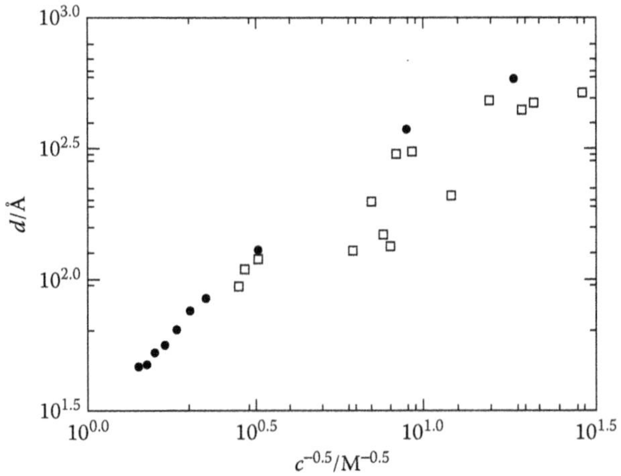

FIGURE 9.7 Clay layer spacings d in vermiculite gels as a function of the external salt concentration c for propylammonium (solid circles) and butylammonium (open squares) vermiculites.

The rate of change of T_c is greater for propylammonium than for butylammonium: 50 K per log unit compared with 14 K per log unit. This still corresponds to a change in surface potential of only a few millivolts, so the qualitative behavior is the same as in the butylammonium system, namely one of approximately constant surface potential with respect to electrolyte concentration. The quantitative difference between the two slopes was important for our purposes in that it enabled us to make $d = 43.6$ Å gels under easily controllable conditions at $c = 0.5$ M, $T = 4$°C.

The crystalline phase of the n-propylammonium Eucatex used had a d-value of 18.3 Å [16]. Twenty flakes of samples, prepared by prolonged soaking in a molar solution of H-chain propylammonium ions (as the chloride) in heavy water, were placed in a 0.5 M solution of the same and left to swell for five days. The swollen samples were contained in a flat-plate can made from a null-coherent-scattering Ti/Zr alloy, as in the previous experiments on the LAD. The sample container was mounted on a rotating closed-cycle refrigerator, and the temperature was maintained at 4°C. Isotopic labeling of the counterions was conducted *in situ* by replacing $C_3H_7ND_3^+$ with $C_3D_7ND_3^+$ in the swelling solution. The samples were aligned with respect to the neutron beam and a chosen detector bank so that the c-axis of the gel was parallel to the scattering vector Q [34]. With this orientation, the measured coherent neutron scattering intensity, $I(Q)$, consists of the $(00l)$ Bragg reflections.

The LAD diffractometer uses neutrons with wavelengths of 0.1 to 6 Å. It has 14 detector banks arranged around the sample at scattering angles between 5 and 150° in 2θ. Two separate runs were conducted for each sample, corresponding to scattering into the detector banks at 5 and 35°. By joining these two data sets, we achieved a total scattering vector range of $0.1\ \text{Å}^{-1} < Q < 12\ \text{Å}^{-1}$. The lower limit allowed us to measure the intensity of the (001) Bragg reflection (at $Q = 0.144\ \text{Å}^{-1}$), while the upper limit provided us with a real-space resolution of 0.5 Å, much higher than on D16. The raw data were corrected for absorption and multiple scattering as well as for scattering due to the background, sample container, and soaking solution. The normalized coherent neutron scattering intensities are presented for both the H-salt and D-salt samples in Figure 9.8. The (001) to (003) peaks are clearly visible, and their positions show that both samples have a regular d-value of 43.6 Å. At higher Q-values, the Bragg peaks are not individually resolved but combine to give rise to broader features.

The normalized neutron scattering intensities, $I(Q)$, are related to the neutron-scattering density profiles normal to the clay sheets, $\rho(z)$, via the crystallographic structure factor, $S(Q)$, as

$$I(Q) = M(Q)S(Q)S^*(Q) \tag{9.1}$$

and

$$S(Q) = \int_0^d \rho(z)[\cos(Qz) + i\sin(Qz)]dz \tag{9.2}$$

where d is the clay-layer spacing, z is the position on the clay c-axis, and $M(Q)$ is a Lorentzian that takes into account the mosaic spread of the sample [27]. Applying this

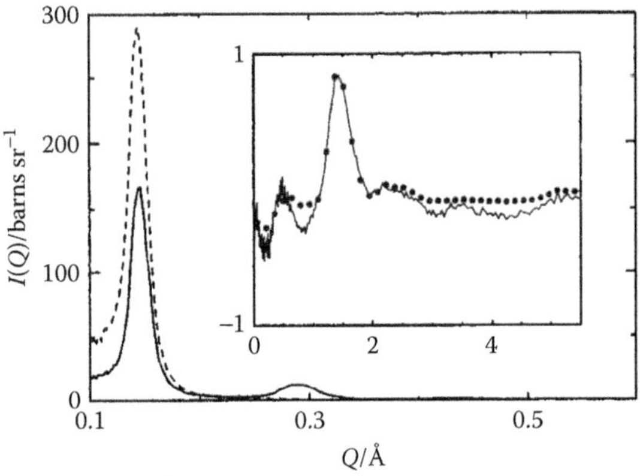

FIGURE 9.8 Neutron scattering intensity $I(Q)$ for propylammonium vermiculite gels with a d-value of 43.6 Å. Data are presented as a function of counterion isotope: dotted line, $C_3H_7ND_3^+$; solid line $C_3D_7ND_3^+$. The main figure illustrates how isotope substitution creates a large difference in the (001) and (002) Bragg peak intensities. The (003) peak was also individually resolved but cannot be seen in this figure owing to scale limitations. The inset shows the data for the $C_3D_7ND_3^+$ sample (solid line), illustrating how the Bragg peaks are not individually resolved at higher Q. The inset also shows the calculated peak intensities for the $C_3D_7ND_3^+$ sample (circles), illustrating that the fitted profiles shown in Figure 9.9 are consistent with the experimental data.

essentially crystallographic description is a novel treatment that is valid along the c-axis. It encompasses the periodically repeated liquid layers present in the gels and has the advantage that it produces a single particle distribution function, $\rho(z)$, rather than a spatially averaged pair distribution function. We were therefore able to locate the water molecules and counterions unambiguously with respect to their distance from the clay surfaces.

Neutron scattering density profiles were obtained simultaneously for both isotopic compositions using an inverse Monte Carlo refinement [27], as follows. Two arrays were set up, $\rho(z)_1$ and $\rho(z)_2$, each containing 150 movable "particles." Every particle is associated with a Gaussian distribution of neutron-scattering intensity; the coherent neutron scattering lengths of the main atomic species present are given in Table 9.2. The two arrays were combined to produce scattering density profiles representing the hydrogenated and deuterated samples, $\rho(z)_H$ and $\rho(z)_D$, as

$$\begin{aligned} \rho(z)_H &= \rho(z)_1 + b_H \rho(z)_2 \\ \rho(z)_D &= \rho(z)_1 + b_D \rho(z)_2 \end{aligned} \tag{9.3}$$

where b_H and b_D are the scattering lengths of hydrogen and deuterium, respectively (see Table 9.2). The density profile $\rho(z)_2$ then gives the location of the substituted alkyl protons: the "difference profile." The profile $\rho(z)_1$ gives the location of all

TABLE 9.2
Coherent Neutron Scattering Lengths *b*

Nucleus	$b/10^{-14}$ m
^{1}H	−0.3374
^{2}H (D)	0.6671
C	0.6651
^{15}N	0.937
^{14}N	0.644
O	0.5805
Mg	0.55
Al	0.345
Si	0.4107
Fe	0.993

atoms common to both samples (clay layers, water molecules, and the unlabeled portion of the counterions): the "common profile."

The starting configuration for the refinement was the clay layer structure derived from X-ray diffraction and chemical analysis [23]. All other particles were distributed evenly over the system. The positions of all particles, including those in the clay layer, were then sampled by an inverse Monte Carlo routine. This weighs acceptance of particle moves by comparing the experimental $I(Q)$ with the one calculated from the model profiles, $\rho(z)_H$ and $\rho(z)_D$, using Equation 9.1 and Equation 9.2. The final profiles, averaged over 100,000 equilibrated configurations, are presented in Figure 9.9. The average R-factors for the final fits were <1% (Figure 9.8, inset). In the common profile, the sharp peaks in the regions 0 to 3.3 Å and 40.3 to 43.6 Å are due to the atoms in the crystalline clay layers. As shown in Table 9.3, the positions and areas of these peaks are in excellent agreement with the X-ray data and chemical analysis.

Moving into the interlayer region, we observe strong common peaks at 6.6 and 9.6 Å (and correspondingly at 37.0 and 34.0 Å). A much weaker pair of peaks is also seen at 12.6 and 31.0 Å. The difference profile has a broad peak across the interlayer region. This has a maximum scattering density around the midplane (at z = 21.8 Å) and an oscillatory component giving rise to smaller maxima at 11.2 and 32.4 Å. Although this component is weak, the smaller maximum (compared with the one at the midplane) in the ion distribution accords well with the previous observation that butylammonium ions sit at about 11 to 14 Å from the center of the plates in the gel phase. The areas of all the peaks are listed in Table 9.3.

We conclude from the main peak in the difference profile that the majority of the alkyl groups are broadly distributed around the center of the interlayer region. This is consistent with the slight dip observed across the center of the common profile, arising from the partial exclusion of D_2O molecules by counterions. It should be noted, however, that the partial volume of the counterions [35], although higher

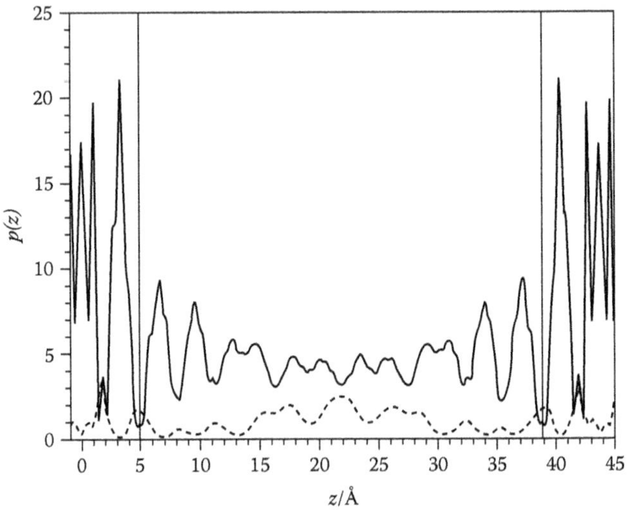

FIGURE 9.9 Neutron scattering density profiles $\rho(z)$ normal to the clay layers for n-propylammonium vermiculite gels with a *d*-value of 43.6 Å. To highlight the contrast arising from isotopic substitution, we present two profiles. The solid line shows the contribution from all species that are not isotopically labeled; the dashed line shows the contribution from the substituted alkyl hydrogen atoms, plus ongoing exchange of H/D in the clay layers. This profile has been expanded by 100% along the $\rho(z)$ axis. The lines at $z = 4.8$ Å and $z = 38.8$ Å indicate the van der Waals surfaces of the clay layers. The common profile has peaks at 6.6 and 9.6 Å (and 37.0 and 34.0 Å), indicating that there are two layers of water molecules associated with each surface. The difference profile shows that the majority of the counterions are broadly distributed around the midplane of the interlayer region.

TABLE 9.3
Analysis of Peak Areas in the Fitted Density Profiles for 43.6 Å n-Propylammonium Vermiculite Gels

Peak Position (Å) [a]	Assignment of Chemical Species to Each Peak	Profile Where Peak Occurs	Peak Area (10^{-14m})	Chemical Equivalent [b]
0.0–0.64	O_h cations	Common	1.58	O_h cations
0.64–1.46	O (clay)	Common	1.75	3 O
1.46–2.1	D (clay)	Common	0.384	0.6 D
2.1–4.8	T_h cations, O (clay), H_2O	Common	5.24	T_h cations, 3 O, 1.4 D_2O
0.9–3.3	H/D (clay)	Difference	0.35	0.6 H/D
4.8–8.1	D_2O	Common	3.7	1.93 D_2O
3.3–6.7	H/D (D_2O)	Difference	0.33	0.6 H/D
8.1–11.1	D_2O	Common	3.5	1.81 D_2O
11.1–32.5	D_2O, C_3-ND_3	Common	21.2	0.65 C_3-ND_3, 9.5 D_2O
13.4–30.2	H/D (alkyl)	Difference	1.06	0.61 H_7

[a] Limits of integration (Å).
[b] Number of atoms corresponding to the peak areas.

than in the D16 experiments, is still only about 15% of the total volume of the region over which they are spread. There is therefore no evidence to suggest significant aggregation of the propyl groups.

Having established that the majority of the counterions are located around the center of the interlayer, we attribute the peaks at 6.6 and 9.7 Å in the common profile to two layers of water molecules associated with each surface. This result is in excellent agreement with the conclusions reached via the spatially averaged pair distribution functions used to analyze the vanilla ripples from the butylammonium samples in Chapter 8. The position of the third pair of peaks in the common profile is consistent with a less well-defined third layer formed either by water molecules or by the ND_3^+ headgroups of the counterions. The peak at 11.2 Å in the difference profile suggests that a small proportion of the ND_3^+ headgroups are also incorporated into the second layer of water molecules, giving an oscillatory component to the counterion distribution. Figure 9.10 shows a molecular graphics snapshot of the system, illustrating the two layers of partially oriented water molecules and the maximum counterion density at the center of the interlayer region. Although exciting as our first picture of the complete structure of the interlayer solution between two colloidal particles, it should be noted that this single snapshot, containing only 12 propylammonium counterions, cannot reproduce all the features of the averaged scattering density profiles.

The peaks in the clay layer region of the difference profile are attributed to ongoing replacement of hydrogen by deuterium within the clay layer during the four days between collection of data from the three samples. (We also studied a sample in which the D_2O was substituted by H_2O, in a less successful experiment.) The peak at 1.7 Å is consistent with exchange of protons on hydroxyl groups in the clay layer, while the peak at 4.9 Å is consistent with the exchange of protons strongly adsorbed to the hexagonal cavities of the clay surface. While no data are available on the propylammonium vermiculite, the timescale involved should be comparable with rates observed in other mica-like minerals [36].

Does Figure 9.10 represent another success for the hypothesis of the "electric rope" forwarded in Chapter 6? It certainly looks at first blush as if counterions are shared by the two plates and may be electrically binding them. But the answer to our question is, "No, not really," because the salt concentration is too high for us to realistically expect mean field theory to apply. For $c = 0.5$ M, the Debye length is between 4 and 5 Å, the size of only one or two water molecules. Obviously we would not expect a primitive model, treating the solvent as a dielectric continuum, to be valid when the electrostatic screening length is less than the diameter of two solvent molecules. Nevertheless, we may still be surprised at how completely Figure 9.10 undermines the classical Stern layer picture of counterion adsorption [37], in which immobile counterions are attached to the clay layers. Our results show that, in contrast to traditional models, the region within 6 Å of the clay surface consists principally of partially ordered layers of water molecules. This structural model of the interfacial water is consistent with the vanilla ripples [31, 32] and previous studies of hydrated surfaces, including surface X-ray scattering [38] and surface force apparatus (SFA) experiments [39]. For example, SFA measurements of the interaction between two mica surfaces immersed in aqueous solutions show

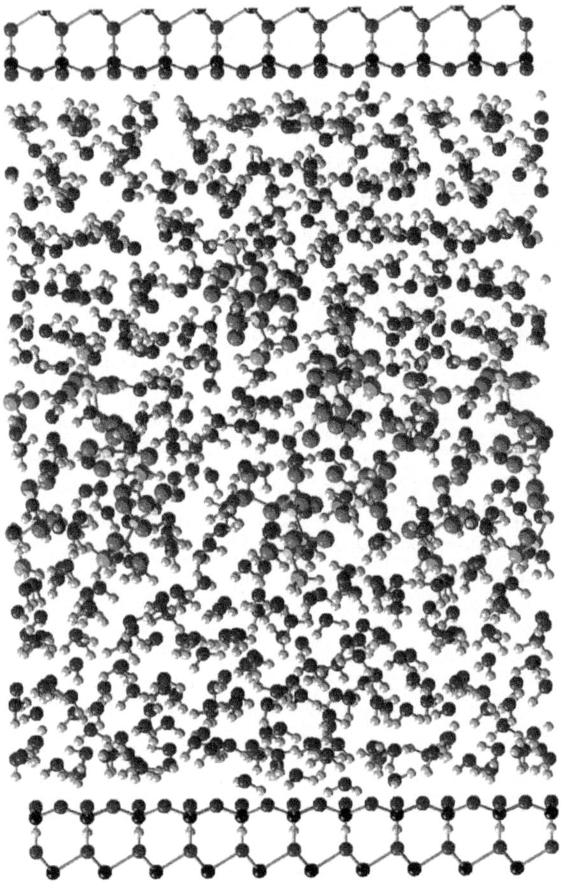

FIGURE 9.10 Molecular graphics snapshot of a $d = 43.6$ Å propylammonium vermiculite gel, illustrating how the majority of the counterions are separated from the charged clay surfaces by two layers of partially oriented water molecules. The atoms are color-coded as follows: labeled hydrogen/deuterium (red); oxygen (blue); hydrogen (white); carbon (yellow); nitrogen (green); and silicon, magnesium, aluminum (black). See color insert following page 76.

that at separations of less than 15 Å there is an oscillatory component to the surface–surface force. Our results suggest that these maxima in the surface–surface force can be attributed to the disruption of discrete layers of water molecules at the mica surfaces. These short-range forces are beyond the scope of mean field theory.

REFERENCES

1. Walker, G.F., *Nature,* 187, 312, 1960.
2. Hunter, R.J., *Foundations of Colloid Science,* Clarendon Press, Oxford, 1987.
3. Sogami, I. and Ise, N., *J. Chem. Phys.,* 81, 6320, 1984.
4. Smalley, M.V., *Mol. Phys.,* 71, 1251, 1990.
5. Smalley, M.V., *Langmuir,* 10, 2884, 1994.

6. Williams, G.D., Moody, K.R., Smalley, M.V., and King, S.M., *Clays Clay Min.,* 42, 614, 1994.
7. Braganza, L.F., Crawford, R.J., Smalley, M.V., and Thomas, R.K., *Clays Clay Min.,* 38, 90, 1990.
8. Smalley, M.V., Thomas, R.K., Braganza, L.F., and Matsuo, T., *Clays Clay Min.,* 37, 474, 1989.
9. Derjaguin, B.V. and Landau, L., *Acta Physicochimica,* 14, 633, 1941.
10. Verwey, E.J.W. and Overbeek, J.Th.G., *Theory of the Stability of Lyophobic Colloids,* Elsevier, Amsterdam, 1948.
11. Sogami, I., Shinohara, T., and Smalley, M.V., *Mol. Phys.,* 74, 599, 1991.
12. Sogami, I., Shinohara, T., and Smalley, M.V., *Mol. Phys.,* 76, 1, 1992.
13. Shinohara, T., Smalley, M.V., and Sogami, I.S., *Mol. Phys.,* 101, 1883, 2003.
14. Skipper, N.T., Soper, A.K., and Smalley, M.V., *J. Phys. Chem.,* 98, 942, 1994.
15. Skipper, N.T., Smalley, M.V., Williams, G.D., Soper, A.K., and Thompson, C.H., *J. Phys. Chem.,* 99, 14201, 1995.
16. Williams, G.D., Skipper, N.T., Smalley, M.V., Soper, A.K., and King, S.M., *Faraday Discuss.,* 104, 295, 1996.
17. Williams, G.D., Skipper, N.T., and Smalley, M.V., *Physica B,* 234, 375, 1997.
18. Williams, G.D., Soper, A.K., Skipper, N.T., and Smalley, M.V., *J. Phys. Chem. B,* 102, 8945, 1998.
19. Ochaia, E-I., *Bio-Inorganic Chemistry: An Introduction,* Allyn and Bacon, Boston, 1977.
20. Fiabane, A.M. and Williams, D.R., *The Principles of Bio-Inorganic Chemistry,* Chemical Society Monographs, Vol. 31, Chemical Society, London, 1977.
21. Hay, R.W., *Bio-Inorganic Chemistry,* Ellis Horwood, Chichester, U.K., 1984.
22. Hewish, N.A., Neilson, G.W., and Enderby, J.E., *Nature,* 297, 138, 1982.
23. Newman, A.C.D., *Chemistry of Clays and Clay Minerals,* Mineralogical Society, London, 1987.
24. Grim, R.E., *Applied Clay Mineralogy,* McGraw-Hill, New York, 1960.
25. Brindley, G.W. and Brown, G., *Crystal Structures of Clay Minerals and Their X-ray Identification,* Mineralogical Society, London, 1980.
26. Howells, W.S., Internal Report RL-80-017, Rutherford Appleton Laboratory, Chilton, Didcot, U.K., 1980.
27. Skipper, N.T., Soper, A.K., and McConnell, J.D.C., *J. Chem. Phys.,* 94, 5751, 1991.
28. Garrett, W.G. and Walker, G.F., *Clays Clay Min.,* 9, 557, 1962.
29. Slade, P.G., Stone, P.A., and Radoslovich, E.W., *Clays Clay Min.,* 33, 51, 1985.
30. Boek, E.S., Coveney, P.V., and Skipper, N.T., *J. Am. Chem. Soc.,* 117, 12608, 1995.
31. Swenson, J., Smalley, M.V., Thomas, R.K., Crawford, R.J., and Braganza, L.F. *Langmuir,* 13, 6654, 1997.
32. Swenson, J., Smalley, M.V., Thomas, R.K., and Crawford, R.J., *J. Phys. Chem. B,* 102, 5823, 1998.
33. Murayama, M., *Science,* 153, 145, 1966.
34. Soper, A.K., Howells, W.S., and Hannon, A.C., *Analysis of Time-of-Flight Data from Liquid and Amorphous Samples,* CLRC Publications, Rutherford Appleton Laboratory, Didcot, U.K., 1989.
35. Desnoyers, J.E. and Arel, M., *Can. J. Chem.,* 45, 359, 1967.
36. Vennemann, T.W. and O'Niel, J.R., *Geochim. Cosmochim. Acta,* 60, 2437, 1996.
37. Stern, O., Z. *Elektrochem.,* 30, 508, 1924.
38. Toney, M.F., Howard, J.N., Richer, J., Borges, G.L., Gordon, J.G., Melroy, O.R., Wiesler, D.G., Yee, D., and Sorenson, L.B., *Surface Sci.,* 335, 326, 1995.
39. Israelachvili, J.N. and Pashley, R.M., *Nature,* 306, 249, 1983.

10 Freezing Experiments on n-Butylammonium Vermiculite Gels

This will be our final chapter on the three-component clay-salt-water system. All the experiments described to date have taken place in liquid water at temperatures above 0°C. What happens when we freeze the solvent? After our excursion in pursuit of the complete ion distribution, we return to our model system, the n-butylammonium Eucatex vermiculite system, to provide an answer. As a continuation of our extensive $\{r, c, T\}$ investigation of the three-component clay-salt-water system, we used neutron and X-ray diffraction experiments to investigate freezing cycles between −5 and +5°C for the model system consisting of butylammonium vermiculite, butylammonium chloride and water [1]. We hope that our results will be useful in understanding the weathering of rocks in freezing cycles, an important problem in geology and physical geography.

We discovered that when the gels are frozen, the colloidal state collapsed into the crystalline phase, with its interlayer spacing of 19.4 Å. This phenomenon was observed throughout a wide range of clay concentrations r and salt concentrations c. The phase transition was observed to be reversible, the gel phase always being recovered upon warming through the freezing point of water. As usual, the disappearance of the gel phase was monitored by small-angle neutron scattering, and the appearance of the crystalline phase, together with ice peaks, was observed by scattering at higher Q.

By one of those odd coincidences, it snowed in Didcot, U.K., on the night we mapped out the $\{r, c\}$ freezing behavior of the system on the LOQ instrument at ISIS. Three salt concentrations, $c = 0.01$ M, $c = 0.03$ M and $c = 0.1$ M and three sol concentrations, $r = 0.01$, $r = 0.1$ and $r = 0.3$, were studied, with three samples used at each of the $\{r, c\}$ points to allow for sample-to-sample variability. The average d-values obtained in the gel phase at $T = +5$°C are shown in Table 10.1, and the patterns obtained at −5 and +5°C for one of the $c = 0.1$ M, $r = 0.3$ samples are shown in Figure 10.1.

It is clear from the low Q data in the top half of Figure 10.1, that the pronounced gel peak (corresponding to an interlayer spacing of approximately 75 Å) observed at $T = +5$°C has completely disappeared at −5°C. The gel has collapsed, just as we observed with respect to increases in temperature in Chapter 1. In a manner of speaking, we could say that the results given in Chapter 1 show that the gel has an upper critical solution temperature (UCST), so the results in Figure 10.1 show that the gel also has a lower critical solution temperature (LCST). We labeled the former

TABLE 10.1
Average *d*-Values (Å) Observed in the LOQ Experiments as a Function of the Sol Concentration *r* and the Salt Concentration *c*, at *T* = +5°C

	Sol Concentration		
Salt Concentration, c (M)	$r = 0.01$	$r = 0.1$	$r = 0.3$
0.01	330	250	110
0.03	240	170	100
0.1	130	110	75

T_c, and we now label the latter T_f. By the time of these experiments, LOQ also had a high-angle bank, and by tilting the sample appropriately, we were able to see the first-order peak of the d = 19.4 Å crystalline phase at $T = -5$°C, as shown in the bottom half of Figure 10.1.

In this case, I have deliberately chosen the highest r, highest c sample (d = 75 Å, in the regime of pastes in disoriented clays) as an example of the phase transition because, in practical problems, the higher clay concentrations will be more important than the dilute sols we have used for theoretical comparisons. The essential feature of the freezing experiments on LOQ was that the behavior shown in Figure 10.1 was observed at all the $\{r, c\}$ points studied.

It was impossible to identify the formation of ice in the LOQ experiments because the maximum Q-value of the instrument, even with the high-angle bank, lies below the Q-value of the first Bragg peak of ice. This was performed in a single experiment at c = 0.1 M, r = 0.1 on LAD (liquids and amorphous materials diffractometer). The normalized structure factor of the gel at +5°C is shown in the Q-range between 1 and 6 Å^{-1} in Figure 10.2a. We note that under these conditions, the first-order Bragg peak of the d = 110 Å gel phase lies well below the low Q limit of LAD, so in this Q-range we see only the diffuse scattering from the gel that has been the subject of the last two chapters. The pattern changed completely at −5°C, as shown in Figure 10.2b.

In Figure 10.2b, the intense peak at Q = 0.324 Å^{-1} (note the difference in intensity scales between parts (a) and (b) of the figure) is the first-order peak of the crystalline phase identified in the high-angle bank data in the LOQ experiments, and the even more intense peak at Q = 1.7 Å^{-1} is the first Bragg peak of ice. The full diffraction pattern obtained was compared with those from known ice structures [2] and was found to correspond to the ordinary hexagonal ice phase one would expect to obtain upon cooling a simple salt solution or pure water through its freezing point. As far as it was possible to tell from the 10°C temperature jump employed, the ice peaks appeared *simultaneously* with the appearance of the clay crystalline peak. This strongly suggests that the clay gel phase is thermodynamically unstable in the solid solvent. Having determined from the higher Q-range of the LAD instrument that ordinary ice is formed in the "double" phase transition of the clay and solvent

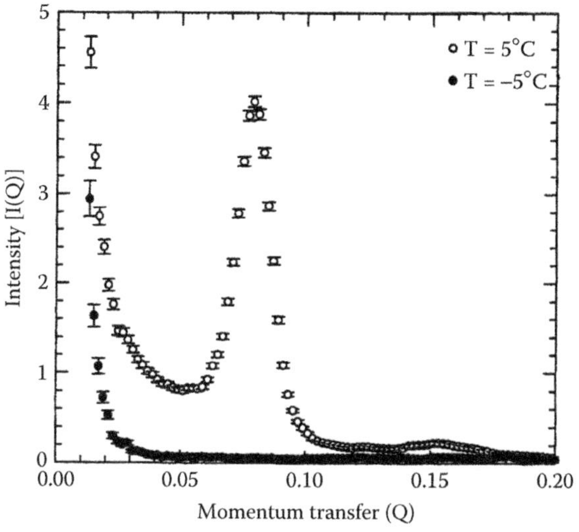

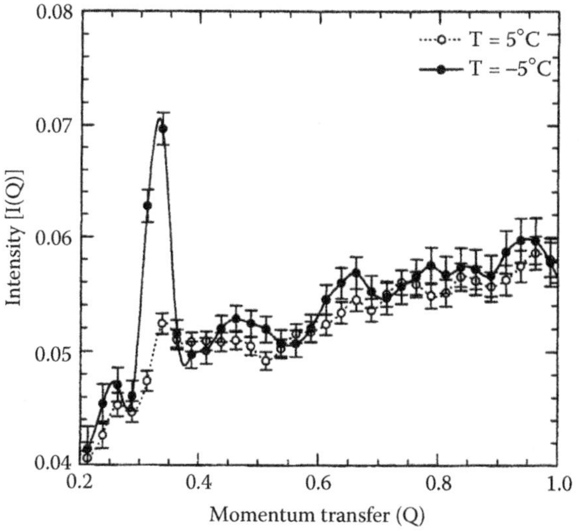

FIGURE 10.1 Top figure (a): LOQ low-angle bank data at $r = 0.3$, $c = 0.1$ M. The pattern obtained at $T = +5°C$ is shown by the open circles and that at $T = -5°C$ by the closed circles. Bottom figure (b): LOQ high-angle bank data at $r = 0.3$, $c = 0.1$ M. The pattern obtained at $T = +5°C$ is shown by the open circles and that at $T = -5°C$ by the closed circles.

systems, the sharpness and reversibility of the transition were then investigated in laboratory X-ray experiments. Before describing the X-ray experiments, I should note that the high penetrating power of neutrons means that they sample the entire bulk of the material, so we can be sure that surface effects are not playing any significant role in the double transition.

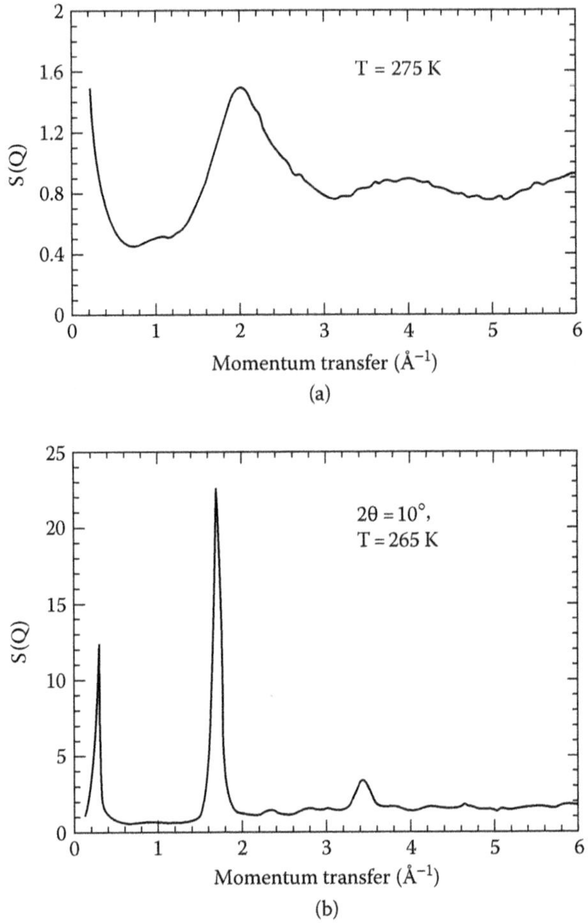

FIGURE 10.2 LAD data at r = 0.1, c = 0.1 M. The low-temperature gel-tactoid phase transition was discovered to occur simultaneously with the formation of ice. Top figure (a) shows the structure factor of an r = 0.1, c = 0.1 M butylammonium vermiculite gel at T = +5°C, and bottom figure (b) shows the pattern obtained at T = −5°C. In (b) the gel has collapsed locally to its tactoid phase (which produces a peak at Q = 0.324 Å^{-1}), and the solution has crystallized at the same temperature.

The X-ray measurements were carried out using a Phillips Xpert θ-θ X-ray diffractometer. In this geometry, both the X-ray source and detector arm subtend an angle θ to the free surface of the sample. The normal operating conditions were 50 kV and 30 mA. An X-ray beam produced by a molybdenum target that passes through a Zr filter placed over the divergence slit was used for all the experiments. The beam consists of two emission lines at 0.7096 and 0.7136 Å and was treated as one unresolvable wavelength at 0.7107 Å. A programmable slit system was used to control the dimensions of the incident and scattered beams. The widths of the divergence and antiscatter slits were set as automatic and controlled by changing the illuminating length (generally 2 mm). The height of the receiving slit was 0.1 mm. The samples were mounted on a temperature-controlled disk-shaped sample holder, which was connected to a cryostat, and the level of the sample surface was adjusted before the inner and outer sample chambers were fixed to the cryostat. The data were collected over the 2θ range from 0.5° to 30°, which corresponds to the Q-range 0.05 to 4.5 Å^{-1}, with a step size of 0.1°. X-rays scattered by the samples were recorded on a two-dimensional area detector and analyzed using the software PC-APD.

The θ-θ X-ray diffractometer used was ideal for our investigation, as we could see both the transition from the gel phase to the crystalline phase of the clay (at high salt and sol concentrations) and the formation of ice. The clay volume fraction was held constant at $r = 0.1$ for two reasons; first, to obtain a d-value in the gel phase that would be observable within the Q-range of the instrument and, second, because the sample container of the X-ray instrument was too small to conveniently study low volume fractions of clay. We already know from the LOQ experiments that the freezing transition behavior is not strongly affected by r in the range between 0.01 to 0.3, or between 1 and 30% clay, so it is natural to choose somewhere in the middle of this range, and $r = 0.1$ results could be compared directly with the LAD data.

With more beam time available in the laboratory, we were able to go through the phase transition in much smaller temperature steps. Before the measurements were taken, the system was kept at each temperature for at least half an hour to ensure that thermal equilibrium had been achieved. The diffraction patterns obtained from a sample with $r = 0.1$, $c = 0.1$ M (the same conditions as for the LAD experiment) are shown in Figure 10.3a at five temperatures between −2 and 2°C, in 1°C steps.

The $I(Q)$ against Q plot at 2°C (the dotted line in Figure 10.3a) seems to show a weak peak at $Q = 0.18$ Å^{-1}, corresponding to a real-space correlation length of 35 Å. In view of the LOQ result that $d = 110$ Å at $r = 0.1$, $c = 0.1$ M, we interpret this peak as the weak third-order Bragg peak of the gel phase, with the lower order Bragg peaks outside the low Q-range of the instrument. At 1°C (solid line) the overall intensity of the scattering in the Q-range 0.1 to 0.4 Å^{-1} increases and then decreases again at 0°C (open circles). The peak also becomes less distinct and seems to shift toward lower Q, indicating an expansion of the gel phase as freezing is approached. What is definitely clear is that there is no peak at 0.32 Å^{-1}, showing the absence of crystalline clay, down to 0°C. The neutron experiments at −5 and 5°C had been performed in heavy water, but from the point of view of the practical utility of the more accurate determination of T_f, the lab experiments were carried out in H_2O, so 0°C was the freezing point of the solvent in this case. At −1°C, the diffraction pattern from the gel phase collapsed entirely, and the Bragg peak typical of the crystalline phase appeared at $Q = 0.32$ Å^{-1}. The second Bragg peak of the $d = 19.4$ Å phase is also clearly visible at $Q = 0.64$ Å^{-1}. Further scans below −1°C were identical to that at −1°C, showing that the phase change is complete within 1°C of temperature and an interval of one hour.

The phase transition was found to be reversible by cycling the temperature through the freezing point between −3 and +3°C, and the transition was always complete within a degree of the freezing point. At two temperatures just above and

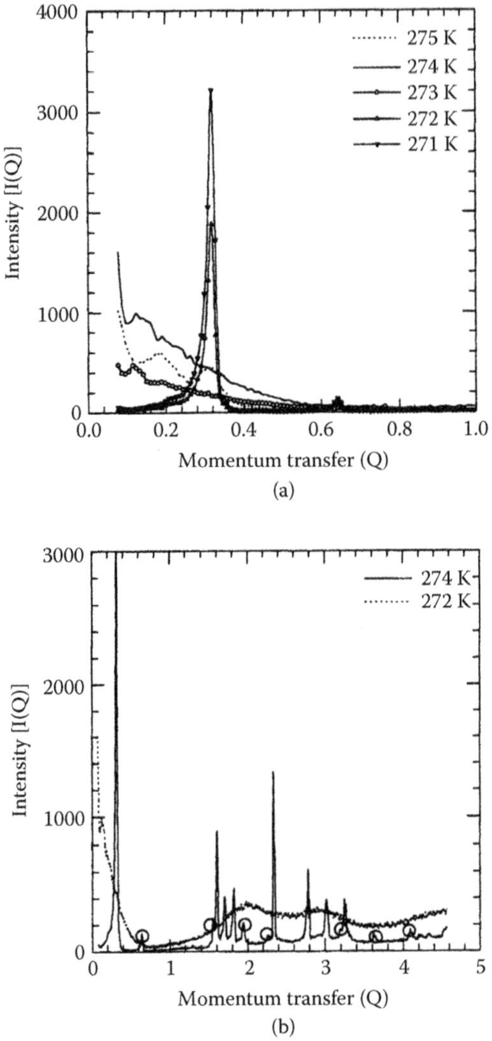

FIGURE 10.3 X-ray data at $r = 0.1$, $c = 0.1$ M. Top figure (a): in the Q-range up to 1.0 Å^{-1}, the appearance of the peaks at $Q = 0.32$ Å^{-1} and $Q = 0.64$ Å^{-1} at $T = -1$°C shows the formation of tactoids between 0 and −1°C. Bottom figure (b): the patterns at $T = -1$°C and $T = +1$°C have been extended to $Q = 4.5$ Å^{-1} to show the appearance of the ice peaks at −1°C. The clay peaks are marked by open circles.

below the freezing temperature (at +1 and −1°C), the sample was scanned over a higher Q-range, up to 4.5 Å^{-1}. The results are shown in Figure 10.3b. It is clear that the scattering pattern at −1°C is very different from that at +1°C, with sharp Bragg peaks appearing in the higher Q-region between 1.5 and 4.5 Å^{-1}. The peaks marked with an open circle (o) are the higher order Bragg peaks arising from the crystalline clay layers, and the other peaks are from the crystalline ice. For comparison, diffraction patterns were obtained for pure water. It was evident that the Bragg peak positions for ice were the same in both cases, again indicating that the structure of ordinary ice [2] is not perturbed by the presence of the clay and the salt.

I should note that the limit of sensitivity of these experiments restricts us to saying that the phase transitions are simultaneous only in the sense that they both occur between −1 and 0°C (in either direction). More subtle variations, of the order of ±0.1°C, would not have been detectable. According to Debye-Hückel theory [3], the depression of the freezing point of pure water is 0.18°C in a 0.1 M uni-univalent electrolyte solution. We would expect the clay to cause a further small depression in the freezing point, as discussed below. Within these limits, the temperature where both the freezing transition and the gel-crystalline phase transition occur is the same in our model clay colloid system, and it can be concluded to be the ordinary freezing point of the soaking solution.

A completely different set of phenomena are observed when the samples are frozen rapidly. If we quench a sample in liquid water at a rate of, say, 1000 K/sec, it is likely that we will not see the thermodynamic phase transitions described above. Indeed, it is an underlying assumption of electron microscopy experiments on aqueous systems that we should more or less preserve the room-temperature structure by rapid freezing. We carried out an electron microscopy study of n-butylammonium vermiculites at CCTR Cryotech, York, U.K. [4].

We prepared three sets of samples, all in large excesses of soaking solution ($r < 0.01$), at three electrolyte concentrations, $c = 0.001$ M, $c = 0.01$ M and $c = 0.1$ M. The samples had been prepared by soaking for several weeks at 7°C to ensure that full equilibrium swelling had been achieved [4]. When the gels were studied, they had been out of the fridge for quite a while, standing at a room temperature of 21 to 22°C. At this temperature, the $c = 0.001$ M ($T_c = 45$°C) and $c = 0.01$ M ($T_c = 33$°C) samples remained in their osmotically swollen gel phase, but the $c = 0.1$ M samples had collapsed into their crystalline phase. One of the replicas we obtained from the crystalline phase is shown in Figure 10.4.

The three thick diagonal white lines running across the center and upper left of Figure 10.4, marked by the large arrows, probably represent packets of vermiculite layers. At the magnification employed, the 2 nm spacing in the crystalline regions would not be resolvable, and packets of layers with such a spacing would appear continuous. If this interpretation is correct, then the thickness of the three bands suggests that they each contain about ten layers. We are then observing directly the "tactoids" described by Kleijn and Oster [5], who use this nomenclature to describe a structure in which groups of vermiculite layers with an interlayer spacing of ≅1 nm are separated by large aqueous regions with spacings in the range 100 to 1000 nm (or 0.1 to 1 μm). In keeping with this description, the packets of 10 or so layers in Figure 10.4 are separated by aqueous regions of a thickness on the order of a micron that contain hardly any vermiculite at all. The situation is illustrated schematically in Figure 10.5.

In Figure 10.5, we see that the reversible phase transition we have been seeing with the neutrons is between states that look like Figure 10.5b and Figure 10.5c in real space, and we should obviously interpret the "crystalline" ($d = 1.94$ nm) phase

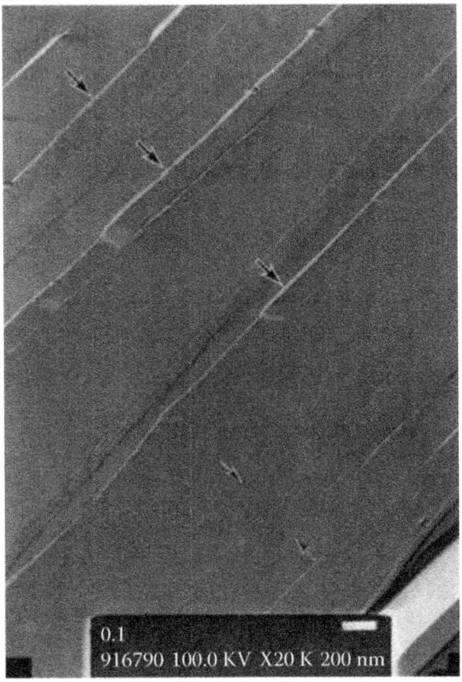

FIGURE 10.4 Electron micrograph of a swollen n-butylammonium vermiculite prepared in a 0.1 M solution of n-butylammonium chloride. The scale is defined by the white bar, which represents 200 nm. The three large arrows indicate packets of vermiculite layers, or tactoids. In the region between the two smaller arrows, the layers appear to be individually separated.

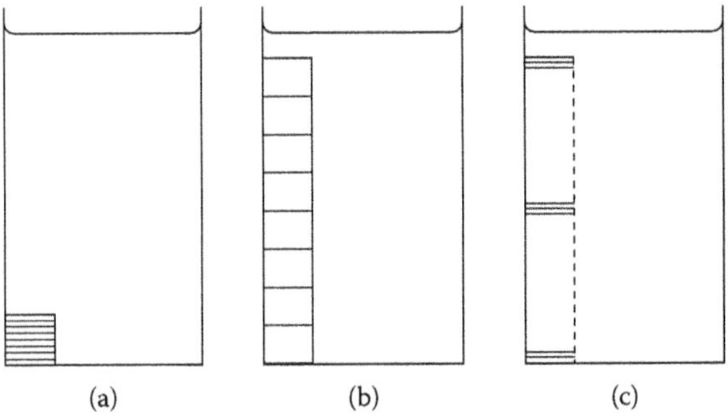

FIGURE 10.5 Schematic illustration of the swelling of n-butylammonium vermiculite in a 0.1 M n-butylammonium chloride solution; (a) represents the n-butylammonium vermiculite crystal ($d \cong 2$ nm) prior to swelling, (b) the gel ($d \cong 12$ nm) formed by a homogeneous sixfold expansion in the range $0°C < T < 14°C$, and (c) the tactoid formed when the gel collapses at $T < 0°C$ or $T > 14°C$. In (c), the dashed line represents the fact that the tactoid structure occupies roughly the same volume as the gel structure.

as the "tactoid" phase of Figure 10.5c. Such a structure is not surprising in view of the silvery sheen that accompanies the collapsed phase. Any regularity of the spacing between the crystalline regions in the tactoid phase, beyond the resolution of the neutron scattering experiments, would give interference effects in the visible light range. Of course, we recognize that Figure 10.5c is an oversimplification; in the bottom right-hand corner of Figure 10.4, between the two smaller arrows, there appears to be a region where all the layers are still individually separated, with spacings of the order of tens of nanometers. Little is known about the kinetics of the aggregation process, so it could be that the bottom right corner represents a region in which vermiculite layers were coagulating before the process was stopped by the rapid freezing. We should also bear in mind that when we look at a picture like Figure 10.4, we are actually looking at a piece of platinum, and the preparation of this platinum replica from an aqueous gel or tactoid structure involves many steps, as described below.

First, thin slices of the samples, whether in the gel ($c = 0.01$ M and $c = 0.001$ M) or tactoid ($c = 0.1$ M) phases, were cleaved from the gel (or tactoid) stacks by cutting along the basal planes with a razor blade. The cutting was performed underneath the aqueous soaking solution to ensure that the samples were maintained at their equilibrium room-temperature condition, and the accuracy of the cutting was ensured by performing it under a low-power (×10) microscope, which made the orientation of the gel stacks clearly visible. Slivers of a thickness of less than 1 mm had to be cut because of the geometry of the sample holder, which consisted of two cylindrical brass rivets of internal diameter 1.2 mm. The rivets were joined end-to-end, with a total length of 6 mm. The thin gel slices cut under the microscope fitted easily into these brass rivets and, after loading into the cells, the dead volume was immediately filled with the aqueous n-butylammonium chloride solution that had been in equilibrium with the gels. The length of the slivers was approximately 4 to 5 mm, so that a part of the gel was always across the junction between the two rivets.

To minimize contamination of the gels by any metal ions that might leach out of the brass into the sample, the rivet pair was rapidly frozen as soon as it had been loaded with a sample. The rivet was plunged into liquid propane, which cooled the sample at a rate on the order of magnitude of 1000 K/sec [6]. The rapidity of the temperature drop in the freeze-fracture replica technique employed is a crucial feature, since we wished as far as possible to maintain the room-temperature structure of the gels. Pairs of frozen rivets were then transferred to the freeze-fracture apparatus (Leybold Heraeus Biotech 2005). The specimen temperature was raised to −140°C, and it was then fractured across its long axis by means of an electromagnet. The vermiculite layers in the gel slivers were loaded parallel to the long axis of the rivets, so the fracturing was perpendicular to the vermiculite layers, exposing a view of the edge of a gel stack.

After fracturing, platinum was evaporated from a carbon crucible at an angle of 45° onto the surface of the exposed plane to form a thin film (1.5 to 2.0 nm thick). Mechanically, this film was not very stable, so a further layer of carbon was evaporated onto the platinum perpendicular to the fracture surface, as a thick layer (15 to 20 nm), to give support to the metal film. After this procedure the rivets were removed from the apparatus and the original sample was dissolved off the platinum replica. This was performed using HF, which does not affect the platinum surface but which dissolves clay very effectively.

A peculiar feature of this step of the replica preparation was that the 0.1 M samples, which were in their tactoid phase, simply floated off the platinum upon being dipped in water or dilute HF solutions, whereas the 0.01 and 0.001 M samples required much more concentrated HF solutions for their removal. We started off by using a drop of concentrated (≈10 M) HF to remove the gel samples from the replicas, and the vermiculite was removed successfully. However, the replicas curled up like miniature rugs, so the HF solution used was diluted as far as possible while still removing the vermiculite. It is difficult to estimate the effective HF concentration used because the single drop added is immediately diluted in the aqueous sample, but it was probably in the region of 0.1 M. Three replicas were prepared at each electrolyte concentration, and it was noted in each case that the HF solution was necessary to remove the gel samples, whereas the tactoid samples could be floated off in water. We did not understand this great difference in adhesion properties, though it must be related to the surface properties of the two different structures shown in Figure 10.5b and Figure 10.5c. In any event, some excellent replicas were also obtained from the gel phase by this method, as shown below.

The replicas were scanned at random, and at least six micrographs were taken at each electrolyte concentration. The best micrograph obtained, for a $c = 0.01$ M sample, is shown in Figure 10.6. It clearly shows the lamellar structure of the gel, with the vermiculite layers running horizontally across the micrograph in this case. Because there is overwhelming evidence from our neutron scattering studies that the vermiculite layers are individually separated under these conditions, it is natural to interpret the horizontal lines in Figure 10.6 as the edges of individual layers. Using this interpretation, the interlayer spacing was determined by drawing lines across the photographs perpendicular to the vermiculite layers and then counting the number of layers in a given interval, using the known magnification of the instrument. Five people made independent assessments of the layer spacings, and we were in broad agreement. One example of our method of estimating the layer spacing has been drawn in Figure 10.6. In this case, 42 spacings were counted in an interval of 2.13 μm, corresponding to a mean spacing of 51 nm.

Even for an individual micrograph, this method did not give a unique result because of the substantial occurrence of bifurcation of the layers. One clear example has been marked by the large arrow in Figure 10.6. The two smaller arrows draw attention to the fact that such bifurcations necessarily lead to an uncertainty in the estimated layer spacing: the region between these two arrows, counted as two layers in the example estimation, would be counted as one layer if the perpendicular line were displaced parallel to itself by a small distance in either direction. Although 42 intervals have been marked off within 2.13 μm, such parallel displacements can lead to any number within the range from 40 to 45, corresponding to average spacings of 53 nm and 47 nm, respectively. There was also sample-to-sample variability from one micrograph to another; in this case, $d = 50 \pm 5$ nm was the average result of the five independent estimates of the spacing for this individual micrograph.

All of the six micrographs obtained from samples prepared in 0.01 M soaking solutions gave clear lamellar regions in which over 30 layers could be counted, and

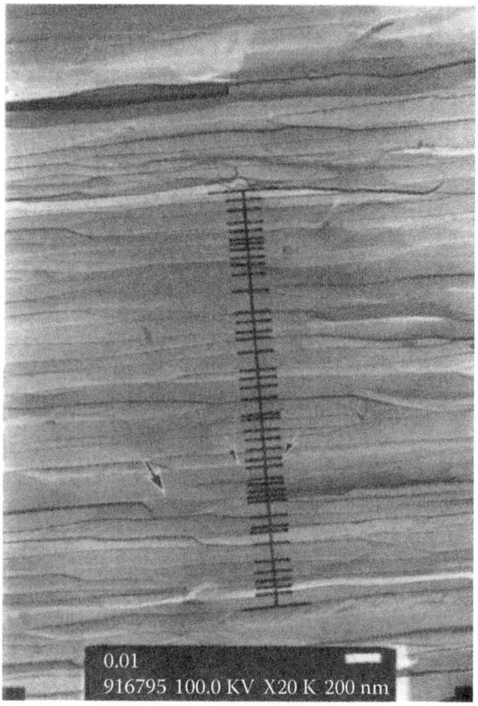

FIGURE 10.6 Electron micrograph of an n-butylammonium vermiculite gel prepared in a 0.01 M n-butylammonium chloride solution. The scale is defined by the white bar, which represents 200 nm. The method of determining the interlayer spacing is indicated by the line drawn perpendicular to the vermiculite layers. The large arrow points out one example of a bifurcation, and the two smaller arrows draw attention to two neighboring bifurcations that affect the estimation of the layer spacing.

the average results for the other five cases were 62, 65, 48, 55 and 60 nm. The unweighted average of the six results at this concentration was 57 nm, and the weighted average (weighted over the number of plates counted in each case) was 53 nm. A reasonable conclusion from these data seems to be that the *d*-value is 55 ± 10 nm for $c = 0.01$ M.

A similar method was used to estimate the average spacing in the $c = 0.001$ M samples. An example micrograph, at the same magnification as that used in Figure 10.4, is shown in Figure 10.7a, which is an example of the poorer quality of the micrographs obtained at $c = 0.001$ M. While it is straightforward to estimate that $d = 108$ nm by drawing a line perpendicular to the vermiculite layers across the lamellar region at the top of the micrograph (16 layer spacings can be counted in an interval of 1.73 μm), it is clear that there are some imperfections in the structure that cannot be interpreted in terms of a well-spaced array of monodisperse parallel layers. The three arrows draw attention to such a disordered region. However, even in such a case, the gel as a whole retains a strong net orientation. This can be seen

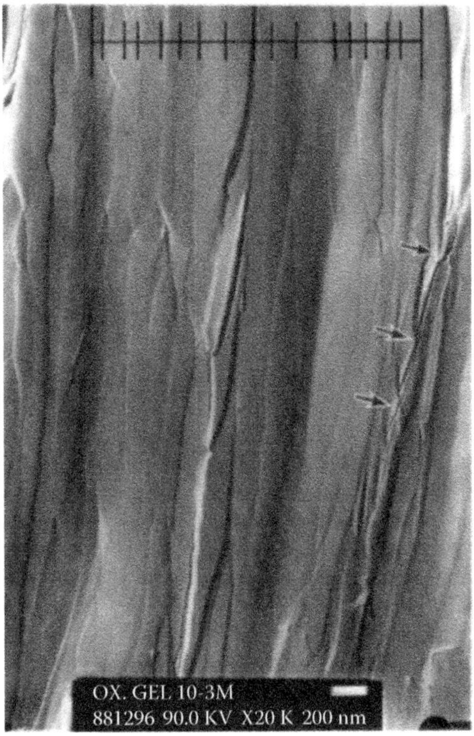

(a)

FIGURE 10.7 (a) Electron micrograph of an n-butylammonium vermiculite gel prepared in a 0.001 M n-butylammonium chloride solution. The scale is defined by the white bar, which represents 200 nm. The line drawn perpendicular to the vermiculite layers across the top of the micrograph was used to estimate the layer spacing. The three arrows indicate a disordered region that cannot be interpreted in terms of monodisperse parallel layers. (b) Electron micrograph of the same gel at a lower magnification. The white bar represents 2 μm. In spite of local inhomogeneities, the structure is clearly lamellar over the region shown, which extends over 20 μm perpendicular to the vermiculite layers.

in Figure 10.7b, which depicts the structure at a lower magnification. The observation of a lamellar region extending over 20 μm, containing hundreds of individual vermiculite layers, shows that the electron microscopy method is well adapted to observing the gel structure over a range of several microns. This real-space image makes it obvious why we see such strong interference effects from the gel phase in the Fourier space probed by neutrons.

The data for the $c = 0.001$ M samples were less reproducible than those for $c = 0.01$ M. Of the eight micrographs obtained, two did not give clearly defined lamellar regions at all, and two of the others showed substantial contamination by ice crystals. For each individual micrograph, the d-values obtained by the five different analyses were closely grouped (in the example case, all the investigators

(b)

FIGURE 10.7 (Continued).

independently chose the region at the top of the micrograph and obtained results between 105 nm and 110 nm), but there was substantial sample-to-sample variability. The six results obtained were 78, 125, 54, 74, 108 and 64 nm, the average and the weighted average being 84 and 83 nm, respectively. Despite the wide spread of these results, it seems reasonable to deduce that the *d*-value is 85 ± 20 nm at $c = 0.001$ M.

It is noteworthy that the *d*-values obtained from the electron microscopy experiments are definitely different from those obtained by *in situ* neutron scattering studies of the gel phase, being much higher for both the $c = 0.001$ M and $c = 0.01$ M cases. As we saw in Chapters 1 and 3, the spacings reported for these electrolyte concentrations were 60 nm and 30 nm, respectively, which have to be compared with the present results. The electron microscopy results are greater by a factor between one and a half and two, the discrepancy being outside the range of experimental error.

The neutron results are certain to be more accurate for three reasons. First, the relative expansion of the crystals upon swelling has also been measured macroscopically and agrees with the neutron results to the limit of experimental error [7]. Second, the neutron results have much greater statistical certainty. A single neutron scattering experiment, with a beam width of several millimeters, samples several thousand vermiculite layers, and this type of measurement has been repeated many times with consistent results. In contrast, the present method has sampled, in total, only a few hundred vermiculite layers. Third, and most important, the neutron scattering measurements provide a direct *in situ* study of a gel in solution, without the need for the preparation of a replica.

Bearing these considerations in mind, there are only two plausible explanations we can give for the discrepancy. The first possibility is that the gels were not fractured exactly perpendicularly to the layers. In this case, the apparent spacing would be increased by a factor $1/\cos\phi$ relative to the true d-value, where ϕ is the angle between the fracture plane and that perpendicular to the layers. For values of 10, 20 and 30°, this factor is equal to 1.02, 1.06 and 1.15, respectively, so even if we made a very inaccurate fracture, the geometric correction would not account for the large expansion observed. It may be a contributory factor, but it cannot be the dominant one. The second possibility is that the gels expand during the rapid freezing process. Weak corroborative evidence for this explanation was found in the fact that ice crystals were detected in several of the micrographs, showing that the structure at room temperature could not have been maintained. It was also noted that, in the micrographs displaying the ice crystals, the spacing tended to be greater than average. Although there are insufficient statistics to prove this point, it does seem to indicate that freezing expands the structure.

We can rule out the possibility that the effect is due to the n-butylammonium ions. An electron micrograph of the supernatant fluid, taken under the same freeze-fracture conditions, was completely featureless. This is consistent with the thermodynamic behavior of simple n-butylammonium salt solutions; their enthalpy of solution is nearly equal to zero [8] and their partial molar volumes are nearly independent of concentration [9], implying that there are no special ion-solvent effects in the system. The necessary conclusion is that the cooling rate used in our experiments, of the order of 10^3 K/sec, was too low to prevent a major reorganization of the microstructure. This could have serious repercussions for data taken from electron microscopy studies of biological systems, which are necessarily aqueous macroionic systems.

We have now finished our experimental studies of the three-component clay-salt-water system. The remaining three chapters will be devoted to the corresponding four-component system with added polymer. Before leaving the three-component system, let us finally return to the thermodynamic phase transitions observed in the neutron and X-ray scattering experiments. A crucial feature of the discussion is that simple n-butylammonium salt solutions are well represented as ideal ions [8, 9]. The dissolution of such salts into water gives colligative properties such as depression of freezing point and elevation of boiling point that can be calculated from Debye-Hückel theory [3]. The thermodynamics of such two-component solutions are well understood. In adding a clay with a counterion identical to that of the simple salt solution, we create a three-component system whose thermodynamic properties have been less thoroughly investigated. Let us imagine holding c fixed at some convenient value, say $c = 0.01$ M, and consider the addition of clay to the salt solution as a pseudobinary system. In the sense that we can view the dispersal of the clay into the aqueous medium (i.e., gel formation) as analogous to the dissolution of salt into water, we would expect to see phenomena typical of the two-component salt-water system, like depression of freezing point, brought about by "dispersal" of the clay layers into the liquid solvent (water). It is clear from the neutron and X-ray experiments that the solid solute (i.e., the clay) does not dissolve in the solid solvent (ice) and that the solid solute does dissolve (i.e., gel) into the liquid solvent. The explanation for all of these phenomena is clearly a thermodynamic one; the secondary minimum state is thermodynamically stable over a wide range of $\{r, c, T\}$ conditions between a lower critical solution temperature (LCST) T_f ($\cong$ 0°C) and an upper critical solution temperature (UCST) T_c ($\cong$ 10 to 40°C for our usual range of salt concentrations).

The importance of the butylammonium vermiculite system as a model system for charged colloids lies mainly in the identification of the crystalline, or tactoid, phase with the primary minimum state ($d \approx 1$ nm) and the identification of the gel phase with the secondary minimum state ($d \approx 10$ to 100 nm) [10, 11]. I have previously noted *ad nauseam* that the reversibility, sharpness and reproducibility of the phase transition at the UCST show that the transition is a true thermodynamic one. All the same considerations clearly apply to the phase transition at the LCST described in this chapter [1]. In the former case, it was possible to directly measure the enthalpy change at the UCST, which was found to be approximately 5 J/g of crystalline vermiculite. It would be very difficult to measure the enthalpy change at the LCST directly because of the simultaneous occurrence of the freezing transition of the water, whose enthalpy change is three orders of magnitude greater per gram of water [12]. Nevertheless, from the diffraction experiments described here, the structure of the clay system changes in a very similar way at both phase transition temperatures.

We do have to be careful in the way we apply the definition of a phase to the n-butylammonium vermiculite system. According to Gibbs [13], a phase is any homogeneous and physically distinct part of a system that is separated from other parts of the system by definite boundary surfaces. Because the gel can be lifted out of the supernatant fluid on a spatula, it clearly justifies description as a phase in the latter sense, but it is inhomogeneous on the nanometer-to-micron (colloidal) length scale. It can only be defined as homogeneous on the macroscopic length scale. The same considerations apply to the tactoid phase.

I have also previously noted *ad nauseam* that the thermodynamic nature of the phase transitions in this system have profound implications for the theory of colloid stability. On the other hand, there are massive practical applications in stabilizing colloids against flocculation with the temperature [14]. For example, sedimentation problems in lakes, the role of drilling mud used in the petroleum industry, and the role of clay lining in landfill barriers at waste disposal sites are mainly affected by the phase behavior. Our results may be of indirect use in these problems as well as impacting directly on the behavior of rocks in freezing and thawing cycles in geological weathering processes. However, in many practical applications, especially in soil science, polymers are present in the aqueous clays. So far, we have looked only at charge stabilization of colloids. We now turn to look at what the model vermiculite system with added polymers can tell us about the other great class of colloids, the polymer-stabilized systems [14].

REFERENCES

1. Hatharasinghe, H.L.M., Smalley, M.V., Swenson, J., Hannon, A.C., and King, S.M., *Langmuir,* 16, 5562, 2000.
2. Hahn, T., Ed., *International Tables for Crystallography,* 4th ed, Kluver, Dordrecht/ Boston/London, 1995.
3. Debye, P.J.W. and Hückel, E., *Physik. Zeits.,* 24, 183, 1923.
4. McCarney, J. and Smalley, M.V., *Clay Min.,* 30, 187, 1995.
5. Kleijn, W.B. and Oster, J.D., *Clays Clay Min.,* 30, 383, 1982.
6. Robards, A.W. and Crosby, P., *Cryo-Letters,* 4, 23, 1983.
7. Braganza, L.F., Crawford, R.J., Smalley, M.V., and Thomas, R.K., *Clays Clay Min.,* 38, 90, 1990.
8. Krishnan, C.V. and Friedman, H.L., *J. Phys. Chem.,* 74, 3900, 1979.
9. Desnoyers, J.E. and Arel, M., *Can. J. Chem.,* 45, 359, 1967.
10. Smalley, M.V., *Mol. Phys.,* 71, 1251, 1990.
11. Smalley, M.V., *Langmuir,* 10, 2884, 1994.
12. Harrison, R.D., Ed., *Book of Data,* 2nd ed., Nuffield, Harmondsworth, Middlesex, U.K., 1973.
13. Gibbs, J.W., *Collected Works,* Yale University Press, New Haven, CT, 1948.
14. Everett, D.H., *Basic Principles of Colloid Science,* Royal Society of Chemistry, London, 1988.

11 The Effect of Adding Polymers to the Vermiculite System

Two books published in 1979 pointed out a need for experimental advances in two very different areas of polymer science. From the polymer physics point of view, de Gennes [1], predicting that a segregated model would apply in two-dimensional situations, hoped that future experiments using chains trapped in lamellar systems would be able to probe how the degree of chain interpenetration depends on dimensionality. The interlayer region in a lamellar clay gel is an obvious candidate for such a trapping experiment. From a more practical point of view, the importance and potential of the four-component clay-polymer-salt-water system in agricultural and industrial applications could hardly be overstated, being the central problem of soil science. As pointed out by Theng [2], "Progress in our understanding of this system is at present limited more by experimental than theoretical inadequacies." Although some progress has been made during the quarter of a century since these remarks, there remains a strong element of truth in this statement. We hope that the study described in this and the remaining two chapters will contribute significantly to improving the experimental situation by describing an investigation into the effect of adding polymers to our model clay system, the n-butylammonium vermiculites.

Adding polymers to a well-characterized clay colloid is potentially interesting both from the point of view of clay science and that of polymer science, but care is needed because the problem is a many-variable one. We have seen in chapters 1 to 10 that the three-component clay-salt-water system has been studied as a function of the set of five variables $\{r, c, T, P, p\}$ (sol concentration, salt concentration, temperature, hydrostatic pressure and applied uniaxial stress, respectively). Adding polymer to the system adds three new variables to be considered: the chemical nature of the polymer x, the molecular weight of the polymer M and the volume fraction of the polymer in the condensed matter system v. The variable x represents the composition of the side groups, the block structure, the stereoregularity and other features of the polymer. The variable M represents the number-average or weight-average molecular weight of the polymer and its polydispersity. The variable v is defined as the ratio of the polymer volume to the total volume occupied by the polymer, clay, salt and water.

Adding the new set of variables $\{x, M, v\}$ presents us with a total of eight variables to be considered, creating an enormous phase space to be investigated. We therefore proceeded cautiously, eliminating six of the variables in our first study, as follows. First, attention was restricted to the unstressed systems: namely, all the experiments were carried out at $P = 1$ atm and $p = 0$. Second, attention was restricted to the case $r = 0.01$ (the clay occupied 1% of the condensed matter system, as a volume fraction), $c = 0.1$ M; these two concentration variables determining the composition of the three-component clay-salt-water system were held fixed. Third, attention was restricted to x = PVME (poly(vinyl methyl ether)) and M = 18,000; a particular polymer was chosen for the study. The experiments were therefore reduced to studying the $\{v, T\}$ behavior of the system, a two-variable problem [3].

We note that the PVME-water system itself has an interesting $\{v, T\}$ phase diagram of LCST (lower critical solution temperature) type [4], but that the polymer and water were miscible over the temperature range we studied. We have seen before that $T_c = 14°C$ for the $r = 0.1$, $c = 0.1$ M system with no added polymer, so the range of interest was $8°C < T < 20°C$ to see whether or not the addition of polymer would change the phase transition temperature between the gel and tactoid states illustrated in Figure 10.5. For a PVME sample with a weight average molecular weight of 98,000, the LCST between PVME and water occurs at 33.0°C in the v range between 0 and 0.1 [4]. For the lower molecular weight PVME sample used, the LCST is shifted to higher temperatures, well above the range $8°C < T < 20°C$. In addition to the effect on T_c, the other obvious question to be answered was how the addition of polymer would affect the d-value in the gel phase.

The PVME had been synthesized by cationic polymerization in toluene at −78°C with boron trifluoride etherate as the initiator. The number average molecular weight (M_n) of the PVME, measured by gel permeation chromatography (GPC) in chloroform, was 27,000 in terms of the polystyrene equivalent, and the polydispersity ratio (M_w/M_n, where M_w is the weight-average molecular weight) was 1.40. We later used much more highly monodisperse polymers, with $M_w/M_n < 1.05$. Experiments on PVME samples with narrow molecular weight distributions ($M_w/M_n < 1.1$) in the molecular weight range between 500 and 4000 have shown that the absolute value of M_n obtained by ^1H NMR is equal to 0.67 times the polystyrene equivalent value obtained by GPC. If we assume that this factor holds for our sample, although it may depend on the GPC conditions (such as temperature, kind of column used and solvent) and the polymer structure (such as stereoregularity), then the true value of M_n is 18,000, corresponding to a degree of polymerization of 310.

Solutions of the required volume fraction of PVME were prepared by dissolving a known mass of the polymer ($\rho = 1.03$ g/cm^3) in a known volume of a 0.1 M n-butylammonium chloride solution, itself prepared by dissolving a known mass of n-butylammonium chloride in D_2O. The clay crystals were prepared as described previously [5]. After weighing, a single vermiculite crystal was placed in a quartz sample cell of dimensions $1 \times 1 \times 4.5$ cm, and an appropriate amount of the polymer solution (typically 2.5 cm^3) was added to prepare an $r = 0.01$ sample. The sample cells were identical to those used in the experiments on the LOQ instrument (ISIS, Didcot, U.K.) described in Chapter 5. As usual, the cells were sealed with parafilm and allowed to stand at 7°C for two weeks prior to the neutron scattering experiments to ensure that equilibrium swelling had been achieved [5].

The neutron-diffraction experiments were carried out at the JRR-3M Research Reactor, Japan Atomic Energy Research Institute (JAERI), Tokai-mura, using the

SANS-J instrument. A monochromatic beam of neutrons of wavelength $\lambda = 6$ Å was used, and the incident beam was collimated by passage through a 5 mm diameter circular aperture. The samples were mounted on a temperature-controlled four-position sample changer under the control of the instrument computer. Neutrons scattered by the gel samples were recorded on a two-dimensional area detector, with 128 × 128 pixels, situated 1.5 m behind the samples and covering the approximate Q-range between 0.02 and 0.2 Å^{-1}, corresponding to real-space correlation lengths in the range between 30 and 300 Å. We took samples with no added polymer ($v = 0$) as control samples, knowing that these should have a spacing of approximately 120 Å and give rise to diffraction effects within this range.

Two typical scattering patterns from gel samples are shown in Figure 11.1, which represents the raw data observed on the detector as intensity contours plotted as a function of the pixel row and column numbers. The geometry of the experiments was as shown in Figure 5.1a for the LOQ experiments on the corresponding three-component system, with the clay layers parallel to the ground. As described previously [5], the sample runs were corrected for background scattering so that the scattering patterns analyzed arose purely from the gels. The scattering pattern consists of two lobes of intensity above and below the plane of the layers in the gel. As shown in Figure 11.1a, for the pure aqueous sample (the control sample with no PVME added), almost all of the scattering falls within two cones of 60° azimuthal width, as observed previously [5]. The scattering from a PVME-added sample is shown in Figure 11.1b. The first remarkable feature of this pattern is that the cones of scattering have become narrower than those observed in the pure aqueous system, with almost all of the scattering now lying within a cone of azimuthal angle 50° rather than 60°. This effect was observed

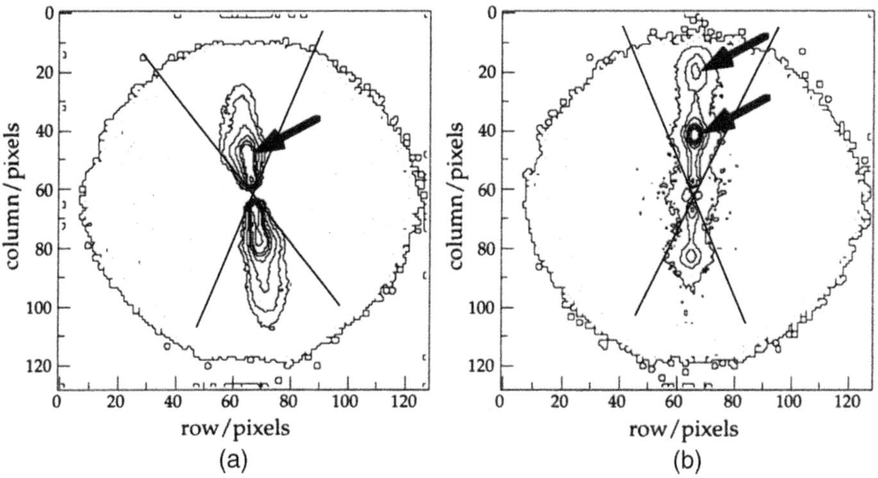

FIGURE 11.1 Contour plots of typical neutron-scattering patterns for $r = 0.01$ and $c = 0.1$ M gels at $T = 8$°C. (Figure legend shows the pixel codes.) (a) No added polymer; the arrow marks the first-order diffraction maximum at $Q_{max} = 0.5$ nm^{-1}. (b) With 1% PVME. The arrows mark the first maximum at $Q_{max} = 0.7$ nm^{-1} and the second maximum at $Q_{max} = 1.4$ nm^{-1}.

for all the polymer-added samples studied. Because the width of the cone is a measure of the mosaic spread of the clay plates, this shows that

1. The addition of polymer causes the clay plates to become more strongly aligned (more parallel) in the gel phase.

The existence of the intensity maxima within the cones of scattering have been our experimental cornerstone for measuring the well-defined separation between the clay plates. The second remarkable feature of a comparison between parts (a) and (b) of Figure 11.1 is that the diffraction pattern from the polymer-added sample is much sharper; it exhibits a more pronounced first-order diffraction maximum and a strong second-order diffraction maximum, which is rare for a pure aqueous sample [5]. This effect was also observed for all the PVME-added samples and means that

2. The addition of polymer causes the clay plates to be more regularly spaced in the gel phase.

It is also clear that the first-order diffraction maximum has shifted to higher Q as compared with the pure aqueous $r = 0.01$, $c = 0.1$ M sample. This was also a general effect and means that

3. The addition of polymer causes a decrease in the interlayer spacing between the clay plates in the gel phase.

To quantify this effect, the scattering patterns were radially summed within the appropriate azimuthal cones. One example of such a radial summation, for a sample containing 2% PVME ($v = 0.02$, $r = 0.01$, $c = 0.1$ M) is shown by the solid line in Figure 11.2, where the intensity has been given on a logarithmic scale.

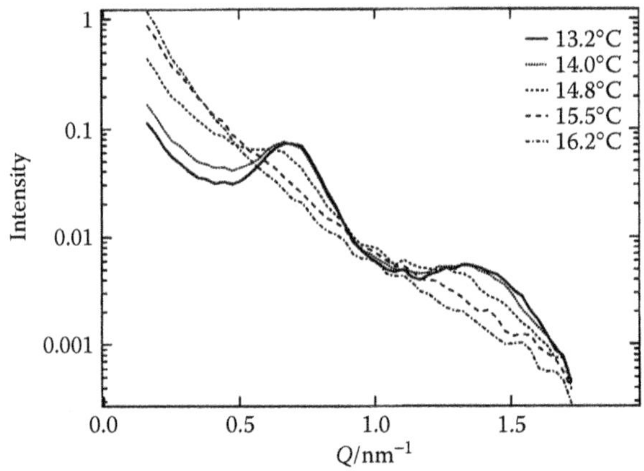

FIGURE 11.2 $I(Q)$ (arbitrary units) vs. Q plots obtained for one sample at $r = 0.01$, $c = 0.1$ M, $v = 0.02$ (2% PVME). Note the logarithmic intensity scale. The solid line shows the scan obtained at $T = 13.2$°C, and the dots, short dashes, long dashes and dot-dashes show those obtained at $T = 14.0$, 14.8, 15.5 and 16.2°C, respectively. The phase-transition temperature $T_c = 15 \pm 1$°C.

The solid line in Figure 11.2 shows a strong first-order diffraction maximum at $Q_{max} = 0.07$ Å^{-1} and a weaker second-order diffraction maximum at $Q_{max} = 0.14$ Å^{-1}, corresponding to a d-value of approximately 90 Å. Identical diffraction traces (not shown in Figure 11.2) were obtained at 1°C intervals between 8 and 13°C. At 14°C, there is a slight change in the pattern, followed by a profound change in the range between 14 and 16°C. This corresponds to the phase transition between the states illustrated in Figure 10.5b and Figure 10.5c. The phase transition temperature in this case is $T_c = 15 \pm 1$°C. The phase transition was always observed at this temperature, irrespective of the volume fraction of PVME in the system. We therefore conclude that

4. The addition of polymer has no effect on the phase transition temperature between the gel and tactoid phases of the clay system.

Between four and eight samples were studied at seven volume fractions up to $v = 0.04$, and the d-values are plotted as a function of v in Figure 11.3. The variation of the average d-value seems to be well represented by the exponential fit shown by the dotted line in Figure 11.3 as

$$d = 79 + 39\exp(-1.0v) \tag{11.1}$$

where d is expressed in Å and v is expressed as a percentage. It is highly suggestive that the d-value depends on polymer concentration in the same way as it depends on uniaxial stress, indicating that the polymers apply an effective stress to the system. We will exploit this similar functional dependence in detail in the next chapter.

By decreasing the sample-to-detector distance to 1.5 m and the wavelength of the incident beam to 3.57 Å, we were able to catch the peak at $Q = 0.324$ Å^{-1},

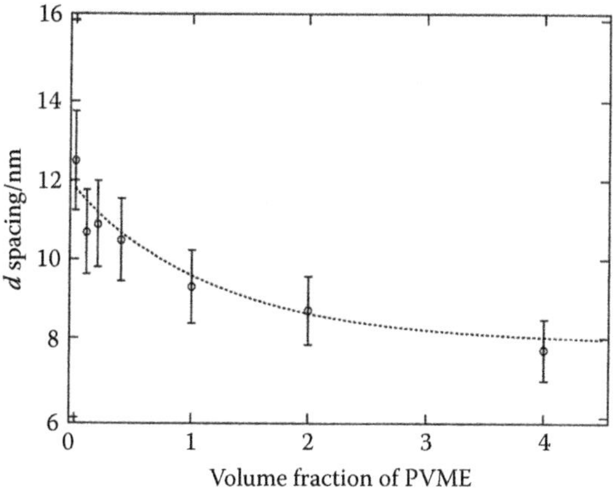

FIGURE 11.3 The average d-value (nm) as a function of the PVME volume fraction (%) for $r = 0.01$, $c = 0.1$ M, $T = 8$°C.

indicative of the 19.4 Å phase, at $T = 16°C$, for both the pure aqueous and polymer-added samples [6]. This identity means that

5. The spacing between the clay plates in the tactoid phase is not affected by the added polymer when the gels are collapsed by an increase in temperature.

When we submitted our results (items 1 to 5) to *Langmuir*, one of the reviewers described them as "mind-boggling." We agreed, and then we started our journey into the vast phase space of the polymer-added system. We continued to study unstressed samples (this will be the case for all the polymer-added samples studied in this book) and restricted our attention to the case $r = 0.01$ [6]. The five variables $\{c, T, x, M, v\}$ still contain myriad possibilities, limited only by the requirement that the polymers be water-soluble. To find our way around the maze of possibilities, one of our guiding principles was the "size" of the polymer molecules used. Brandrup and Immergut [7] give the end-to-end distance l of a PVME chain in solution as

$$l = 900M^{1/2} \times 10^{-4}\,\mathrm{nm}$$

Inserting $M = 18{,}000$ into his formula gives $l \cong 120$ Å, and dividing this number by $\sqrt{6}$ gives the radius of gyration $R_g \cong 50$ Å. Because the d-value of 120 Å in the pure aqueous system is composed of a clay plate of approximate thickness 10 Å and an interlayer spacing of approximate thickness 110 Å, it would seem that the first polymer we have studied should be able to "fit" into the interlayer region in the gel. Indeed, if we think of the end-to-end distance as giving an approximate size for the polymer molecule, then we had roughly matched this size to the size of the interlayer spacing, both being about 100 Å.

Although size matters, the first obvious question was whether or not the effects would be observed for other polymers. We saw similar effects for various uncharged polymers, including the PEO (polyethylene oxide) that later became our model system [6]. Using a PEO sample that fulfilled the approximate matching condition between polymer size and interlayer spacing, we obtained remarkably similar $I(Q)$ vs. Q plots for PEO-added and PVME-added samples (with the concentration variables fixed at $r = 0.01$, $c = 0.1$ M and $v = 0.01$, at $T = 8°C$), as shown in Figure 11.4. In both cases, the polymer-added samples had d-values of 90 Å. Results 1 to 3 above were clearly reproducible between the two systems, as shown in Figure 11.4 by the comparison with the scattering from a pure aqueous sample in the same experiment. The phase-transition temperatures were the same, too, and results 4 and 5 were thus also reproduced [6]. PEO and PVME have widely different chemical structures, so the results depend on a more general property of polymer physics rather than the chemical nature x of the polymer. When you are going to study a many-variable problem, you have to make some decisions. I decided to focus attention on x = PEO. PEO was an obvious candidate because (a) it is widely available as monodisperse samples of various molecular weights, (b) it is easy to handle and dissolve in water, and (c) the PEO-salt-water system is one of the best characterized polymer-salt-water systems [8–10].

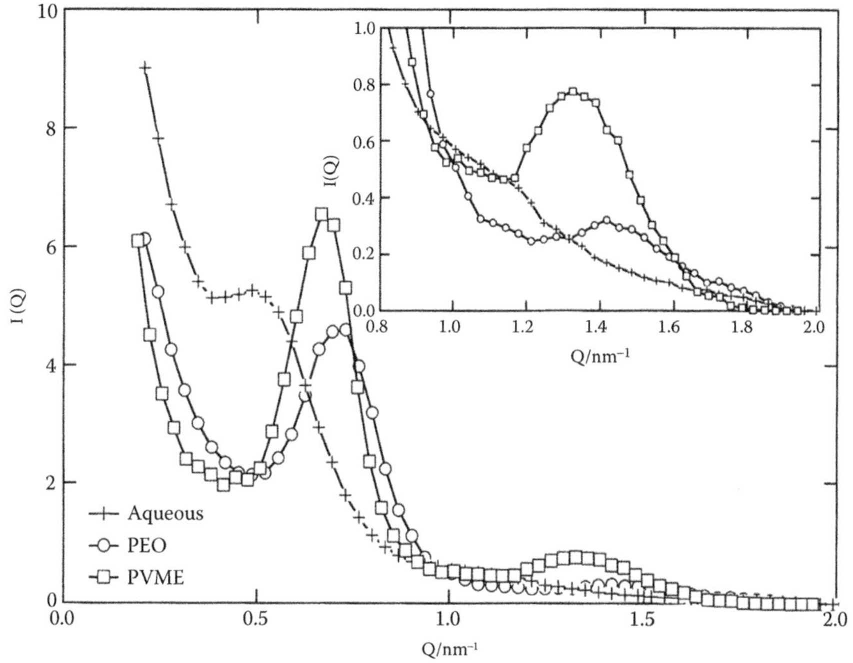

FIGURE 11.4 $I(Q)$ (arbitrary units) vs. Q plots obtained for $r = 0.01$ and $c = 0.1$-M gels at $T = 8°C$. The crosses, squares and circles denote the scattering from the pure aqueous system and those with $v = 0.01$ (1%) PVME and $v = 0.01$ (1%) PEO, respectively. The insets are the scattering on a magnified (×5) intensity scale, showing the clear second-order diffraction maxima for the polymer-added samples.

In starting systematic work on the vermiculite-PEO-salt-water system [11], we again proceeded cautiously. The agreement in d-values for the $r = 0.01$, $c = 0.1$ M, $v = 0.01$ PVME gels (at $T = 8°C$) and their $x =$ PEO counterparts could have been a fluke. It was obviously necessary to repeat the measurement of the d vs. v curve for the PEO-added system. We chose a polymer with a molecular weight $M = 18{,}000$, eliminating M from the comparison between the PVME-added and $x =$ PEO systems. The sample-to-sample variability study on the new model vermiculite-PEO system at $v = 0.02$ is shown in Figure 11.5, where the scattering patterns observed on the two-dimensional multidetector have been reduced to plots of $I(Q)$ vs. Q, as described previously [5]. It is clear that the scattering is reproducible between gels prepared from different pieces of vermiculite. All four samples exhibit a sharp first-order diffraction peak at $Q_{max} = 0.083$ Å^{-1}, with a weaker second-order diffraction peak (shown in the insets in Figure 11.5) at twice this Q-value. The d-value at this volume fraction is therefore equal to 76 Å.

The d-values obtained as a function of the polymer volume fraction v for the PEO-added system are shown in Figure 11.6. The values are remarkably similar to the corresponding curve for the PVME-added system shown in Figure 11.3, giving rise

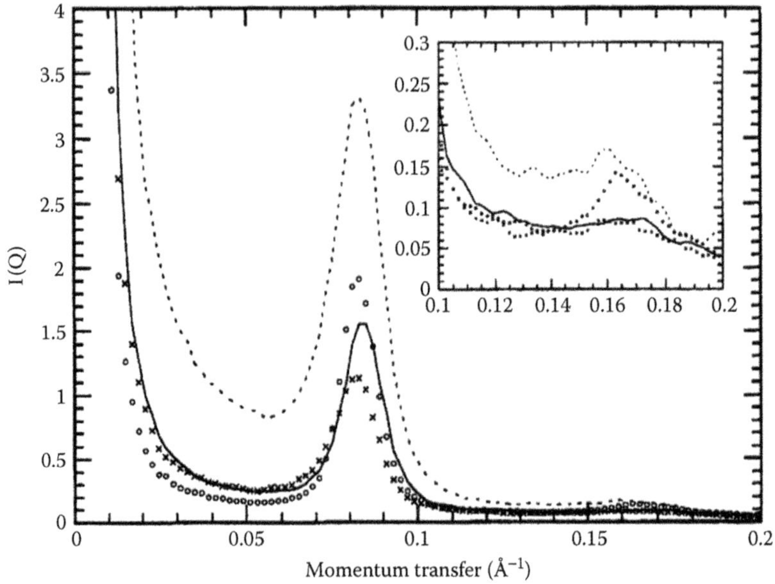

FIGURE 11.5 The scattering patterns obtained from four different samples at $r = 0.01$, $c = 0.1$ M, $v = 0.02$ (PEO), and $T = 7°C$. The insets show the weak second-order peak on an expanded scale.

to a similar exponential fit. This proves experimentally that two chemically different neutral polymers bring about a similar contraction of the gel phase. The effect of PEO on the phase transition temperature T_c between the gel phase and tactoid phase of the system was also the same as that observed for PVME; namely, there was no effect at

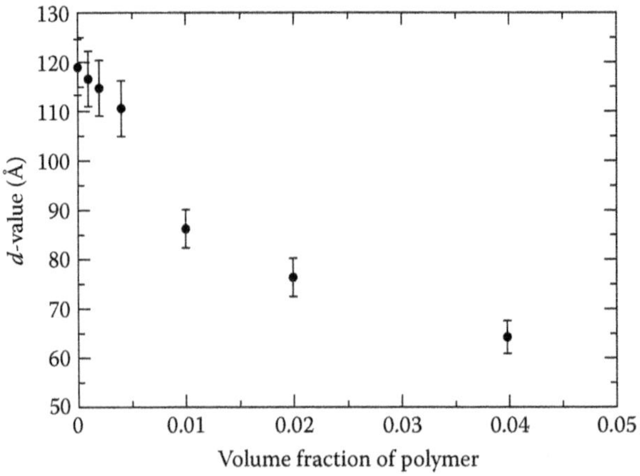

FIGURE 11.6 The average d-value (Å) as a function of the PEO volume fraction at $r = 0.01$, $c = 0.1$ M, and $T = 7°C$.

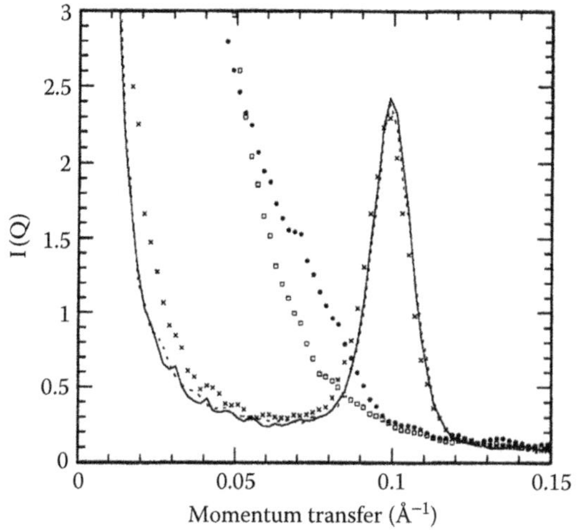

FIGURE 11.7 $I(Q)$ vs. Q plots obtained for one sample at $r = 0.01$, $c = 0.1$ M, $v = 0.04$ (PEO). The solid line shows the scan obtained at $T = 7$°C, and the dashes, crosses, open squares, and solid circles show those obtained at 10, 12, 14, and 16°C, respectively. The phase transition temperature $T_c = 13 \pm 1$°C.

all at volume fractions up to 4%. The diffraction traces obtained as a function of temperature for the 4% PEO sample are shown in Figure 11.7. It is clear that the gel peak disappears between 12 and 14°C, corresponding to $T_c = 13 \pm 1$°C. The same effect was observed for all the samples studied.

In Tokai-mura, we made a first attempt to determine the effect of varying the salt concentration c in PEO solutions at $c = 0.01$ M, and we found that the addition of a polymer with a molecular weight M of about 9,000 had had no discernible effect on the corresponding spacing of $d \cong 300$ Å in the $r = 0.01$ pure aqueous system [6]. We also investigated at $c = 0.01$ M with the $M = 18{,}000$ polymer on LOQ [11]. We used, in addition to the control samples at $v = 0$, three volume fractions at $v = 0.004$, $v = 0.01$ and $v = 0.04$. Three typical traces are shown in Figure 11.8. The weak diffraction effects in the region of $Q \cong 0.02$ Å^{-1}, corresponding to $d \cong$ 300 Å, are typical of $c = 0.01$ M gels without added polymer; in this case, there is no sharpening of the pattern and no contraction of the gel interlayer spacing.

The radius of gyration of a PEO molecule with $M = 18{,}000$ is approximately the same as that of a PVME molecule of the same molecular weight, namely $R_g \cong$ 50 Å [11]. This corresponds to a mean end-to-end distance of $l = 120$ Å. It is thus possible that the difference in behavior between the $c = 0.1$ M and $c = 0.01$ M cases is that in the former case the polymer molecules are able to form bridges between the clay plates, whereas in the latter the polymer molecules are too small to be able to do this. Of course, with 400 monomer units, a fully extended chain could have a length of 1000 Å or more, but such a stretched configuration is so entropically unfavorable with respect to a random walk of the polymer segments that such configurations will make a negligible contribution to the thermodynamic free energy

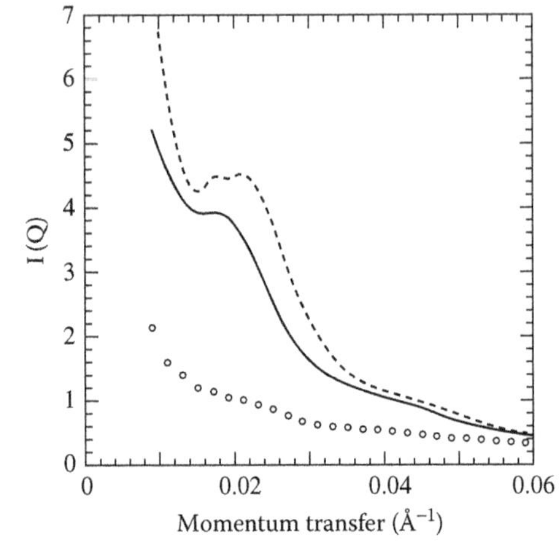

FIGURE 11.8 Scattering patterns obtained from gels prepared in 0.01 M n-butylammonium chloride solutions at T = 20°C. The dashes, solid line and circles correspond to 0.4, 1 and 4% added PEO, respectively.

of the system. The statistically favored configuration of the polymer is the random walk, and the studies described in the following chapter, when we vary M, will definitely indicate that the effective size of the polymer molecules vis-à-vis the size of the clay lamellar spacing is the end-to-end distance, equal to 120 Å in this case. We already suggested from our preliminary studies that bridging flocculation was a strong candidate to explain the phenomena [11], but before we jump to conclusions, let us consider various possible mechanisms.

There are four obvious possibilities for the state of a polymer molecule in and around a clay gel:

Site A: adsorbed onto the surface of a single plate
Site B: as free chains inside the gel phase
Site C: as free chains in the supernatant fluid surrounding the gel
Site D: as bridges between the vermiculite layers, adsorbed onto the surfaces of two neighboring layers

It was possible to estimate the relative contribution of site C by direct chemical analysis of the supernatant fluid, and such measurements are described in Chapter 12. We found that the concentration of the PEO inside the gel (sites A, B and D combined) was about one-half the concentration outside the gel [12]. The partial exclusion of polymer molecules from the gel phase must lead to some osmotic pressure due to the excess molecules in the supernatant fluid.

It is well known that the presence of an excess of nonadsorbed polymer can result in flocculation of colloidal particles by the so-called depletion flocculation mechanism [13]. In the clay-PEO system, the excess PEO molecules in the supernatant fluid would exert an osmotic pressure on the gel, and the effect should be similar to that of applying an external pressure to the gel; indeed, because the pressure is applied from the fluid to the gel (not to the condensed matter system as a whole), the effect should be similar to that of uniaxial stress, which we studied in detail in Chapter 3. As we saw there, uniaxial stress sharpens up the diffraction patterns and decreases the *d*-values, just as we have seen here as effects of polymer addition. An osmotic pressure mechanism could describe all the results we have seen so far.

Other mechanisms are also possible. Inside the gel, we might expect that A and D would be the most favored sites. Adsorption isotherms for neutral polymers on the more widely studied montmorillonite clays invariably show high affinity of the polymer for the surface [2], and the conventional wisdom [14] is that because a large number of solvent molecules must be desorbed to accommodate a single polymer molecule, the translational entropy so gained by the system provides a strong driving force for polymer adsorption. This driving force outweighs the loss of conformational entropy of the polymer and favors A- and D-type adsorption compared with site B, where conformational entropy loss for the polymer inside the gel would not be compensated by solvent entropy gain. Both A- and D-type adsorption could lead to displacement of small ions from near the surface, leading to an increase in the effective surface charge, which would again produce a decrease in the *d*-value and an increase in the strength of binding between the plates [15]. Type D adsorption implies another possibility, that of bridging flocculation [13]. This phenomenon is exploited in water purification, where the addition of a few parts per million of a high molecular weight polyacrylamide leads to flocculation of the remaining particulate matter in the water. In this scenario, it is supposed that the two ends of a polymer chain adsorb on separate particles and draw them together. Such a drawing force could also explain our results. Yet another possible mechanism is that the addition of polymer to the system could perturb the distribution of salt between the gel and the supernatant fluid [5], again affecting the plate–plate interaction. Before considering the $\{r, c, v, M, T\}$ variations that unraveled these possible mechanisms, I wish to pause to consider the area of clay surface available for polymer adsorption.

To estimate the degree of polymer adsorption, we first calculate the total layer surface area of the vermiculite in the crystalline state. The formula for the dry sodium Eucatex sample considered in Chapter 1 can be abbreviated as

$$\text{Layer}^{1.3-}\ 1.3\text{Na}^{+} \tag{11.2}$$

where the unit cell weight of 807 is composed of 777 from the vermiculite layer and 30 from the 1.3 charge balancing interlayer sodium cations. The dry n-butylammonium Eucatex therefore has the formula

$$\text{Layer}^{1.3-}\ 1.3\text{C}_4\text{H}_9\text{NH}_3^{+} \tag{11.3}$$

and the contribution of 96 from the 1.3 n-butylammonium ions increases the unit cell weight to 873. We recall that the dry n-butylammonium Eucatex has a *d*-value of 1.49 nm and that the wet n-butylammonium Eucatex has a *d*-value of 1.94 nm. Although the in-plane dimensions of the unit cell are slightly dependent on the type of isomorphous substitutions in the lattice, it seems reasonable to take the standard value of 0.515×0.89 nm^2 for the surface area of the unit cell [16]. This determines the number of water molecules in the wet crystalline state via the density: the volume of the unit cell is 0.89 nm^3, and this must contain 996 a.u. to reproduce the observed density $\rho = 1.86$ g cm^{-3}. The water therefore contributes 123 a.u. per unit cell, and the formula for the wet n-butylammonium Eucatex is

$$\text{Layer}^{1.3-}\ 1.3\{C_4H_9NH_3 \cdot 5.3H_2O)\}^+ \tag{11.4}$$

You may think this is a crude calculation, but we have an independent test of the formula yielded here because of our fit to the scattering-length density along the z-axis of the mineral, described in Chapter 9 [17]. This gave approximately five water molecules per counterion, in excellent agreement with the formula.

The calculation of the surface area is now straightforward. Each cell has a surface area of 0.515×0.89 nm^2 on each side, and 996 g of the clay contains 6.02×10^{23} cells (Avogadro's number). Thus, the total surface area of 1 g of n-butylammonium vermiculite is

$$\left(\frac{1}{996}\right) \times 6.02 \times 10^{23} \times 0.515 \times 0.89 \times 2\,\text{nm}^2\text{g}^{-1} = 554\,\text{m}^2\text{g}^{-1} \tag{11.5}$$

which corresponds to 1030 m^2/cm^3 of clay. It is unlikely that the third figure in this number is truly significant, so we take the surface area of 1 cm^3 of vermiculite to be 1000 m^2, a convenient number for considering the polymer adsorption.

The calculation of the area occupied by 1 cm^3 of adsorbed polymer is less clear-cut. The true segment-density profile of a polymer in contact with a clay surface is supposed to consist of trains, loops and tails [14]. We will see what it really looks like in Chapter 13. Because we have no means of knowing *a priori* what proportion of the segments will be in the trains occupying the surface, we turn to empirical observations of the H-type (high affinity) adsorption isotherms for polymer adsorption on clay surfaces [2]. These refer to the site A adsorption in our description, as the measurements are generally performed for individually dispersed clay layers. We can therefore gain a rough upper estimate for type A adsorption from the data collected by Theng [2] for the plateau value of poly(vinyl alcohol) (M = 70,000) adsorption on clay surfaces. This was found to be remarkably similar for many 2:1-type layer silicates, ranging from 1.1 to 1.3 mg m^{-2}. (One milligram per square meter seems to be a rule of thumb accepted in the world of polymers at interfaces.) This corresponds to about 1 cm^3 of polymer absorbed into 1 cm^3 of vermiculite in our case, or complete coverage at $v = 0.01$ for H-type adsorption at $r = 0.01$. It would seem that type A adsorption should lead to saturation of the surface at equal volume fractions of polymer and clay, but at $r = 0.01$, the contraction of the d-value continues up to at least $v = 0.04$. This throws doubt on the hypothesis that type A adsorption underlies the phenomena and points to the likelihood that type D adsorption, where the polymer forms "bridges" between adjacent layers, must be dominant within the gel.

We finally remark on the insensitivity of T_c to the addition of neutral polymers and its sister result that the c-axis d-value of 19.4 Å in the tactoid regions in samples with added polymer is equal to that in the pure aqueous system. We later found that the intensity pattern of the scattering from the tactoids was identical in both cases [12], proving that the polymer molecules, although trapped between the clay layers in the gel phase, are expelled from the crystalline regions. Of course, whether we have type A or type D adsorption inside the gel, this implies that adsorbed trains of polymer segments can desorb from the clay surfaces more rapidly than the clay plates can rearrange themselves. Given the relative masses of the macroions (which are of macroscopic or mesoscopic dimensions in the x–y plane) and the PEO molecules, this is not a surprising result. It suggests that we are again, as with the three-component system, looking at equilibrium effects and that the clay–polymer gels have a well-defined equilibrium structure [12].

REFERENCES

1. de Gennes, P.G., *Scaling Concepts in Polymer Physics,* Cornell University Press, Ithaca, NY, 1979.
2. Theng, B.K.G., *Formation and Properties of Clay-Polymer Complexes,* Elsevier, Amsterdam, 1979.
3. Jinnai, H., Smalley, M.V., Hashimoto, T., and Koizumi, S., *Langmuir,* 12, 1199, 1996.
4. Tanaka, H., *Phys. Rev. Lett.,* 70, 53, 1993.
5. Williams, G.D., Moody, K.R., Smalley, M.V., and King, S.M., *Clays Clay Min.,* 42, 614, 1994.
6. Smalley, M.V., Jinnai, H., Hashimoto, T., and Koizumi, S., *Clays Clay Min.,* 45, 745, 1997.
7. Brandrup, J. and Immergut, E.H., *Polymer Handbook,* Wiley, New York, 1989.
8. Briscoe, B., Luckham, P., and Zhu, S., *Macromolecules,* 29, 6208, 1996.
9. Crowther, N.J. and Eagland, D.J., *J. Chem. Soc., Faraday Trans.,* 92, 1859, 1996.
10. Polverari, M. and van den Ven, Th.G.M., *J. Chem. Phys.,* 100, 13687, 1996.
11. Hatharasinghe, H.L.M., Smalley, M.V., Swenson, J., Williams, G.D., Heenan, R.K., and King, S.M., *J. Phys. Chem. B,* 102, 6804, 1998.
12. Smalley, M.V., Hatharasinghe, H.L.M., Osborne, I., Swenson, J., and King, S.M., *Langmuir,* 17, 3800, 2001.
13. Everett, D.H., *Basic Principles of Colloid Science,* Royal Society of Chemistry, London, 1988.
14. Fleer, G.J., Cohen Stuart, M.A., Scheutjens, J.M.H.M., Cosgrove, T., and Vincent, B., *Polymers at Interfaces,* Chapman and Hall, London, 1993.
15. Sogami, I.S., Shinohara, T., and Smalley, M.V., *Mol. Phys.,* 76, 1, 1992.
16. van Olphen, H., *An Introduction to Clay Colloid Chemistry,* 2nd ed., Wiley, New York, 1977.
17. Williams, G.D., Skipper, N.T., Smalley, M.V., Soper, A.K., and King, S.M., *Faraday Discuss.,* 104, 295, 1996.

12 The Mechanism and Strength of Polymer-Bridging Flocculation

There are two basic mechanisms for the stabilization of colloids: charge stabilization and steric stabilization. We had a good look at the former in Chapters 1 to 10. The latter is due to the adsorption of polymers onto the surface of colloidal particles, and we started to look in the previous chapter at what happens when PEO is added to our model colloid. In this chapter we will describe much more systematic and widespread studies of the four-component clay-polymer-salt-water system and propose a new model of polymer bridging flocculation [1]. This phenomenon, where high molecular weight polymers adsorb on separate particles and draw them together, has many industrial applications in a large variety of areas, such as in the preparation of paints and papers, in the stabilization of drilling fluids, and in water purification, where the flocs are used for the removal of unwanted particles. However, despite its widespread occurrence, the physical origin of polymer-bridging flocculation has been poorly understood.

The mechanism at work in the vermiculite-PEO system was elucidated by measuring d-values in the gel phase on the LOQ instrument (ISIS, Didcot, U.K.) as a function of systematic variation of $\{r, c, v, M\}$ (sol concentration, salt concentration, volume fraction of the polymer in the condensed matter system, molecular weight of the polymer). The crucial breakthrough was when we varied M with the concentration variables (and, of course, $p = 0$, $P = 1$ atm, $T = 5°C$, and $x =$ PEO) fixed. The obvious first point to choose was $r = 0.01$, $c = 0.1$ M, $v = 0.02$, where we had already observed a contraction from $d = 120$ Å in the pure aqueous system (this result is henceforth described as $d_0 = 120$ Å, with a subscript zero to distinguish the system with no added polymer) to $d = 76$ Å with 2% PEO of a molecular weight of 18,000 (see Figure 11.5). Initially [2], we chose seven molecular weights in the range between 1,000 and 2 million, as shown in Table 12.1.

To show the data in a form suitable for analyzing the mechanism of the contraction, the molecular weight M has been converted into a mean end-to-end distance of the polymer. Using data given in reference [3], we can derive the approximate empirical relation

$$l = \frac{M^{0.57}}{1.84} \tag{12.1}$$

between the mean end-to-end distance l (in Å) and molecular weight M of PEO dissolved in water. Thus, the chosen molecular weights correspond to mean end-to-end distances

TABLE 12.1
Molecular Weights *M* and Mean End-to-End Distances *l* of the PEO Molecules

Molecular Weight, *M* (a.u.)	Mean End-to-End Distance, *l* (Å)
1,000	28
4,000	61
18,000	140
75,000	330
300,000	720
1,000,000	1,400
2,000,000	2,100

in the range 28 Å < l < 2100 Å. Equation 12.1 is an empirical relationship that corresponds approximately to that expected for a self-avoiding random walk of the polymer segments in the solution. We will see what the polymers between the clay plates look like in the next chapter. For now we note that Table 12.1 gives only approximate values for the "size" of the polymer molecules.

Selected results from our study of molecular-weight variations at $r = 0.01$, $c = 0.1$ M, $v = 0.02$ in the gel phase at $T = 5$°C are shown in Figure 12.1. There is a dramatic

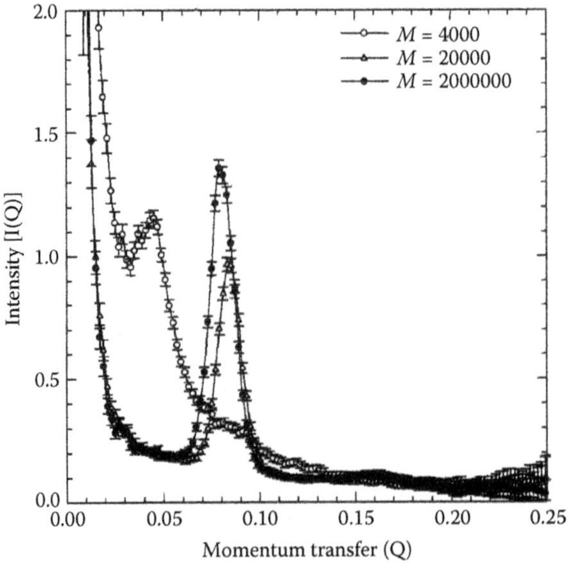

FIGURE 12.1 Molecular weight variations at $r = 0.01$, $c = 0.1$ M, $v = 0.02$ in the gel phase at $T = 5$°C. The open circles, triangles and closed circles show the patterns for $M = 4{,}000$, $M = 20{,}000$ and $M = 2$ million, respectively.

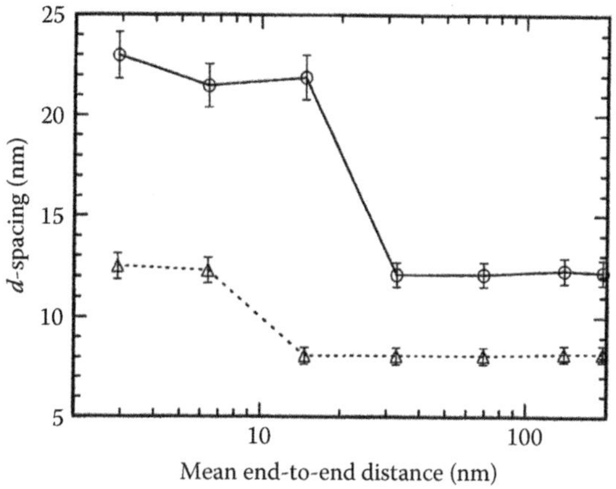

FIGURE 12.2 The d-value as a function of the mean end-to-end distance of the polymer in 0.1 M (triangles) and 0.03 M (open circles) n-butylammonium chloride solutions. The volume fraction of the polymer was 2% in both cases. The lines between the data points are drawn as a guide for the eye.

change in behavior between the two lower molecular weights, M = 4,000 and M = 18,000, but hardly any change at all when M is changed by two orders of magnitude thereafter, from approximately 20,000 to 2 million. In all three cases, clear diffraction effects are observed, with d-values of 130 Å for the M = 4,000 polymer and d = 75 Å for the higher molecular weight polymers. The d-values obtained for all the polymers studied at c = 0.1m are shown by the lower set of points in Figure 12.2.

The results in Figure 12.2 effectively rule out the depletion flocculation mechanism for the contraction between the clay layers upon polymer addition. As we increase the molecular weight from 20,000 to 2 million (×100) at constant volume fraction (2%), the number of polymer molecules in the system decreases by a factor of 100. Osmotic pressure is a colligative property by definition, so it likewise drops by a factor of the order of magnitude of 100 over this interval, but the d-value remains constant. It is therefore not making a significant contribution to the contraction mechanism even though there is an excess of polymer molecules in the supernatant fluid. Instead, the discontinuous change in behavior at a molecular weight corresponding to a polymer mean end-to-end distance roughly equal to the interlayer spacing strongly suggests an important role for the bridging flocculation mechanism. Below this critical value for the molecular weight, which we label M_b (the molecular weight at which bridging occurs), the "ordinary" clay layer spacing d_0 is observed; above it, the contracted value is seen. In the former case, the polymer would be too small to bridge between the clay plates without severe loss of conformational entropy; in the latter, bridging can occur.

One of the major advantages of using the n-butylammonium vermiculite system for polymer adsorption experiments is that we can use the salt concentration to control d_0. Our preliminary results at $c = 0.03$ M (and $r = 0.01$, $v = 0.02$) are also shown in Figure 12.2. For a larger value of d_0, the bridging molecular weight increases, this time lying in the range $18{,}000 < M_b < 75{,}000$. It was obviously desirable to determine M_b as a function of d_0 more accurately. One of the major advantages of using PEO as the added polymer is that it can easily be obtained as monodisperse samples of a wide range of molecular weights, giving us almost continuous control of the polymer end-to-end distance as a variable via M_b. At $c = 0.03$ M, $v = 0.02$, $r = 0.01$, two new M values were studied, $M = 30{,}000$ and $M = 59{,}000$. In both cases, all four samples gave clear gel peaks with contracted spacings in the bridging regime. The bridging molecular weight was therefore found to be in the range $18{,}000 < M_b < 30{,}000$, corresponding to mean end-to-end polymer distances l between 140 Å and 190 Å. Because $d_0 = 190$ Å in this case, an approximate matching condition for d_0 and l is again suggested, with $1.0l < d_0 < 1.4l$ as the experimental result. We likewise refined our range for the crossover at $c = 0.1$ M, $v = 0.02$ by varying M as 7,000, 9,000 and 12,000. In this case, all 12 samples gave clear diffraction effects with d in the range between 70 and 75 Å in the bridging regime. The bridging molecular weight was therefore found to be in the range $4{,}000 < M_b < 7{,}000$, corresponding to mean end-to-end polymer distances l between 60 and 85 Å. Because $d_0 = 110$ Å in this case, the crossover occurs between l values in the range between $1.3l$ and $1.8l$. Combining this with the range obtained at $c = 0.1$ M, we find that the crossover actually occurs at separations somewhat greater than the mean end-to-end distance, at around $1.3l$ to $1.4l$.

In the previous paragraph we remarked on a couple of the advantages of the vermiculite-PEO system for studying the mechanisms of polymer interactions. Another big advantage is that the force-distance curves between the clay layers are known for a wide range of salt concentrations (and therefore d_0), as we saw in Chapter 3. This means we can map any contractions due to added polymer onto an effective drawing force, thereby obtaining quantitative information about the interaction. We have already shown the d vs. v curve obtained at $c = 0.1$ M, $r = 0.01$ (see Figure 11.6), which could be mapped onto the corresponding d vs. p uniaxial stress curve [4]. Before we do this, let us consider what happens when we measure the d vs. v curves at $c = 0.03$ M and $c = 0.01$ M. Obviously our mapping will be the more accurate if we measure the contractions with added polymer over as wide a range as possible.

At $c = 0.01$ M, the interlayer spacing $d_0 = 330$ Å, so obviously a high molecular weight polymer is required, with the matching condition $d_0 \approx l$ occurring for $M = 75{,}000$. In fact, M_b was estimated to be about 75,000 in this case [1], but a molecular weight of 2 million was chosen for the systematic study. We shall see later that this data was vital to test our quantitative model of bridging flocculation, but it was a difficult experiment for two reasons. First, the pure aqueous d-value of $d_0 = 330$ Å gives rise to a first-order diffraction effect at $Q \cong 0.02$ Å^{-1}, near the limit of the Q-range of LOQ. Second, there is greater sample-to-sample variability for the dilute gels [5], so clear-cut results are more difficult to obtain, and some contradictions between different data sets were found. However, the fact that the polymers sharpen up the scattering patterns and reduce the variability in the d-values [6] made the measurements feasible, as shown in Figure 12.3. Part (a) of Figure 12.3 shows our

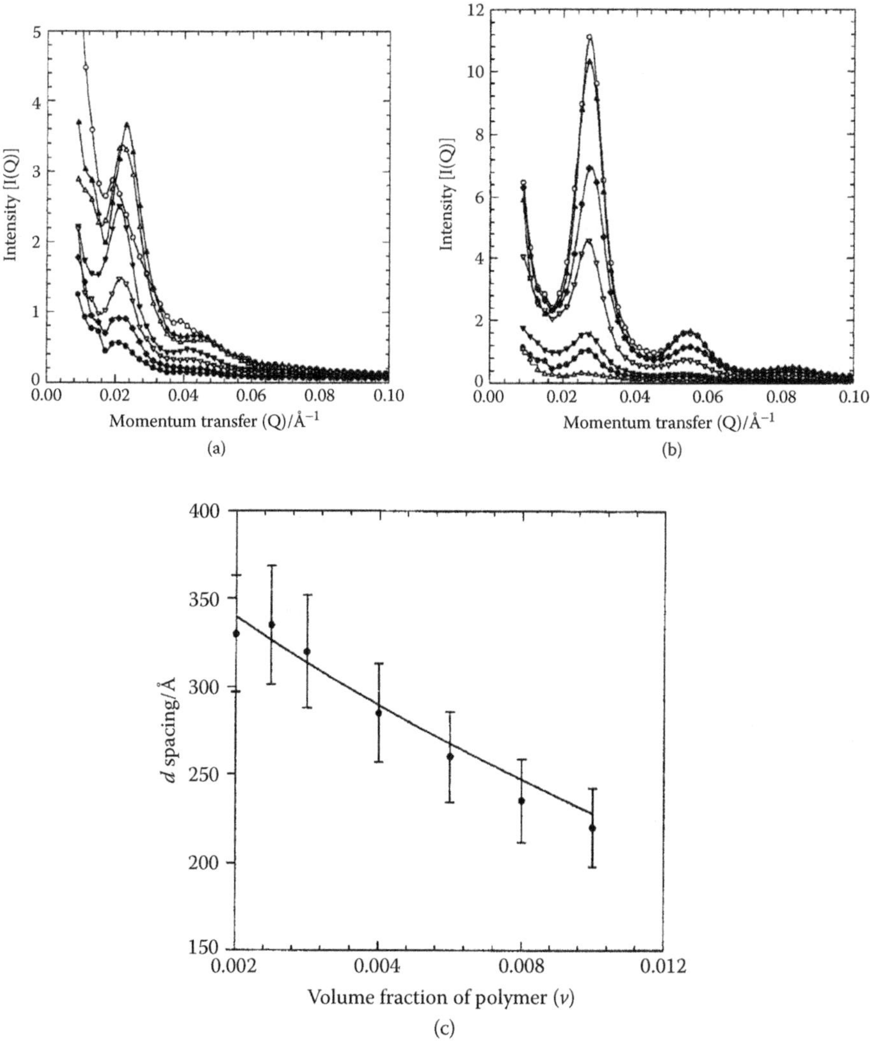

FIGURE 12.3 Parts (a) and (b) show sample-to-sample variability studies, the different symbols denoting patterns obtained from different pieces of gel under identical conditions. (a) Diffraction traces obtained from seven samples in the gel phase at T = 5°C. The PEO molecular weight was 2 million, and the concentration variables were fixed as r = 0.01, c = 0.01 M, v = 0.004. (b) Diffraction traces obtained at v = 0.008, with the other variables as defined in part (a). (c) Plot of the d-values obtained as a function of v at M = 2 million, r = 0.01, c = 0.01 M, T = 5°C.

sample-to-sample variability study at v = 0.004, and (b) shows it at v = 0.008. These were easily our best data sets for single d vs. v measurements at c = 0.01 M, r = 0.01. For lower volume fractions, much greater variability was seen, as shown in Table 12.2. Nevertheless, the pure aqueous separation d_0 = 330 Å was observed

TABLE 12.2
***d*-Values Obtained as a Function of *v* for Two Separate Experiments at *r* = 0.01, *c* = 0.01 M, *T* = 5°C, *M* = 2 million**

v	*d* (observed) (Å)	*d* (observed) (Å)	*d* (average) (Å)
0 (pure aqueous)	—	—	330
0.001	350, 380, 270	—	330
0.002	275, 325	350	320
0.004	295, 320, 270	280, 280, 280, 280	285
0.006	—	260, 260, 260, 260	260
0.008	235, 230, 230	235, 235, 235	235
0.01	—	220, 220, 220	220
0.015	Crowded	Crowded, 195	195
0.02	—	Crowded	Crowded

for the control samples, and the complete *d* vs. *v* curve at $c = 0.01$ M has good statistics. It is shown in Figure 12.3c.

A noteworthy feature of Figure 12.3c is that there is no data for volume fractions greater than $v = 0.01$. Samples with higher volume fractions (apart from one anomalous sample at $v = 0.015$) did not give any Bragg peaks at all, and these are described as "crowded" in Table 12.2. We coined this expression because we believe that the disappearance of the Bragg peaks occurs as a result of the interlayer becoming too crowded to accommodate further bridges as the amount of polymer in the system is increased. The volume fraction at which crowding occurs, v_c, depends strongly on the electrolyte concentration *c* (and hence d_0). The values for v_c obtained at $c = 0.03$ M and $c = 0.1$ M were about 4% and 9%, respectively, as shown in Table 12.3. The crowding transition with respect to volume fraction was as sharp as the bridging transition with respect to molecular weight. The volume fractions of the bridging polymers were varied in small steps, as shown in Figure 12.4. In Figure 12.4a, it is obvious that the well-defined *d*-value suddenly collapses between

TABLE 12.3
Crowding Volume Fractions v_c, Crossover Molecular Weights M_b, and Crossover End-to-End Distances l_b as Functions of the Salt Concentration *c* and Pure Aqueous *d*-Value d_0

c (M)	$d_0 - 10$ (Å)	v_c	M_b	l_b (Å)
0.01	320	0.01	$M_b < 75{,}000$	$l_b < 330$
0.03	180	0.04	$18{,}000 < M_b < 30{,}000$	$140 < l_b < 190$
0.1	110	0.09	$4{,}000 < M_b < 7{,}000$	$60 < l_b < 85$

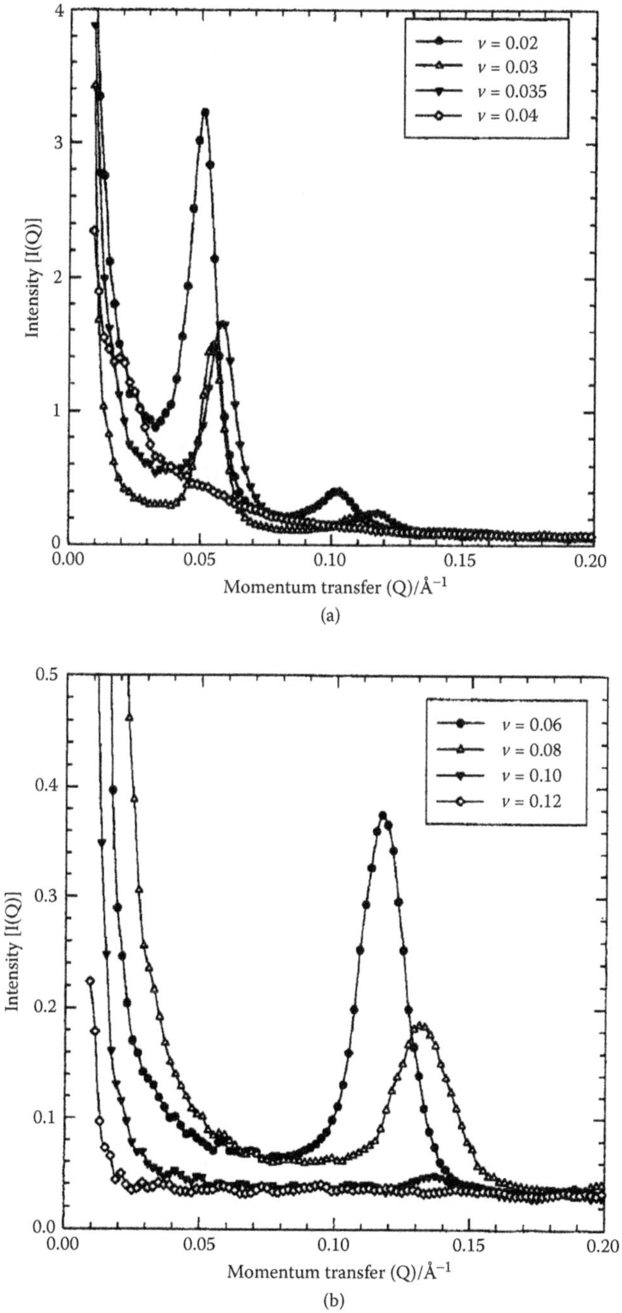

FIGURE 12.4 (a) Example diffraction traces at $r = 0.01$, $c = 0.03$ M, $T = 5°C$, $M = 75{,}000$ (bridging) with v varied in the range between 0.02 and 0.04. The crowding volume fraction v_c is between $v = 0.035$ and $v = 0.04$. (b) Example diffraction traces at $r = 0.01$, $c = 0.1$ M, $T = 5°C$ for $M = 18{,}000$ (bridging) with v varied in the range between 0.06 and 0.12. In this case, $v_c = 0.09 \pm 0.01$.

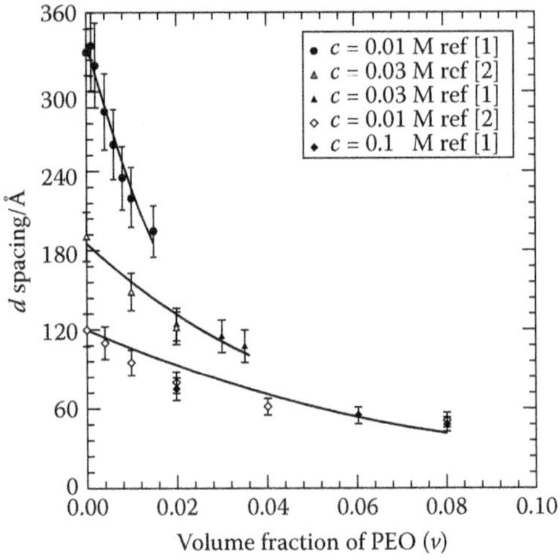

FIGURE 12.5 Plot of d-values obtained as a function of v at $r = 0.01$, $T = 5°C$ for the three salt concentrations $c = 0.01$ M, $c = 0.03$ M, and $c = 0.1$ M. The open symbols correspond to the results in reference [2], and the closed symbols are the results in reference [1]. The lines between the points are drawn as a guide for the eye.

$v = 0.035$ and $v = 0.4$; in Figure 12.4b the transition occurs between $v = 0.08$ and $v = 0.10$. The data points at $v = 0.06$ and $v = 0.08$ enabled us to extend our previously observed d vs. v curve at $c = 0.1$ M, and the effect of lower volume fractions than $v = 0.02$ at $c = 0.03$ M were also measured, giving the complete data set shown in Figure 12.5. Of course, all three curves terminate at v_c. We will return to the interpretation of crowding in the following chapter, when we have looked at the polymer structure. The most urgent problem is to find a theory that will explain the data in Figure 12.5.

Before looking for a quantitative interpretation, we need to know the concentration of the PEO molecules in the gel phase. As we saw in Chapter 5, the salt concentration inside the gel is lower than that in the supernatant fluid by a constant factor of 2.8 [5]. In Chapter 5, we also saw that it was much easier to measure the distribution of the salt between the two phases when they have roughly equal volumes, and that it is easy to arrange for this experimentally by choosing $r = 0.1$, $c = 0.1$ M. Likewise with the polymer distribution, it is more convenient to work at higher sol concentrations, and our main study will be at $r = 0.1$, $c = 0.1$ M. This means we have an intermediate step to take in linking our neutron scattering results to macroscopic measurements of v_i, the volume fraction of PEO inside the gel. Namely, we have to perform neutron scattering measurements at $r = 0.1$ with added PEO. Of course, in doing this, we are not going to try to cover the massive $\{c, v, M\}$ space we have covered at $r = 0.01$. Instead, we restrict ourselves to a couple of core studies to see if the essential phenomena are the same.

Because we had extensive data at $r = 0.01$, $c = 0.1$ M and $M = 18{,}000$ (bridging), the main study of the effect of sol concentration on the swelling was made under these

TABLE 12.4
Average *d*-Values Obtained at Higher Sol Concentrations, with $c = 0.1$ M, $T = 5°C$, $M = 18{,}000$

v	*d* (average) (Å) at $r = 0.1$	*d* (average) (Å) at $r = 0.3$
0.01	77	70
0.02	69	—
0.04	58	52
0.08	46	—
0.12	Crowded	—
0.16	—	Crowded

$\{c, M\}$ conditions, with the two high sol concentrations $r = 0.1$ and $r = 0.3$ studied for several volume fractions v in the range between 0 and 0.12. For the batch used in these experiments, the control on the pure aqueous $v = 0$ system gave $d_0 = 130$ Å at $r = 0.01$ (slightly above the average value), $d_0 = 110$ Å at $r = 0.1$ and $d_0 = 75$ Å at $r = 0.3$. Typical results of the effect of PEO addition are shown in Figure 12.6, and the average *d*-values obtained from the traces are given in Table 12.4. The contraction of the gel phase at $r = 0.1$ is similar to that observed at $r = 0.01$, with the overall *d*-values lower. It is noteworthy that the Bragg peak disappears at a similar volume fraction, $v \cong 0.10$, at $r = 0.1$ as it does at $r = 0.01$. The crowding volume fraction therefore seems to be fairly insensitive with respect to *r*. At $r = 0.3$, our grid of points was too coarse to be able to say more than $v_c = 0.10 \pm 0.06$, a compatible result. The important comparison is between the $r = 0.01$ and $r = 0.1$ results, which are plotted in Figure 12.7a. This shows that the *d* vs. *v* curves are approximately parallel in the two cases, showing that the same mechanism is at work. The magnitude of the contraction is also what is expected from the effect of increasing *r* at fixed *c* in the three-component clay-salt-water system [5]. It is also instructive to plot *d* as a function of *r* at a fixed polymer volume fraction *v*, as shown in Figure 12.7b. The polymer clearly exerts a similar drawing force throughout the two-phase region.

The concentration of the polymer in the supernatant fluids was measured by gel permeation chromatography (GPC) with refractive index detection. Measurement of the refractive index of a polymer solution is a widely used method for the determination of polymer concentrations, which we here express as volume fractions. After being in contact with the vermiculite for two weeks, the supernatant fluids containing PEO can be contaminated by other particles, for example, silica particles dissolving from the edges of the clay, that affect the refractive index. Taking advantage of the narrow molecular weight distribution of the PEO molecules ($M_w/M_n = 1.02$), we were able to use GPC to obtain a pure solution of the polymer for analysis by differential refractometry.

The experiments were carried out at Unilever Research Port Sunlight Laboratory (Wirral, U.K.) using a Hewlett-Packard HP1100 modular liquid chromatograph connected to a Water 410 differential refractometer. A Tosoh TSK GMPW column and 0.3 M $NaNO_3$ were the respective stationary and mobile phases used for the separation [7]. The instrument provides automated sample injection and precise

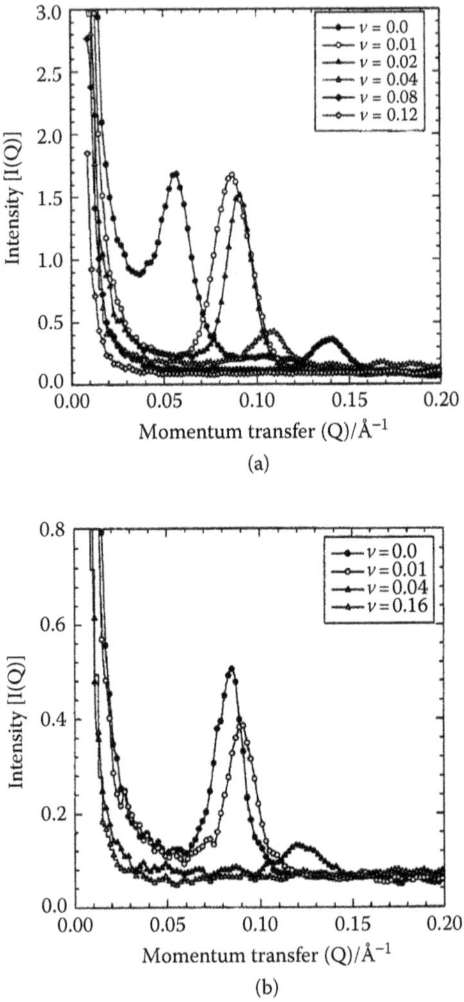

FIGURE 12.6 Typical diffraction traces observed at higher volume fractions, with $c = 0.1$ M, $T = 5°C$, $M = 18{,}000$; (a) $r = 0.1$ and (b) $r = 0.3$.

high-pressure pumping with computer-assisted data treatment. The polymer concentration in each sample was determined by means of an external calibration using the same source of PEO, purchased from Scientific Polymer Products Ltd. (Church Stretton, U.K.) and used without further purification. In each set of measurements under fixed $\{r, c, v, M\}$ conditions, two samples of the PEO solution from the supernatant fluid and two samples of the original soaking solution were analyzed. For these experiments, samples of a total volume of typically 5.0 cm^3 were prepared at a fixed clay concentration of $r = 0.1$ in glass jars sealed with Parafilm (see Figure 5.4). The jars were allowed to stand at 7°C for two weeks prior to the analyses.

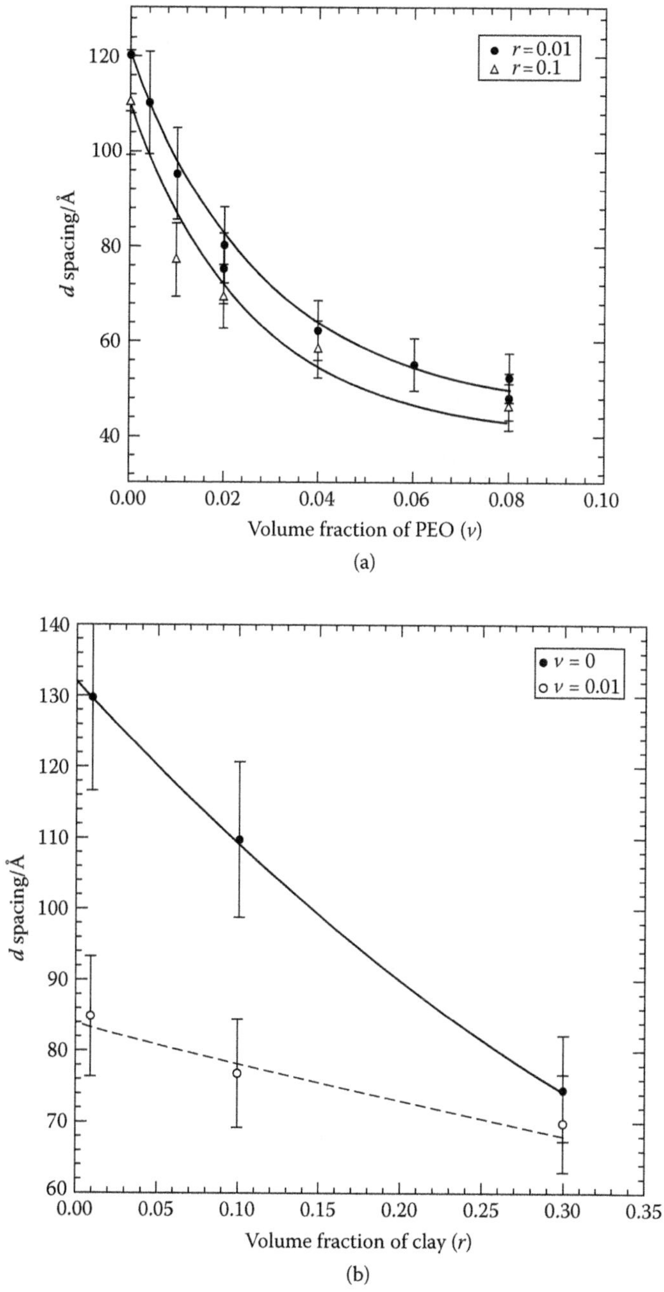

FIGURE 12.7 (a) Plot of d-values as a function of v for the two sol concentrations $r = 0.01$ (upper curve) and $r = 0.1$ (lower curve). (b) Plot of d-values as a function of r for the two polymer volume fractions $v = 0$ (upper curve) and $v = 0.01$ (lower curve). The lines between the points are drawn as a guide for the eye.

To make an independent analysis of the PEO concentrations in the supernatant fluids, a chemical method was employed in certain cases for samples prepared in the same way as those used in the GPC analysis. This method is based on the ability of PEO to form precipitates with large anions [8]. The precipitation was performed using phosphomolybdic acid ($H_3Mo_{10}PO_{32}\cdot 24H_2O$), barium chloride ($BaCl_2\cdot 2H_2O$), and hydrochloric acid, all purchased from Aldrich Chemicals and used without further purification. The reagent was prepared by dissolving 0.5 g of $H_3Mo_{10}PO_{32}\cdot 24H_2O$, 0.5 g of $BaCl_2\cdot 2H_2O$, and 1.5 ml of concentrated HCl in 250 ml of distilled water. After precipitation by adding 1 ml of the reagent to 1 ml of the PEO sample in a clean, dry centrifuge tube, the tube was shaken gently and stored at 20°C for 15 min before being centrifuged for 5 min at 10,000 rpm to separate the complex from its supernatant fluid. This supernatant fluid was then analyzed for excess phosphomolybdic acid by diluting 1 ml of the fluid up to 50 ml in a volumetric flask and determining the absorbance at 216 nm [8]. As with the GPC analysis, calibrations were performed using solutions containing known concentrations of PEO, and the polymer volume fractions were obtained using the calibration graph.

The main aim was to determine the distribution of PEO molecules between the gel and the supernatant fluid at $r = 0.1$, $c = 0.1$ M, $T = 5$°C for $M = 18{,}000$ (bridging) and polymer volume fractions in the range between $v = 0$ and $v = 0.12$. The corresponding neutron diffraction traces are shown in Figure 12.6a. In comparing these structural analyses with an independent analysis of the concentration of the PEO in the supernatant fluid, we established the following protocol in preparing the samples.

1. The salt was always dissolved first into the heavy water. We use molarity for the salt concentration variable c and note that at the highest concentration studied, $c = 0.1$ M, the volume fraction of salt is approximately 1%.
2. The polymer was then dissolved into the solution such that, for example, in preparing a $v = 0.12$ sample, 12 cm^3 of PEO was added to 100 cm^3 of the salt solution. The actual volume fraction of polymer in the standard solution v_{add} was $v/(1 + v)$, or 0.11 in this case.
3. The solution containing the polymer and salt was added to the vermiculite clay crystals such that in preparing an $r = 0.1$ sample, 100 cm^3 of solution was added to 10 cm^3 of clay. The global volume fraction of polymer in the condensed matter system v_{global} was $v_{add}/(1 + r)$, or

$$v_{global} = \frac{v}{(1+v)(1+r)} \tag{12.2}$$

with the concentration variables v and r experimentally controlled as above. We note that the global volume includes the volume of the clay plates ($d \approx 10$ Å), which can be regarded as inaccessible to the polymers. In keeping with Equation 12.2, the global volume fraction r_{global} of clay in the system is given by

$$r_{global} = \frac{r}{1+r} \tag{12.3}$$

or $r_{global} = 0.091$ for the unswollen crystals.

TABLE 12.5
Adsorption Isotherm Results at $r = 0.1$, $c = 0.1$ M, $T = 5°C$, $M = 18{,}000$ (bridging)

First Set of GPC Results

v	v_{ex}	r_{gel}	v_i	f
0.01	0.012	0.36	0.0046	0.46
0.02	0.024	0.32	0.0061	0.31
0.04	0.044	0.28	0.014	0.35
0.08	0.081	0.21	0.018	0.22
0.12	0.106	0.21	0.081	0.67

Second Set of GPC Results

v	v_{ex}	r_{gel}	v_i	f
0.06	0.059	0.24	0.034	0.56
0.07	0.067	0.22	0.041	0.59
0.08	0.077	0.21	0.038	0.47
0.09	0.083	0.21	0.056	0.62
0.10	0.089	0.21	0.074	0.74
0.11	0.099	0.21	0.075	0.68
0.12	0.105	0.21	0.086	0.71

Note: The average value of the fractionation factor f is equal to 0.45 in the bridging regime.

In analyzing the supernatant fluid, we measured the volume fraction v_{ex} in the fluid external to the clay gel. The results of the GPC analysis at $c = 0.1$ M, $M = 18{,}000$ (bridging) are given in Table 12.5. To determine the volume fraction of polymer v_{gel} inside the gel from v_{global} and v_{ex}, we need to know the relative volume occupied by the two phases. Because of errors due to adhesion of the sticky fluids to the outside of the clay gel, it is better to use the neutron data to determine the gel volume because we know that the microscopic and macroscopic expansions match well. If we let x be the factor by which the crystals have expanded, then $x = d_{gel}$ (Å)/19.4, and the final volume fraction of clay in the condensed matter system is $r_{gel} = xr_{global} = 0.091x$. The results are given in the third column of Table 12.5. Although it was not possible to obtain the d-values for $v \geq 0.09$ from the neutron data in the crowded regime, the macroscopic expansions observed in these cases were similar to that at $v = 0.08$, so we can estimate that $r_{gel} = 0.21$ in these cases also. The volume fraction of polymer in the gel phase is calculated as

$$v_{gel} = \frac{v_{global} - v_{ex}(1 - r_{gel})}{r_{gel}} \quad (12.4)$$

and the volume fraction of polymer v_i in the fluid inside the gel is calculated from this quantity by assuming that the clay plates exclude PEO molecules. The results are given in the fourth column of Table 12.5. Finally, the fifth column of Table 12.5 gives the polymer fractionation factor f defined by $f = v_i/v$. In the bridging regime up to $v = 0.09$, the average value obtained for f was 0.45. As with the salt fractionation factor s, there is a wide spread of values, with individual results lying between $f = 0.22$ and $f = 0.62$. We remarked previously in Chapter 5 that such macroscopic measurements are more susceptible to sample-to-sample variability, creating a random variation in the results. However, there was no evidence for any systematic variation of f with respect to v, and we conclude that f is roughly constant and equal to 0.45 ± 0.10 in the bridging regime. In the following discussion, this will be written as $v_i = 0.45v$, relating the volume fraction of polymer inside the gel to the experimentally controlled v.

Next, results for a nonbridging polymer at $c = 0.1$ M, $M = 4{,}000$ provide a comparison with the more widely studied case of polymer adsorption at a single interface [9]. The results are given in Table 12.6. In this case, it is clear that the fractionation factor f is approximately constant, with an average value of 0.95. This means that the polymer is roughly equally divided between the two phases.

The usual way of representing polymer adsorption onto clay surfaces is to plot an isotherm showing the amount of polymer adsorbed in grams per gram of clay as a function of the equilibrium concentration of polymer in units of g cm^{-3}. We have to be careful in comparing our results with standard isotherms because we are measuring the total amount of PEO inside the clay. This absorbed mass is not necessarily adsorbed onto the clay surfaces, but may be located in the interlayer solution. To reflect this difference, we have used the unusual nomenclature "absorption isotherm" rather than the usual "adsorption isotherm" in the presentation of the data.

We know from the density that 1 g of clay occupies 0.54 cm^3, so according to our protocol, an $r = 0.1$ sample is prepared by adding 5.4 cm^3 of polymer solution. Because the actual concentration v_a of polymer in this fluid is given by $v_a = v/(1 + v)$, the total

TABLE 12.6
Adsorption Isotherm Results at $r = 0.1$, $c = 0.1$ M, $T = 5°C$, $M = 4000$ (nonbridging)

v	v_{ex}	r_{gel}	v_i	f
0.01	0.0090	0.56	0.0098	0.98
0.02	0.018	0.56	0.019	0.95
0.03	0.027	0.56	0.029	0.95
0.04	0.036	0.56	0.037	0.93
0.05	0.045	0.56	0.046	0.92
0.06	0.055	0.56	0.053	0.89
0.07	0.061	0.56	0.064	0.91
0.08	0.067	0.56	0.074	0.92

TABLE 12.7
Absorption Isotherm Results at $r = 0.1$, $c = 0.1$ M, $T = 5°C$

v	Mass Added (g)	Mass Absorbed (g) $M = 18{,}000$	Mass Absorbed (g) $M = 4{,}000$
0.01	0.060	0.0095	0.034
0.02	0.12	0.011	0.067
0.03	0.18	—	0.099
0.04	0.23	0.021	0.13
0.05	0.29	—	0.16
0.06	0.35	0.044	0.19
0.07	0.40	0.048	0.22
0.08	0.45	0.030	0.26
0.09	0.50	0.060	—
0.10	0.55	0.079	—
0.11	0.60	0.081	—
0.12	0.66	0.092	—

volume of polymer added in such an experiment is $5.4v/(1 + v)$. Because the density of PEO is 1.13 g cm^{-3}, the mass of polymer added per gram of clay is $6.1v/(1 + v)$ grams, as shown in the second column of Table 12.7. The third column of Table 12.7 gives the mass of polymer absorbed (g) per gram of clay for the bridging polymer ($M = 18{,}000$), and the fourth column gives the same quantity for the nonbridging polymer ($M = 4{,}000$). The results have been plotted in Figure 12.8. For both molecular weights, the results seem to be linear to the limit of experimental accuracy, confirming the constant fractionation of polymer in both cases. The much steeper gradient of the $M = 4{,}000$ data reflects the facts that (a) f is approximately twice as great for the nonbridging PEO and (b) the clay gel occupies a much larger proportion of the condensed matter system when the polymer is unable to bridge.

Finally, because the accuracy with which we can determine the strength of the bridging force depends on the accuracy of f, three independent chemical analyses were performed in the bridging regime at $r = 0.1$, $c = 0.1$ M, $M = 18{,}000$ at the v-values 0.01, 0.04, and 0.08. The results of the phosphomolybdic acid titrations were $v_{ex} = 0.012$ at $v = 0.01$, $v_{ex} = 0.044$ at $v = 0.04$, and $v_{ex} = 0.076$ at $v = 0.08$, in close agreement with the GPC results given in Table 12.5. The fractionation factor was therefore confirmed to be $f = 0.45 \pm 0.10$ in the bridging regime.

Given the volume fraction of the polymer inside the gel, we are now able to propose a quantitative model of bridging flocculation. Because Crawford et al. [4] studied the contraction of the interlayer spacing as a function of uniaxial stress for the same system without any added polymer, we are able to convert the observed d-values to effective uniaxial pressures caused by the bridging polymers. If we assume that we have one polymer bridge when the end-to-end polymer distance l (calculated according to Equation 12.1) exactly matches the d-value with the

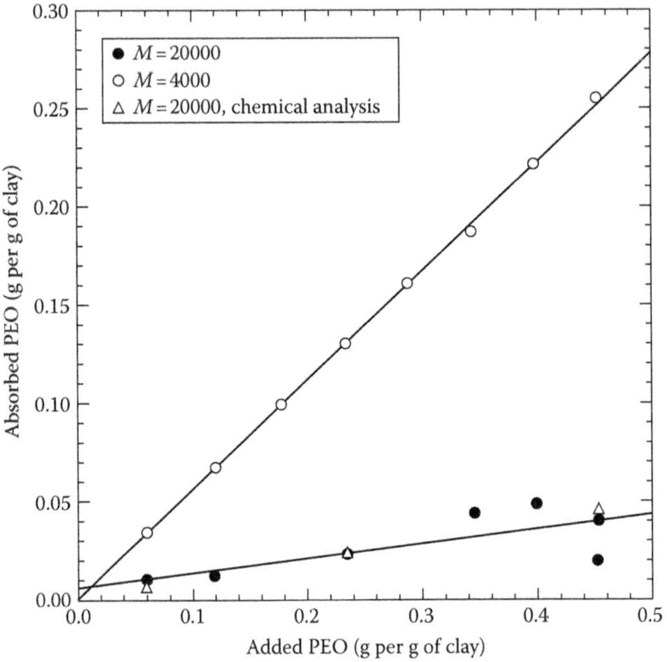

FIGURE 12.8 Absorption isotherm data at $r = 0.1$, $c = 0.1$ M. The closed circles and open triangles show the amount of PEO taken up by the gel for $M = 18{,}000$ (bridging), with the circles and triangles representing GPC analyses and chemical analyses, respectively; the open circles are for $M = 4{,}000$ (nonbridging).

thickness (≈ 10 Å) of the clay plates subtracted, we are able to calculate the total number of bridges per unit area of the clay plates from the observed d-values and polymer volume fractions, as follows. One polymer bridge is obtained when

$$d - 10 \approx l \approx \frac{M_{\min}^{0.57}}{1.84} \tag{12.5}$$

where $M_{\min}$ stands for the lowest molecular weight polymer that is able to bridge. For larger polymer molecules, the number of bridges is proportional to the molecular weight of the polymer, as indicated by the molecular weight independent d-value. If the molecular weight of the polymer is M, then the number N_p of bridges per polymer chain is given by

$$N_p = \frac{M}{M_{\min}} = \frac{M}{[1.84(d-10)]^{1/0.57}} \tag{12.6}$$

The average number of polymer chains per unit volume of solution N_V (μm³) can be obtained from the density $\rho = 1.13$ g/cm³ of PEO and the polymer volume fraction v_i inside the clay gel through the relation

$$N_V = \frac{\rho N_A v_i}{10^{12} M} = \frac{6.8 \times 10^{11} v_i}{M} \tag{12.7}$$

where N_A is Avogadro's number (6.02×10^{23}). If the solution inside the gel initially occupied one unit volume (μm³), then the total available area (μm²) of the clay plates is given by $20{,}000/(d_0 - 10)$, where d_0 stands for the initial d-value, for adsorption onto a single plate. However, because the polymers bridge between two clay plates, the available area of each plate is $10{,}000/(d_0 - 10)$. The average number of polymer chains per unit clay area N_a (μm²) can therefore be expressed as

$$N_a = \frac{(d-10)N_V}{10{,}000} = \frac{6.8 \times 10^7 (d-10) v_i}{M} \tag{12.8}$$

If we combine Equation 12.6 and Equation 12.8, the total number of polymer bridges N per unit clay area (μm²) is given by

$$N = N_a N_p = \frac{2.3 \times 10^7 (d_0 - 10) v_i}{(d_0 - 10)^{1/0.57}} \tag{12.9}$$

In Figure 12.9, we show the relation between the calculated number of bridges/μm² and the measured uniaxial pressures [4] corresponding to the observed d-values. It can be seen that the effective uniaxial pressure caused by the bridging polymers increases linearly (within the experimental errors) with the number of polymer bridges. This means that the average drawing force per polymer bridge must be effectively the same for all the samples studied, independent of the salt concentration, polymer volume fraction, and molecular weight. Using the data points given in Figure 12.9, we can calculate the average drawing force per polymer bridge to be approximately 0.6 pN, which corresponds to the straight line shown in Figure 12.9. The presence of a constant force is consistent with the assumptions made in deriving Equation 12.5 through Equation 12.9.

In Figure 12.9a, we have plotted the data on a log–log scale to emphasize that the linearity of the plot holds over three orders of magnitude in the effective drawing force exerted by the polymer chains. The open symbols show the data points obtained at $c = 0.01$ M and $c = 0.03$ M in reference [1], and the closed symbols show the previously obtained data points [2] at $c = 0.03$ M and $c = 0.1$ M. The excellent agreement confirms the essential feature of the bridging mechanism proposed in reference [2]. To demonstrate the linear relation between the effective uniaxial pressure p and N more convincingly, the data have been replotted on a linear scale in Figure 12.9b. This has the demerit of crushing the $c = 0.01$ M data points together

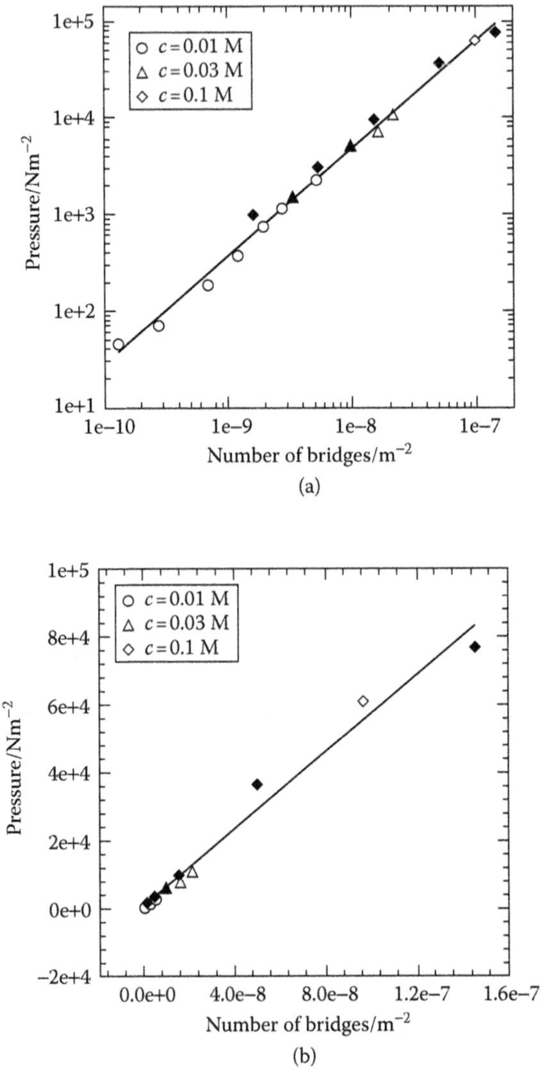

FIGURE 12.9 Effective uniaxial pressure as a function of the calculated number of bridges. The circles, triangles and diamonds correspond to $c = 0.01$ M, $c = 0.03$ M, and $c = 0.1$ M, respectively, with the open symbols showing the data given in reference [1] and the closed symbols showing the data given in reference [2]. The log–log plot in (a) shows that the linear relation between p and N extends over three orders of magnitude, and the linear plot in (b) emphasizes the direct proportionality of p and N. The straight line corresponds to an effective uniaxial force of 0.6 pN per bridge.

close to the origin but the advantage of showing that the linearity is not an artifact of using a logarithmic scale.

Although the linearity of the relationship between the effective stress and the number of polymer bridges confirms the general features of the bridging mechanism, there are three serious sources of error in the absolute magnitude of our value of 0.6 pN as the drawing force per polymer bridge. First, the absolute value of the force is very sensitive to the exact nature of the condition $d_0 = l$ that has been used in the present calculation. If the condition really turns out to be $d_0 \approx 1.3l$, as suggested by the results for the bridging molecular weight at $c = 0.1$ M, the estimated force would be lower by a factor of $1.3^{1.75} = 1.58$. This would imply a value of 0.4 pN per polymer bridge. It is obvious from these considerations that we cannot give the force to better than one significant figure. It is certainly on the order of 1 pN.

The second source of error lies in the considerable sample-to-sample variability in the uniaxial stress data [4]. We saw in Chapter 3 that the global outcome of many experiments was that the intercepts of the pressure–distance curves pointed decisively to a constant surface potential of $\psi_0 = 70$ mV and that the slopes of the curves κ_{in} were described by the relation $\kappa_{in}d_0 = 9$, where d_0 is the zero stress d-value. These relations have been used in the present mapping of the contractions, but other choices would be possible, For example, if we used the global relation $\psi_0 = 70$ mV together with the values for κ_{in} determined from the individual ln p vs. d curves, we would increase the magnitude of the drawing force by a factor of approximately 1.8. This serves to emphasize that the absolute value of the force is very sensitive to the exact form of the pressure–distance curves, which are difficult to measure when you are compressing fragile gels with a quartz plate in the small cell appropriate for neutron diffraction experiments.

The third source of error lies in the accuracy of our value for the polymer fractionation factor f. A value of $f = 0.5$, about 10% higher than the mean value, is well within the bounds of possibility and would provide for 10% more bridges than assumed in the present calculation, leading to a 10% lower drawing force per bridge. This serves to emphasize that the absolute value of the force is also sensitive to the exact form of the isotherms and to the uncertainties of mapping f between different sol and salt concentrations. This all may seem to make our final estimate of the strength of the bridges rather vague, but in a multicomponent system with the sample-to-sample variability inherent in colloid science, we believe it may not be possible to do better than limit the range to 0.6 ± 0.2 pN. Although we have formulated a qualitative mechanism for bridging flocculation, to be described in the next chapter, the calculation of the absolute value of the drawing force constitutes a new challenge in colloid and polymer science.

REFERENCES

1. Smalley, M.V., Hatharasinghe, H.L.M., Osborne, I., Swenson, J., and King, S.M., *Langmuir*, 17, 3800, 2001.
2. Swenson, J., Smalley, M.V., and Hatharasinghe, H.L.M., *Phys. Rev. Lett.*, 81, 5840, 1998.
3. Cohen-Stuart, M.A., Waajen, F.H.W.H., Cosgrove, T., Vincent, S., and Crowley, T.L., *Macromolecules*, 17, 1825, 1984.
4. Crawford, R.J., Smalley, M.V., and Thomas, R.K., *Adv. Colloid Interface Sci.*, 34, 537, 1991.

5. Williams, G.D., Moody, K.R., Smalley, M.V., and King, S.M., *Clays Clay Min.*, 42, 614, 1994.
6. Smalley, M.V., Jinnai, H., Hashimoto, T., and Koizumi, S., *Clays Clay Min.*, 45, 745, 1997.
7. Kato, Y., Matsuda, T., and Hashimoto, T., *J. Chromatogr.*, 332, 39, 1985.
8. Nuysink, J. and Koopal, L.K., *Talanta*, 29, 495, 1982.
9. Fleer, G.J., Cohen Stuart, M.A., Scheutjens, J.M.H.M., Cosgrove, T., and Vincent, B., *Polymers at Interfaces*, Chapman and Hall, London, 1993.

13 The Structure of Bridging Polymers

The clear picture that emerged from our studies on the LOQ instrument (ISIS, Didcot, U.K.) of the vermiculite–PEO system was of polymers forming bridges between the clay plates, with a dramatic change in the behavior of the system when the mean end-to-end distance of the polymer roughly matched the position of the secondary minimum in the electrostatic system. In the bridging regime, the amount of PEO adsorbed inside the gel seems to be such that there is a constant fractionation of polymer between the gel and the supernatant fluid. The only difference between a polymer of molecular weight 20,000 and one of molecular weight 2 million is that the latter forms 100 times as many bridges as the former, provided that both are large enough to bridge. Each bridge provides a drawing force of 0.6 pN.

In Chapter 12 we proposed a model for the drawing force that gave a consistent explanation for all the $\{r, c, v, M\}$(sol concentration, salt concentration, volume fraction of the polymer in the condensed matter system, molecular weight of the polymer) variations in the d-values that we had observed on LOQ [1]. To understand the mechanism of the polymer bridging flocculation, we need to determine the location of the polymer chains in the interlayer solution. I hope you, the reader, do not mind an occasional lighthearted aside. When we determined the dressed macroion structure from the D16 data [2, 3] in Chapter 8, I described the scattering as vanilla ripples, the basic flavor. When we determined the whole interlayer structure from the LAD (liquids and amorphous materials diffractometer) data [4] in Chapter 9, I described the scattering as raspberry ripples, a more exotic flavor. The ripples in this chapter are then definitely of the chocolate and pistachio variety. The chocolate ripples gave the polymer structure by replacing H-PEO with D-PEO to determine the distribution of the polymer segments [5], and the final pistachio ripples gave the ion distribution by replacing H-butylammonium with D-butylammonium in the PEO-added system to determine how the polymer adsorption affects the dressed macroion structure [6]. The combination of the double isotope substitution on both the counterions and the polymer chains gave us a uniquely detailed picture of the interlayer region and enabled us [7] to refine the new mechanism for polymer bridging flocculation proposed in reference [1].

We saw in Chapters 8 and 9 that the large difference in coherent scattering length between H and D ($b_H = -3.74$ fm and $b_D = 6.67$ fm) made possible the determination of the location of the alkylammonium counterions; it also makes possible the determination of the location of the polymer segments. Our first experiments were on the LAD instrument using H-PEO and D-PEO samples (with H-salt and D_2O) [5]. We fixed the concentration variables as $r = 0.05$, $c = 0.1$ M, $v = 0.04$ for experimental convenience. We recall that with 4% PEO in the supernatant fluid, we would expect the volume fraction inside the clay to be about 2%, so only a small amount of the total number of atoms in the system are changed by the isotope substitution. Nevertheless, differences between the H-PEO patterns and the D-PEO patterns are obvious, as shown in Figure 13.1.

Figure 13.1 was obtained on the LAD instrument, more or less exactly as described in Chapter 9. It shows the structure factors $S(Q_z)$ of the two samples with hydrogenous (solid line) and deuterated (dashed line) PEO on three different scales. In the lower Q-range shown in Figure 13.1a we observe the Bragg peaks due to the interplate correlations. Although the two samples were prepared under the same $\{r, c, v\}$ conditions, it is clear that the d-values along the swelling axis were not identical; the hydrogenous and deuterated samples showed strong first-order Bragg peaks at about 0.115 and 0.105 $Å^{-1}$, respectively, and second-order peaks at twice these Q-values. The ratio in intensity between the first- and second-order Bragg peaks is significantly lower for the deuterated sample, which indicates either that the first-order Bragg intensity is suppressed by a higher scattering density in the middle between the clay layers (if this is the case, a large part of the polymer molecules must be located there) or that the deuterated sample shows some kind of a well-defined characteristic distance of approximately half the d-value. As the d-value is about 60 Å, such a characteristic distance would, for instance, be produced if a large fraction of the polymer segments were located at about 15 Å from the center of the clay layers, giving rise to a correlation length of approximately 30 Å. In the intermediate Q-range shown in Figure 13.1b, the first- and second-order Bragg peaks are off scale on both samples. The third and fourth peaks, occurring at Q-values between 0.3 and 0.5 $Å^{-1}$ (corresponding to a length scale between about 10 and 20 Å), contain information about the local structure around the clay plates [2, 3], but they may also be strongly influenced by the third- and fourth-order Bragg peaks, since their positions almost coincide with the expected Q-values for such higher-order Bragg peaks. Independently of the origins of the peaks, it is worth noting that the intensity of the third peak is much higher for the hydrogenous sample. The higher Q-range shown in Figure 13.1c is similar for the two samples and is dominated by correlations within the water molecules. The higher Q data ($Q > 4$ $Å^{-1}$) will not be important for the present analysis of the intermediate-range correlations involving PEO.

Sometimes you get lucky with an experiment using isotope substitution in clay science, and you get two samples with the same d-value. The raspberry ripples obtained on the LAD instrument in Chapter 9 were from H-salt/D-salt-substituted samples that had an identical d-value of 43.6 Å. This enabled us to obtain the complete scattering length density profile for the ions along the z-axis. The analysis of the H-PEO/D-PEO data described here was complicated by the fact that the hydrogenous sample has a clearly lower d-value of about 55 Å, compared with about 60 Å for the deuterated sample. We therefore decided to analyze the data by the pair correlation function technique. This has been described in detail in Chapter 8 in our work on the vanilla ripples obtained on H-salt/D-salt-substituted samples on the D16 long-wavelength neutron diffractometer at the Institut Laue-Langevin (ILL), Grenoble. To avoid large differences between the hydrogenous H-PEO and deuterated D-PEO samples due to the different d-values of the two samples, the $S(Q_z)$ of the

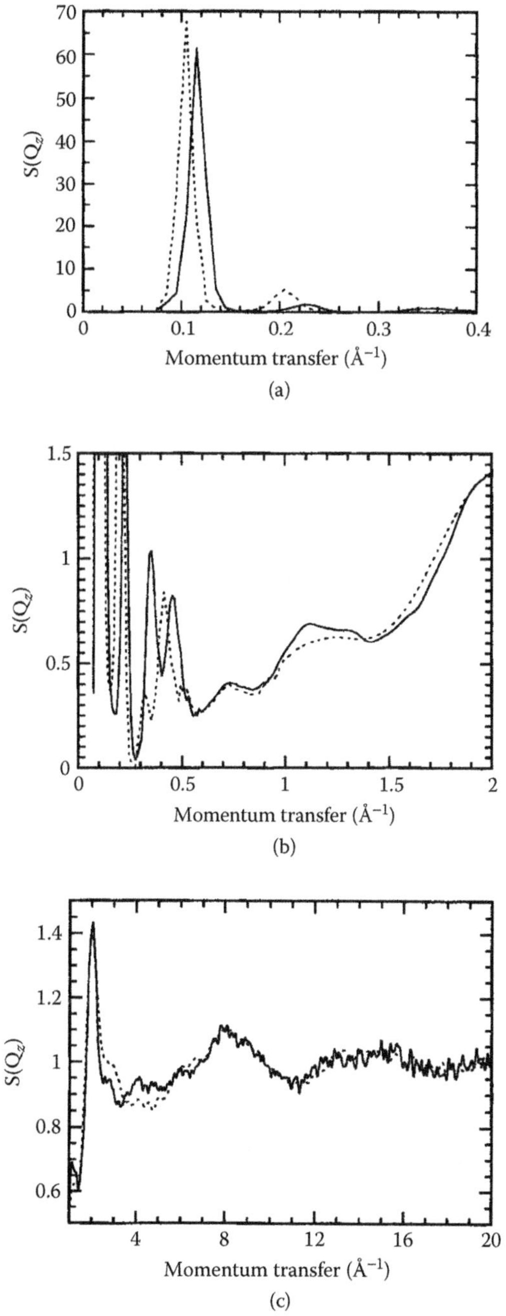

FIGURE 13.1 Structure factors $S(Q_z)$ of the two samples with hydrogenous (solid line) and deuterated (dashed line) PEO. The structure factors are shown on three different scales in (a), (b) and (c) for clarity.

hydrogenous sample was compressed by 8% for $Q < 0.27$ Å^{-1} to get the positions of the first- and second-order Bragg peaks to coincide for the two samples. Thus, we made the assumption that the different *d*-values only affect the positions of the first two Bragg peaks and that the length scale of all the other short- and intermediate-range correlations are the same for the slightly different interlayer spacings. It should be noted that this approximation is by no means perfect, since the third and fourth peaks may be strongly influenced by the third- and fourth-order Bragg peaks. To test the scaling employed, we also performed a scaling of the positions of all the first four peaks ($Q < 0.50$ Å^{-1}) with the 8% scaling factor. The structure factors were Fourier transformed to atomic pair-correlation functions $G(r_z)$ for both types of scaling; reassuringly, we found that the new scaling caused only minor changes in $G(r_z)$ for the hydrogenous sample. Thus, the exact form of the scaling does not affect the argument below.

Figure 13.2 shows the difference function $\Delta G(r_z) = G_D(r_z) - G_H(r_z)$ between the $G(r_z)$ obtained for the deuterated and hydrogenous samples. Thus, $\Delta G(r_z)$ should give us the correlations due to the replacement of H-PEO by D-PEO, which means that the correlations involving PEO should give positive contributions. Moreover, since we use normalized $G(r_z)$, it is obvious that all the remaining correlations involving the clay layers, D_2O and the butylammonium ions are less weighted for the deuterated sample and thus give negative contributions to $\Delta G(r_z)$. It can be seen in Figure 13.2 that $\Delta G(r_z)$ shows at least four relatively sharp peaks in the r_z region 4 to 14 Å and two very broad peaks around 27 to 41 Å and 56 to 70 Å. The four pronounced peaks in the low r_z region are located at approximately 4.5, 8, 11 and 14 Å.

Due to the sharpness of the four peaks in the region $4 < r_z < 14$ Å in $\Delta G(r_z)$, it is evident that the correlations are between well-defined positions. Furthermore, since the intensities of the peaks are clearly above zero, the polymer molecules must be involved in the correlations. Thus, it is very likely that we are observing the

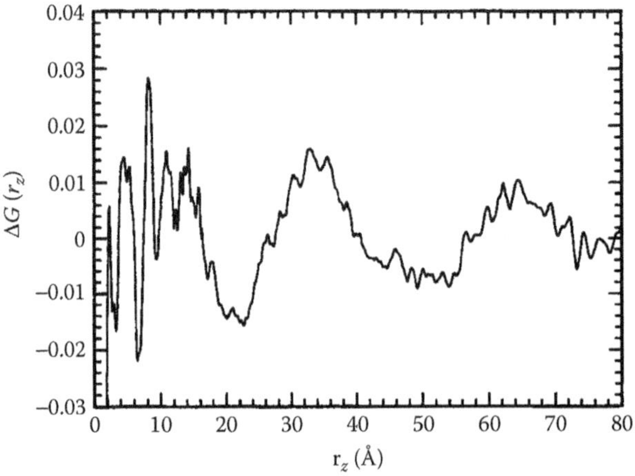

FIGURE 13.2 Difference function $\Delta G(r_z)$ between the atomic pair correlation functions of the deuterated $G_D(r_z)$ and hydrogenous $G_H(r_z)$ samples.

correlations between well-defined atomic layers in the clay plates and polymer segments aligned parallel to the clay layers. A detailed analysis confirms this conjecture [5]. In particular, the peaks located at approximately 11 and 14 Å are mainly due to the correlation between polymer segments adsorbed on either side of the clay plates. The shorter distance is from the correlation between hydrogen atoms bound directly to the surface oxygens (at the same distance as the hydrogen atoms of the first layer of water molecules in the dressed macroion), and the longer distance is then determined by the geometry of the PEO chain, assumed to have a zigzag trans conformation, being the correlation between the hydrogen atoms farther away from the outer oxygen layer. From the intensities of the two peaks at about 11 and 14 Å in Figure 13.2, we estimated that approximately 25% of the PEO segments are stuck to each of the bridged layers for $c = 0.1$ M and $v = 0.04$. Thus, we obtain a picture where the PEO segments are displacing, at least partly, the water molecules immediately adjacent to the clay layers.

Clearly, the results give strong indications that the bridging polymers are adsorbed directly onto the clay layers. This, in combination with the fact that we do not observe any well-defined polymer–clay correlation on a length scale of about 15 Å, means that the lower ratio between the intensities of the first- and second-order Bragg peaks for the deuterated sample than for the hydrogenous sample (see Figure 13.1a) must be due to a suppression of the first-order Bragg intensity of the deuterated sample by the presence of a large fraction of polymer segments located in the middle between the clay layers. These segments give rise to a higher scattering density (compared with the hydrogenous sample) in the middle of the interlayer region, which both decreases the scattering contrast for the first-order Bragg peak and amplifies the second-order Bragg peak. Thus, the broad peak around 27 to 41 Å in Figure 13.2 is interpreted as arising from correlations between polymer segments located in the middle of the interlayer region and clay plates with their neighboring aligned polymer segments. The peak at 56 to 70 Å is basically due to the periodicity along the z-axis, which produces polymer–polymer correlations between polymer segments located at similar interlayer positions around consecutive clay plates. The peak at 27 to 41 Å would have been broader and more symmetric if it had not been partly suppressed by water–water correlations (observed as dips in Figure 13.2) at about 20 and 40 Å. Such an interpretation is consistent with the expected result that the nonadsorbed polymer segments show a relatively unperturbed Gaussian-like distribution centered in the middle between the clay layers, which produces a very low concentration of polymer segments (and therefore a higher concentration of water molecules) just outside the adsorbed polymer segments. This is also consistent with the idea that the structure of the 30 Å thick dressed macroion (consisting of the clay layers, two adsorbed water layers and the butylammonium ions) is also relatively unperturbed by the polymer adsorption, apart from the displacement of some of the water molecules immediately adjacent to the clay layers.

Let us now test this interpretation with the pistachio-flavored ripples with which we conclude our experimental investigation into the four-component clay-polymer-salt-water system [6]. In reference [6], we used simultaneous H/D substitution of both the butylammonium ions and the PEO chains; for each composition, four samples — H-PEO and H-salt, D-PEO and H-salt, D-PEO and D-salt, and H-PEO

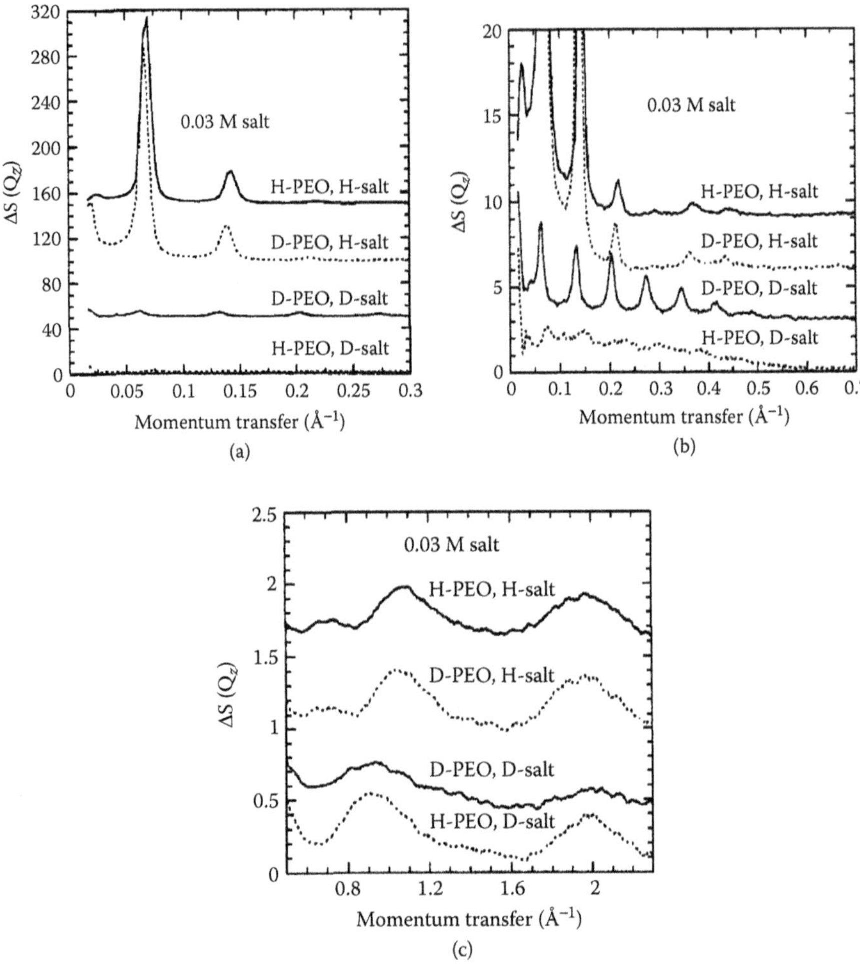

FIGURE 13.3 Difference structure factors $\Delta S(Q_z)$ (obtained from the difference between the measured $S(Q_z)$ of the samples and $S(Q_z)$ of pure D_2O) of the four isotopically different compositions, H-PEO and H-salt, D-PEO and H-salt, D-PEO and D-salt, and H-PEO and D-salt, with salt concentration c = 0.03 M and volume fraction of PEO v = 0.04. The difference structure factors are shifted vertically and shown on three different scales in parts (a), (b) and (c), for clarity.

and D-salt — were prepared. Furthermore, we prepared samples with two different concentrations of n-butylammonium chloride, 0.03 and 0.1 M, to test our model of bridging flocculation. The experiments were carried out on the D16 neutron diffractometer.

Figure 13.3 and Figure 13.4 show difference structure factors $\Delta S(Q_z)$ (parts (a), (b) and (c) of the two figures show different scales) for the four isotope compositions for 0.03 and 0.1 M butylammonium vermiculite gels, respectively. The difference structure factors $\Delta S(Q_z)$ were obtained by subtraction of the $S(Q_z)$

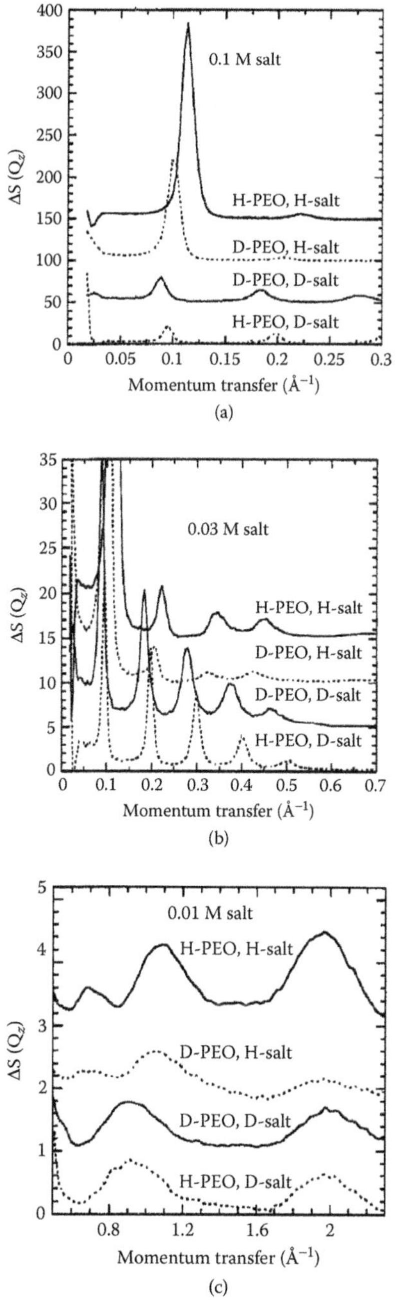

FIGURE 13.4 Difference structure factors $\Delta S(Q_z)$ of the four isotopically different compositions, H-PEO and H-salt, D-PEO and H-salt, D-PEO and D-salt, and H-PEO and D-salt, with salt concentration $c = 0.1$ M and volume fraction of PEO $v = 0.04$. The difference structure factors are shifted vertically and shown on three different scales in parts (a), (b) and (c), for clarity.

of pure D_2O from the measured $S(Q_z)$. From the positions of the Bragg peaks in the lower Q-ranges — shown in Figure 13.3a, Figure 13.3b and in Figure 13.4a, Figure 13.4b — it is clear that the d-value is approximately 90 Å for the 0.03 M salt concentration and about 60 Å in the case of the 0.1 M salt concentration. These are typical results for $r = 0.05$, $v = 0.04$, conditions repeating those of the chocolate ripples. It is also clear that the d-value along the swelling axis was not identical for the four samples at each salt concentration, although they were prepared under the same $\{r, c, v\}$ conditions. Figure 13.3c and Figure 13.4c show the ripples of diffuse scattering caused mainly by the plate–solution and solution–solution correlations. Before we move on to the structural modeling of the data, we can draw some conclusions directly from the results. First, from a comparison of Figure 13.3c with Figure 13.4c, it is evident that the interlayer structure, that is, the distributions of the ethylene oxide segments and the butylammonium ions, must be very similar for the two salt concentrations, despite their different d-values. It is also clear from Figure 13.3b and Figure 13.4b that H/D isotope substitution of the butylammonium chains has a much greater effect on the intensity of the Bragg peaks, as well as on the diffuse scattering pattern, than a corresponding substitution of ethylene oxide segments. This indicates that the butylammonium ions are much more inhomogeneously distributed in the interlayer region, compared with the ethylene oxide segments.

The interlayer structure can in principle be determined from a combined analysis of the diffuse scattering and the intensity variation of the Bragg peaks of the four isotopically different samples. In practice, this is not an easy task when the isotopically different samples have slightly different d-values. To make a proper comparison in real space of the isotopically different samples, we therefore had to scale their d-values to the same value before the Fourier transform was performed. The scaling enabled us to obtain difference pair correlation functions $\Delta G(r_z)$ by subtracting one pair-correlation function $G(r_z)$ from another, without spurious peaks being introduced by the different d-values. Figure 13.5 shows the difference function $\Delta G(r_z) = G_D(r_z) - G_H(r_z)$ between the $G(r_z)$ values obtained for the samples with the deuterated and hydrogenous PEO, respectively. In the case of the 0.1 M salt concentration, the butylammonium ions were deuterated in both samples, whereas the counterions were hydrogenous in the two samples of 0.03 M concentration. It can be seen in Figure 13.5 that, for both salt concentrations, $\Delta G(r_z)$ shows five relatively sharp peaks located at approximately 3, 5.5, 9, 11.5 and 15 Å. This result is very similar to the findings from the LAD data with H-salt at $c = 0.1$ M, although the real-space resolution is lower in this case because of the limited Q-range. Figure 13.5 confirms our interpretation for the 0.1 M salt concentration and shows no indication that the distribution of ethylene oxide segments is significantly different for the 0.03 M salt concentration.

It is interesting to know that the distribution of the ethylene oxide segments is rather independent of the salt concentration, but the point of the pistachio ripples was to investigate how the presence of the polymer affects the distribution of the counterions. To elucidate this in more detail, we employed the modeling technique used to fit the raspberry ripples [4]. We recall that this essentially crystallographic technique is a method for producing a single-particle distribution function, $\rho(z)$,

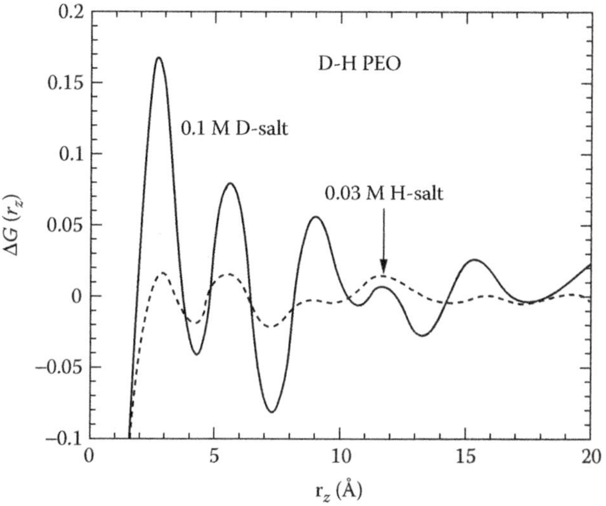

FIGURE 13.5 Difference pair correlation functions $\Delta G(r_z)$ between the atomic pair correlation functions of samples with deuterated and hydrogenous PEO. In the case of the 0.1 M salt concentration (solid line), the butylammonium chains were deuterated in both samples, whereas for the 0.03 M salt concentration (dashed line), both samples contained hydrogenous counterions. The volume fraction of PEO was 4% in all cases.

rather than a spatially averaged pair correlation function, $G(r_z)$. The crystallographic method mainly uses the intensities of the Bragg peaks to determine the structure. Because these Bragg peaks show large sample-to-sample variability (due to better ordering of the clay plates in one sample than another), errors can easily be introduced in this method. It was therefore inappropriate for the determination of the relatively homogeneous distribution of the polymer molecules in the interlayer region. However, H/D substitution of the butylammonium chains causes considerable alterations in the diffraction data, both in the intensity of the Bragg peaks and in the Q-range of the diffuse scattering (see Figure 13.3 and Figure 13.4). Thus, the modeled scattering length density profile along the z-axis, $\rho(z)$, for the butylammonium chains should be at least qualitatively correct. It is shown in Figure 13.6.

Figure 13.6a shows $\rho(z)$ for the composition with a salt concentration of 0.03 M, and the corresponding function for the 0.1 M salt concentration is shown in Figure 13.6b. The centers of the clay plates are located at 0 and 90 Å in Figure 13.6a and at 0 and 62 Å in Figure 13.6b. By comparison of the two figures, it is evident that the distribution of butylammonium chains is qualitatively the same for the two salt concentrations. The most important result is that for both salt concentrations a major part of the butylammonium chains are located in a 4 Å thick layer at a distance of 12 to 16 Å from the center of the clay plates; that is, they are situated just outside the approximately 6 Å thick layer of adsorbed polymer segments and water molecules. This means that the highest concentration of butylammonium chains is located at the same distance from the clay layers as in the corresponding three-component

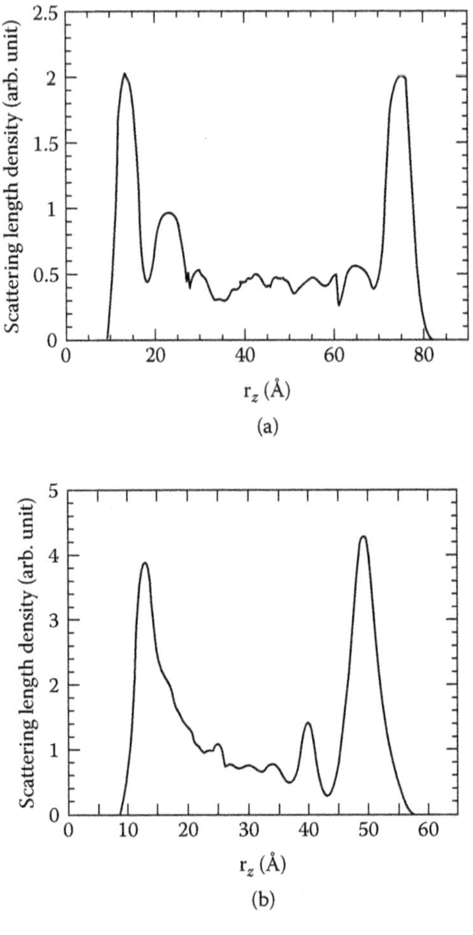

FIGURE 13.6 Scattering length density profiles along the z-axis, $\rho(z)$, of butylammonium chains in the interlayer region as a function of r_z. The salt concentration is 0.03 M in (a) and 0.1 M in (b), and the polymer volume fraction is 4% in both cases. The centers of the clay plates are located at $r_z = 0$ and $r_z = 90$ Å in (a) and at $r_z = 0$ and $r_z = 62$ Å in (b). The results were obtained from structural modeling of the diffraction data shown in Figure 13.3 and Figure 13.4.

system without added polymer [2, 3]. Thus, there is no indication that the presence of the PEO affects the distribution of the butylammonium ions.

In our final study, we found no evidence that PEO alters the dressed macroion structure other than that some of the polymer segments displace water molecules immediately adjacent to the clay plates, bonding directly to them by physical adsorption. Thus, the very inhomogeneous distribution of the counterions remains basically unaffected, and the structurally ordered range around each clay plate is preserved. At first sight, this seems to be a rather surprising result. The substantial

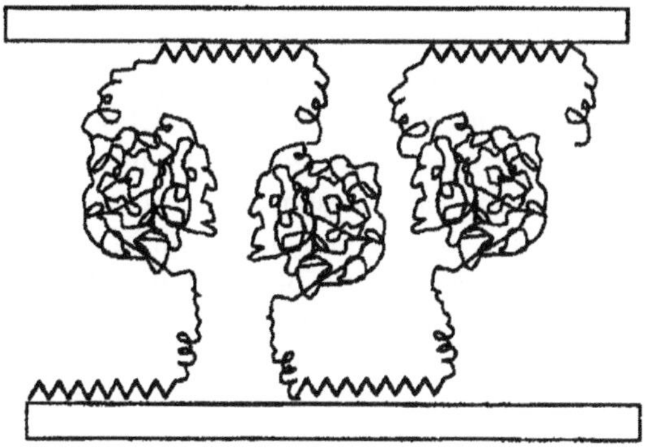

FIGURE 13.7 Schematic illustration of the model for the bridging polymers: (a) shows the case for a polymer that is just large enough to form a single bridge, and (b) is for a much larger polymer, with multiple bridges. The drawing force between the vermiculite plates is the same in both cases.

rearrangement of the first layer of adsorbed water molecules necessary to accommodate the polymer segments might be expected to have a noticeable effect on both the adjacent layer of water molecules and their neighboring counterions. The apparent insensitivity of the counterion layer 11 to 15 Å from the center of the clay plates to the PEO adsorption 6 to 9 Å from the center of the clay plates is probably connected with the insensitivity of the dressed macroion structure to the thermodynamic conditions in the three-component system and raises another challenge to theory.

We end up with a picture of the interlayer structure of a bridging polymer like that illustrated in Figure 13.7. Although Figure 13.7 is only a sketch, I would like to draw your attention to several features that are consistent with all the results on the clay–PEO system described in the last three chapters. First, some PEO segments zigzag across the surface, in the layer directly adjacent to the surface oxygens of the vermiculite layers. Second, there is some kind of random coil, comprising approximately half the segments, in the middle between the vermiculite layers. Third, the region between the surface and the coil is relatively depleted of polymer segments, as represented by the stringy bits connecting these two parts. Fourth, the only difference between bridging polymers of (a) molecular weights just sufficient to bridge and (b) much higher molecular weights, under identical $\{r, c, v\}$ conditions, is that in the former case there is one bridge per molecule and in the latter there are many bridges per molecule, with the number of bridges per unit amount of polymer the same in both cases. Fifth, the polymer chains have been drawn well-separated from each other, to take account of our ideas about how the surface becomes "crowded" at higher polymer volume fractions.

There are some uncertainties in our interpretation of the crowding volume fraction, v_c. The abrupt disappearance of the well-defined Bragg peaks at volume fractions of around 1, 4 and 9% for the cases $d_0 = 330$ Å, $d_0 = 190$ Å and $d_0 = 120$ Å, respectively (see Table 12.3), suggests an important role for the proportion of the surface area of the clay plates occupied by adsorbed polymer segments. To understand this phenomenon, it is necessary to have a model of the polymer segments. Using the known bond lengths and bond angles [8, 9], the length of each monomer in a zigzag trans conformation can be estimated to be about 3.8 Å. Because the density of PEO is 1.13 g/cm^3, the polymer chain with its excluded volume can be approximated by a coil with a diameter of approximately 4.7 Å. We therefore anticipate that the area of a clay plate occupied by a monomer unit will be approximately 3.8×4.7 Å$^2 \approx 18$ Å^2. Now let us consider the 1 μm^3 box inside the clay gel introduced in the calculation of the bridging force in Chapter 12, taking as an example the case $r = 0.1$, $c = 0.1$ M, $v = 0.08$ at the edge of the crowding regime. Using the experimentally determined results $d = 44$ Å and $f = 0.45$, the number of monomer units inside 1 μm^3 of gel is 5.6×10^8, and the available area of clay surfaces is 450 μm^2. To calculate the surface area occupied by the polymer segments, we need a quantitative model of the picture for the polymer segment density profile provided by the chocolate ripples; these give an approximate distribution of the PEO chain inside the gel that has one-quarter of the segments stuck to each of the bridged layers, with the remaining one-half of the segments in the fluid in the interlayer region. We therefore have 2.8×10^8 segments, occupying $2.8 \times 10^8 \times 1.8 \times 10^{-7} \approx$ 50 μm^2. At crowding, approximately 10% of the vermiculite surfaces are therefore covered by PEO segments.

We can obtain an independent estimate of the surface coverage at crowding from the isotherm data displayed in Figure 12.8, which show that for the bridged polymer at $v = 0.08$, approximately 0.05 g of PEO is absorbed per gram of clay. In this (bridging) case, the amount absorbed is also the amount adsorbed, because there are no free chains inside the gel. We have seen that the butylammonium vermiculite has a surface area of approximately 500 m^2 per gram, so we have 0.1 mg of PEO adsorbed per square meter of surface. A general rule of thumb for monolayer adsorption of homopolymers is 1 mg/m^2 [11], so the macroscopic property also corresponds to 10% coverage.

Although the figure of 10% coverage is a rough one, it is clear that the surface is still very open when the crowding transition occurs. One possible interpretation of this result is that we have found evidence for the hypothesis proposed by de Gennes [12] that polymers cannot entangle in two dimensions. In reference [12], "two-dimensional" really refers to a confined geometry in which the confinement distance d is shorter than the mean end-to-end distance l. This is clearly the case studied here. In these circumstances, the large part of each polymer chain that remains in a Gaussian-like distribution in the middle between the clay layers must exclude both other parts of the same chain between different bridges (self-exclusion) and other polymer molecules (mutual exclusion), leading to a wide exclusion zone around each bridge. The idea was included in the sketch of Figure 13.7.

In the previous chapter we obtained a global fit to the d-values with respect to $\{r, c, v, M\}$ variations with a single drawing force per polymer bridge. Combining this knowledge with the structure of the bridging polymers, the mechanism of polymer bridging flocculation can be outlined as follows. The polymer chains diffuse between the clay plates, and each chain adsorbs at both surfaces, provided that the mean end-to-end distance of the polymer is approximately equal to or larger than the d-value. At this point, it is likely that each chain is like a random coil with many bends and that the distance along a chain between the two adsorbed points is much larger than the actual d-value. Thereafter, larger and larger parts of the polymer chains will adsorb at the initial clay–polymer contact points, so the length of the adsorbed train increases at the same time as the chains stretch out between the two clay layers. This will continue until equilibrium occurs between the drawing force, arising from the segmental motions of a "stretched" polymer chain, and the bonding force to the clay plates. Thus, the bonding force determines the strength of each polymer bridge, which then explains why we obtain a constant effective drawing force per bridge.

On several occasions, we felt as if we could see the mechanism in action when we studied the scattering patterns as a function of time, coming into the gel phase from the tactoid phase. We studied the upper phase transition temperature T_c between the gel and tactoid phases under a wide variety of $\{r, c, v, M\}$ conditions and never found any change with respect to polymer addition. Similarly, the tactoids obtained from the polymer-added gels all had a normal d-value of 19.4 Å, showing that polymers did not become trapped by the clay layers when they were collapsed by heating. This means that the bridging polymer chains follow the water molecules in or out of the gel at the phase transition. When we measured the chocolate ripples, we actually used the (001) Bragg peak of the tactoid phase to provide a convenient reference for orienting the samples more accurately than is possible by the rough orientation provided by the confining walls of the sample. We take the example $r = 0.05$, c = 0.1 M, v = 0.04, M = 75,000 studied above, with T_c = 14°C. The sample temperature was held at 20°C while the sample was rocked until the intensity of the first-order Bragg peak of the tactoid phase was maximized. The temperature was then decreased to 7°C to bring the sample into its gel phase for the ripples measurement. Figure 13.8 shows the change of the diffraction pattern as a function of time when the H-PEO sample went through the tactoid-gel phase transition. The measurements cover a period of approximately one hour, with a time difference between consecutive curves of about 10 min. It is evident in Figure 13.8 how the fraction of the sample in the gel phase successively increases at the expense of the tactoid phase,

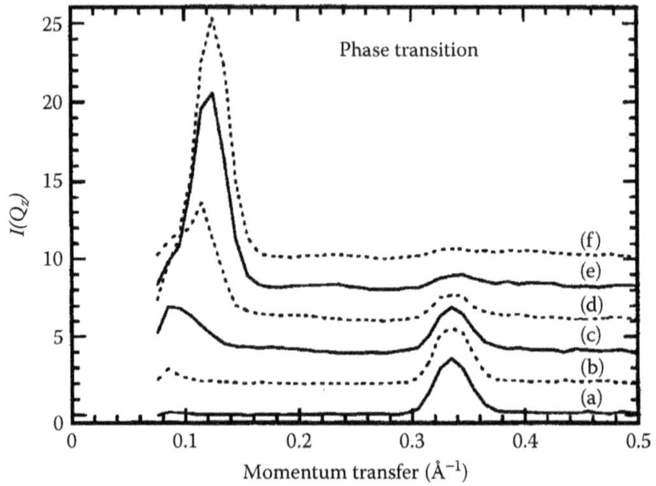

FIGURE 13.8 Change of the diffraction pattern $I(Q_z)$ for the clay-polymer-salt-water sample (c = 0.1 M, M = 75,000, r = 0.05, and v = 0.04) containing hydrogenous PEO when it goes through the tactoid–gel phase transition temperature at about 14°C. The time difference between consecutive curves (figures (a) through (f) are shifted vertically by 2.0 for clarity) is approximately 10 min.

and the interesting feature is that the position of the first-order Bragg peak arising from the interlayer correlation of the gel phase increases from about 0.085 to 0.12 Å^{-1} during the time the whole sample goes through the phase transition. Converting the positions of the first-order Bragg peak to d-values gives a decrease in d-value from approximately 74 to 52 Å during this time. You could easily imagine that we are seeing the polymer pulling the clay layers together. In fact, the kinetics of the bridging mechanism are very clearly illustrated in Figure 13.8, and the bridging structure is clearly an equilibrium one, since polymers can desorb from the surfaces rapidly on the timescale of the plate rearrangements.

Although our value for the drawing force per polymer bridge, 0.6 ± 0.2 pN, necessarily lies within a fairly wide range, the resolution of the measurement is very high. Going back over 20 years, there has been considerable advancement in the understanding of the effect of polymers in modifying surface interactions in colloidal dispersions. In particular, the forces between surfaces with adsorbed PEO layers have been investigated using the mica surface-force apparatus [13–15]. It is difficult to compare the results of our experiments with these for three reasons. First, the weakest forces measured in these experiments were on the order of 10 nN, a force resolution four orders of magnitude lower than ours. Second, the mica force balance measures the complete force–distance curve, whereas we are sensitive only to the force at the equilibrium separation of the surfaces. Third, the PEO is introduced into the system in different ways in the two types of experiment. In the force balance, the mica surfaces are incubated in the presence of polymer for about a day at a macroscopic separation, so that adsorbed layers are formed separately on each surface prior to their approach. In our experiments, the polymer penetrates the clay gel as the interlayer spacing expands from the tactoid phase, so the bridging configuration is achieved before the surfaces become crowded with polymer chains.

The widely invoked model to explain why interparticle separations between colloidal particles decrease when a large polymer is introduced into the system is known as depletion flocculation [16, 17]. This is basically an equilibrium osmotic exclusion model, in which the reduction in d-value is driven by an osmotic pressure of polymer molecules excluded from the interparticle region. We have indeed discovered that the PEO molecules are partially excluded from the gel, but the constancy of the d-value with respect to molecular weight at a fixed volume fraction rules out the depletion flocculation mechanism. This is because osmotic pressure is a colligative property, and the change in the number of polymer molecules at fixed v and variable M would give rise to a varying contraction. Furthermore, there is no obvious way in which the crowding transition could come about via depletion flocculation. Instead, the bridging mechanism gives a coherent explanation of all the available data.

REFERENCES

1. Swenson, J., Smalley, M.V., and Hatharasinghe, H.L.M., *Phys. Rev. Lett.*, 81, 5840, 1998.
2. Swenson, J., Smalley, M.V., Thomas, R.K., Crawford, R.J., and Braganza, L.F., *Langmuir*, 13, 6654, 1997.
3. Swenson, J., Smalley, M.V., Thomas, R.K., and Crawford, R.J., *J. Phys. Chem. B*, 102, 5823, 1998.
4. Williams, G.D., Soper, A.K., Skipper, N.T., and Smalley, M.V., *J. Phys. Chem. B*, 102, 8945, 1998.
5. Swenson, J., Smalley, M.V., and Hatharasinghe, H.L.M., *J. Chem. Phys.*, 110, 9750, 1999.
6. Swenson, J., Smalley, M.V., Hatharasinghe, H.L.M., and Fragneto, G., *Langmuir*, 17, 3813, 2001.
7. Smalley, M.V., Hatharasinghe, H.L.M., Osborne, I., Swenson, J., and King, S.M., *Langmuir*, 17, 3800, 2001.
8. Rosi-Schwartz, B. and Mitchell, G.R., *Polymer*, 35, 5398, 1994.
9. Carlsson, P., Swenson, J., Börjesson, L., Torell, L.M., McGreevy, R.L., and Howells, W.S., *J. Chem. Phys.*, 109, 8719, 1998.
10. Smalley, M.V., Jinnai, H., Hashimoto, T., and Koizumi, S., *Clays Clay Min.*, 45, 745, 1997.
11. Fleer, G.J., Cohen Stuart, M.A., Scheutjens, J.M.H.M., Cosgrove, T., and Vincent, B., *Polymers at Interfaces*, Chapman and Hall, London, 1993.
12. de Gennes, P.G., *Scaling Concepts in Polymer Physics*, Cornell University Press, Ithaca, NY, 1979.
13. Klein, J. and Luckham, P.F., *Nature*, 300, 429, 1982.
14. Klein, J. and Luckham, P.F., *Nature*, 308, 836, 1984.
15. Luckham, P.F. and Klein, J., *J. Chem. Soc., Faraday Trans.*, 86, 1363, 1990.
16. Dickenson, E., *J. Chem. Soc., Faraday Trans.*, 91, 4413, 1995.
17. Jenkins, P. and Snowden, M., *Adv. Colloid Interface Sci.*, 68, 57, 1996.

Index

F

G

H

I

J

K

L

M

N

O

P

Q

R

S

T

U

V

W

X

Y

Index

D

E

F

G

H

I

J

K

L

M

N

O

P

T

U

V

W

X

Y

Printed and bound by CPI Group (UK) Ltd, Croydon, CR0 4YY
23/10/2024
01778227-0010